T0317424

Photovoltaic Power System

Photovoltaic Power System

Modeling, Design, and Control

Weidong Xiao
University of Sydney
Australia

Registered Office(s)
John Wiley & Sons, Inc., 111 River Street, Hoboken, NJ 07030, USA
John Wiley & Sons Ltd, The Atrium, Southern Gate, Chichester, West Sussex, PO19 8SQ, UK

Editorial Office
The Atrium, Southern Gate, Chichester, West Sussex, PO19 8SQ, UK

For details of our global editorial offices, customer services, and more information about Wiley products visit us at www.wiley.com.

Wiley also publishes its books in a variety of electronic formats and by print-on-demand. Some content that appears in standard print versions of this book may not be available in other formats.

Library of Congress Cataloging-in-Publication Data
Names: Xiao, Weidong, 1969- author.
Title: Photovoltaic power system : modeling, design, and control / Weidong
    Xiao.
Description: Hoboken, NJ : John Wiley & Sons, 2017. | Includes
    bibliographical references and index.
Identifiers: LCCN 2016056659 (print) | LCCN 2017001056 (ebook) | ISBN
    9781119280347 (cloth) | ISBN 9781119280361 (pdf) | ISBN 9781119280323
    (epub)
Subjects: LCSH: Photovoltaic power systems.
Classification: LCC TK1087 .X56 2017 (print) | LCC TK1087 (ebook) | DDC
    621.31/244–dc23
LC record available at https://lccn.loc.gov/2016056659

Cover Design: Wiley
Cover Images: (Background) © Alessandro2802/Gettyimages; (Circles: Top right corner to bottom left corner) © chapin31/Gettyimages; © PaulPaladin/Gettyimages; Courtesy of author; © Dwight Smith/Shutterstock

Set in 10/12pt WarnockPro by SPi Global, Chennai, India

10 9 8 7 6 5 4 3 2 1

This book is dedicated to my son, William, and daughter, Emily, with deep love.

Contents

Preface

Photovoltaic (PV) power engineering has attracted significant attention in recent years. This book sets out to fulfil an important need in academia and industry for a comprehensive resource covering modeling, design, simulation, and control of PV power systems. Initially developed to support teaching senior-undergraduate and graduate courses, the work also covers practical design issues, that make it useful for industry practitioners seeking to master the subject through self-study and training. The book provides a smooth transition from fundamental knowledge to advanced subjects of interest to academics and to those working on system improvements in industry. A fundamental knowledge of power electronics and linear control theory is required to benefit fully from this book.

This comprehensive treatment covers fundamental and advanced subjects in technologies, power electronics, and control engineering for PV power systems. Throughout, the description of PV power systems follows a clear framework for each section.

The book is divided into ten chapters. The interrelationship of the chapters is illustrated in Figure 1. The step-by-step introduction of the individual system components and controls for PV power systems is covered in Chapters 4–8. With the support of the system classification and the safety guidelines, which are discussed in Chapters 2 and 3, respectively, the overall system integrations for standalone systems and grid-tied systems are set out in Chapters 9 and 10.

Chapter 1 provides a brief introduction to solar power systems. This includes the clarification of vocabulary which proves integral to the remainder of the book.

Chapters 2 and 3 provide comprehensive classifications of PV power system configurations, in particular grid-tied systems, approached according to the level at which the MPPT is applied, MPPT techniques, power-conditioning topologies, and technologies for battery balancing. The reader is assisted, using clear definitions, to develop an understanding of the latest systems and directions of research and development, which later informs research directions for PV power systems. Reader understanding of relevant safety standards, guidance, and regulations is developed to prevent researchers deviating from standard practice in industry. A system of reference is provided for safe practice in engineering and design. Though the codes and guidelines cited are implemented in the USA and Europe, they are universally applicable and allow all readers to practice PV power engineering in a safe manner. These chapters also cover the certification of PV modules, the safety standards of power interfaces, the system requirements for grid interconnection, and the important means of protection. The main conversion units are the PV-side converters, battery-side converters, and grid-side converters. The

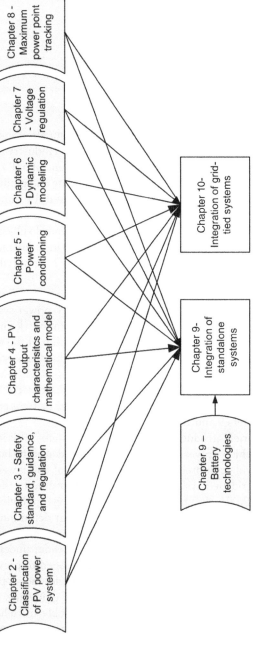

Figure 1 Organization and interconnection of Chapters 2 to 10.

interconnection of the conversion units are the PV link, DC link, and grid link. The PV generation section is divided into the PV source circuits and PV output circuits. Two types of modeling are demonstrated for the reader. Firstly, simulation models that represent a practical system and prove the design concept are discussed in Chapters 4 and 5. Secondly, mathematical models are developed to illustrate the system dynamics used for controller synthesis. Model development and verification are covered in Chapter 6.

Chapter 4 discusses PV output characteristics and mathematical models for simulation and analysis. It builds upon an understanding of PV product datasheets and provides a straightforward approach to building a mathematical model for simulation purposes. When model accuracy becomes the priority, an advanced approach is also provided. The tradeoff between model simplicity and accuracy is extensively considered, discussed, and demonstrated by practical examples.

Chapter 5 provides the information necessary to specify, design, simulate, and evaluate power-conditioning circuits and accessories for PV power applications. The main conversion units are PV-side converters, battery-side converters, and grid-side converters. The important interconnections are the PV link, DC link, and grid link. The description provided comprehensively covers application to all PV power systems. The chapter covers system design, steady-state analysis, and simulation verification. Current modulation for AC grid interconnection is introduced and simulated for design verification. This book places emphasis on computer-aided analysis, design, and evaluation. For universality, all included simulation models are built using the fundamental blocks of Simulink. The analysis reveals the fundamental system dynamics for the purpose of both time-domain simulation and control synthesis. Although provided with software such as Simscape Power Systems, written for power electronics and power systems, this is not used due to the aim to demonstrate the simulation principle and focus on fundamental implementations instead. The control system design, analysis and evaluations are based on the functions of Matlab and Matlab version R2010b. Simulink is used to demonstrate all simulation cases; the same or higher versions of this software can be used to duplicate simulation results or develop results further. Most chapters present practical examples in order to demonstrate designs and verify them, which are presented in case studies. Photographs, diagrams, flowcharts, graphs, equations, and tables are included to provide clear explanations of technical subject matter. Readers can then duplicate results through computer-aided design and analysis, leading to the development and evaluation of new systems.

Chapter 6 focuses on dynamic modeling of PV power systems. The mathematical modeling starts with the state-space averaging, followed by linearization. Dynamic models are developed for the voltage signals at the PV and DC links and the variety of converter topologies is also considered. The dynamics of the developed mathematical models are verified through simulation and comparison. One section is given over to modeling the dynamics of dual active bridges when used as the battery power interface.

Chapter 7 concerns the application of linear control theory. The chapter discusses evaluations of relative stability and system robustness. Voltage regulation for the PV and DC links is introduced and analyzed in depth. Examples and simulations are used to demonstrate the effectiveness of proposed approaches and methodology. Based on the system model, advanced control techniques, such as Affine parameterization, anti-windup, and feedforward implementation, are introduced. The implementation of sensing and digital control is briefly discussed at the end of the chapter.

Chapter 8 focuses on MPPT, which is important and unique to PV power systems. A comprehensive overview is provided, and MPPT algorithms are classified and discussed. The chapter introduces a simple algorithm and develops to consider advanced techniques to improve tracking performance. The simulation and implementation of MPPT techniques are also discussed in this chapter.

Chapter 9 discusses the integration of standalone PV power systems. The chapter enables readers to understand the latest progress and choose suitable battery types for standalone applications through the introduction and comparison of battery characteristics. This proceeds onto a discussion of battery characteristics and models and how they relate to system design and analysis. A new classification is proposed to avoid confusion seen in the earlier literature and provide a clear framework for understanding the methodologies used for battery balancing. A method to integrate the MPPT function with the battery cycle charge is proposed. Examples are given to demonstrate the effectiveness of modeling, design, control, and simulation. Simulation models for the controller and power interface are developed at different levels: short term, medium term, long term, and very long term. A simulation of eight-hour system operation is created, demonstrating the state of MPPT, battery voltage regulation, and variation of state of charge corresponding to solar irradiance and cell temperature.

The final chapter, Chapter 10, addresses the integration of grid-tied PV power systems, including two small-scale single-phase interconnections and one utility-level three-phase system. Examples are given demonstrating the effectiveness of the design, integration, control, and simulation with additional consideration for safety protection. The simulation study is divided into two parts: a short-term simulation that aims to capture the fast transient response of switching dynamics and grid disturbances, and a long-term simulation that illustrates the system operation in response to environmental conditions.

Technical Support

One advantage of this book is that all modeling and simulation for the case studies is based on the basic functions of Simulink and Matlab. The modeling and simulation approach is based on system dynamics, which helps readers to understand the fundamental principle behind various simulation tools. The construction of output models, power interfaces and control, and standalone and grid-tied systems are illustrated in detail. Version R2010b or higher of Matlab and Simulink can be used to duplicate the results or to develop new studies. Other software tools are unnecessary.

University of Sydney   *Weidong Xiao*
Australia

Acknowledgments

I would like to acknowledge the contributions of my former students in the Masdar Institute of Science and Technology (MIST). Dr Yousef Mahmoud initialized the study of PV modeling and simulation while he was a graduate student at MIST. Mr. Po-Hsu Huang continued this development and presented an effective method to solve the nonlinear equations involved and balance model complexity and accuracy. Mr. Edwin Fonkwe raised the idea of distributed PV generation at PV module level. Mr. Omair Khan continued studying fine granularity MPPT and state-of-the-art technologies of gallium nitride power devices for power conditioning. He also proposed the start–stop mechanism, which has been proven to be effective for the hill-climbing-based MPPT.

I received tremendous support from Dr. William G. Dunford, Dr. Antoine Capel, and Dr. Luis Martinez Salamero, during a period of graduate study at the University of British Columbia. In 2010, as a visiting scholar, I spent eight months at the Massachusetts Institute of Technology working with Prof. David Perreault and Prof. James Kirtley. I received great support in the research area of power interfaces for PV power systems. Over the past four years, my former colleague at MIST, Dr Mohamed Elmousi helped me to understand the grid concept when a PV power plant, significant in size, is connected to the network. Working with Dr Yang Du at MIST, I gained significant knowledge regarding the analysis of DC link voltage ripples in single-phase AC systems. I have also been working with my former classmates, Dr Fei Richard Yu and Dr Peng Zhang through collaborative research and joint publications.

Special thanks go to Alpha Technologies Ltd and MSR Innovations Corp, which are based in the beautiful British Columbia, Canada. Both companies provided generous support for my Masters and Doctoral degrees. Working with them, I gained invaluable industrial experience that has helped me to learn and understand the fundamentals of power electronics, and to research and develop practical PV power systems. A former colleague, Mr Tim Roddick, supported me by providing photos and initializing the discussion about building integrated PV.

I would like to thank Ms Ella Mitchell and Ms Nithya Sechin, the editors at John Wiley & Sons, for their professional support for this project through all its phases. Last but not least, I heartily thank all of my family members; I was so focused on writing, that I have spent little time with them over the past 13 months. Without their patience and understanding, this book would have been impossible.

About the companion website

Don't forget to visit the companion website for this book:

www.wiley.com/go/xiao/pvpower

This contains a wealth of valuable material to enhance your learning, including:

- simulation files
- presentation slides.

Scan this QR code to visit the companion website:

1

Introduction

The photovoltaic (PV) effect is the generation of DC electricity from light. Alexandre Edmond Becquerel, a French experimental physicist, discovered the effect in 1839. More recently, scientists have discovered that certain materials, such as silicon, can produce a strong PV effect. In the 1950s, Bell Labs of the USA produced PV cells for space activities. This can be considered as the beginning of the PV power industry. The high cost of PV materials mostly prevented applications elsewhere.

Over the past 20 years, the PV power industry has experienced significant growth. PV power generation has become more and more common. The capacity of PV systems ranges from milliwatts for portable devices such as calculators, to gigawatts for power plants connected to the electricity grid. A grid-connected PV power system can be economically installed, and can be rated as low as just a few hundred watts. The advantages of PV power systems that have led to their rapid growth are:

- green, renewable
- reliability and long lifetime
- advanced manufacturing process
- static, so noise-free operations
- improving efficiency
- decreasing prices
- flexibility of construction
- highly modular nature
- availability of government support and incentives.

Using the latest technologies, the manufacturing of crystalline-based PV cells consumes significant amounts of energy, which prevents further cost reductions. The levelized cost of electricity generated using solar PVs is still high in comparison with conventional generation resources, such as coal, natural gas, and wind, according to a technical report published by the US Department of Energy's National Renewable Energy Laboratory (Stark et al. 2015). The report was based on a study of the USA, Germany, and China. Several large-scale PV power systems were announced in 2016 and projected significantly lower costs, but these must be treated as special cases. The project feasibility and system reliability need to be carefully evaluated until the projects are successfully delivered. It is clear that the price of PV products mostly reflects their quality and reliability. High-quality, certified PV products are usually more expensive than non-certified ones. It is unrealistic to judge a PV power system only on the installation cost since reliable and long-lifetime operations are always expected.

Photovoltaic Power System: Modeling, Design, and Control, First Edition. Weidong Xiao.
© 2017 John Wiley & Sons Ltd. Published 2017 by John Wiley & Sons Ltd.
Companion Website: www.wiley.com/go/xiao/pvpower

Table 1.1 Price schedule of feed-in tariffs in Ontario, Canada.

Type	System capacity (kW)	Price ($/kWh)*
Rooftop	≤10	0.294
	10–100	0.242
	100–500	0.225
Non-rooftop	≤10	0.214
	10–500	0.209

*Canadian dollars.

The feed-in tariff (FIT) is the major driver of the boom in PV power all over the world. The regulatory incentives are different from country to country, but all are designed to accelerate investment in PV-related technologies. One FIT example can be found on the website of the Ontario Power Authority, Canada. Parts of the FIT price schedule are shown in Table 1.1, which covers projects under 500 kW in capacity. It shows that the government contributes significant funds for PV system installations since the listed price is higher than the charge for residential consumption. It should be noted that the listed price is based on the 2016 schedule. Like FITs in most other countries, it is always subject to change. Another disadvantage of PV power systems lies in their low power density, which limits their use mainly to static applications rather than vehicles. Motor vehicles are usually considered as one of the major contributors to air pollution.

1.1 Cell, Module, Panel, String, Subarray, and Array

A PV cell, also commonly called a solar cell, is the fundamental component of a PV power system. A crystalline-based solar cell features a p-n junction, as shown in Figure 1.1. The manufacturing process includes melting, doping, metallization and texturing. The positive and negative sides of the junction form the DC voltage and supply electricity when a load is connected. However, the voltage of a single p-n junction cell is less than 1 V, which is low for most practical applications. Moreover, it is mechanically fragile, and must be laminated and protected for practical use.

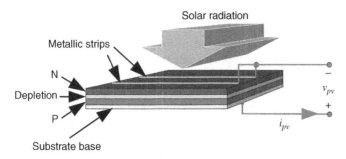

Figure 1.1 Typical crystalline PV cell construction.

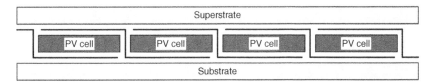

Figure 1.2 Lamination of PV module.

To end users, the basic unit is the PV module or solar panel, which can produce higher voltages and more power than a single cell. A PV module consists of cells that are interconnected and laminated together. The old PV panel was usually designed to match the nominal voltage of batteries, since standalone systems were the beginning of the PV industry. For example, traditional 36-cell PV modules used to be popular for direct charging of batteries with a nominal voltage of 12 V. Nowadays, with the increasing numbers of grid-connected systems and the advances in power-conditioning devices, the number of cells in each PV module is no longer limited to matching the nominal voltages of batteries or loads. The manufacturers are more concerned with cost-effective solutions and supply all different sizes of solar panels: usually incorporating 48, 54, 60, or 72 cells. Solar cables and connectors are usually integrated with the module for straightforward interconnection and installation.

To form a PV panel, crystalline-based PV cells are sandwiched by the superstrate and substrate for protection, as illustrated in Figure 1.2. Tempered glass is commonly used as the superstrate, supporting the module lamination and protecting the fragile cells. Glass also has the same ratio of thermal expansion as a crystalline PV cell, since both are made of silicon. Furthermore, tempered glass is strong and has good transparency, with about 94% light transmission. The glass surface is also textured to reduce light reflections. Metal conductors connect the PV cells from the surface to the bottom for series interconnection. The cells are also protected by an encapsulant, which is a material that surrounds the PV cells between the superstrate and substrate.

Figure 1.3 shows a standard PV module and its internal electrical configuration. It consists of 72 cells in series connection. The cells are divided into three groups, which are termed the submodules. Each submodule includes one bypass diode in parallel connection with 24 interconnected solar cells. The bypass diodes are standard components that are integrated in the crystalline-based PV module. The implementation prevents the destructive effects of hot spots, should there be unbalanced generation among the series-connected cells. The overall electrical connection is configured inside the junction box, which is commonly located at the back of the PV panel. The output cable always indicates the polarity of the positive and negative terminals.

Attention should be given to AC PV modules, often simply called AC modules. Just the same as standard PV modules, an AC module is an environmentally protected unit consisting of interconnected solar cells, junction box, superstrate, substrate, electrical interconnections, and other lamination components. However, the module includes an inverter inside the junction box to produce AC power at the output terminal. The concept of an AC module is the same as the microinverter solution, which converts DC to AC at the PV module level. However, the difference is that the microinverter is an independent unit that is electrically connected to the PV module instead of being fully integrated. Microinverters belong to the class of module-integrated parallel inverters (MIPIs), which will be discussed in Section 2.4.1.

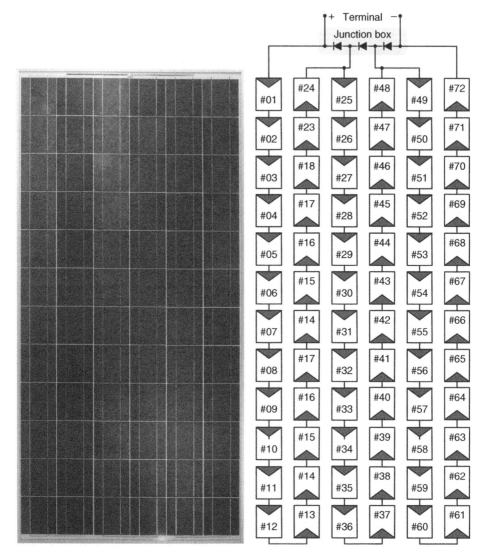

Figure 1.3 72-cell PV module: left, appearance; right, configuration.

It is very important to use the correct terms to describe PV generators: cell, module, panel, string, subarray, and array. Figure 1.4 illustrates how the power capacity is built up from cell level to array level. PV power systems are commonly assembled by configuration of PV modules in series and/or in parallel. The series connection of solar modules in order to stack up the output voltage is commonly referred to as a "string." The parallel connections of PV strings forms an array, in which the power capacity can be built up to the levels of hundreds, thousands, or even millions of watts. In large-scale PV power systems, an array is divided into multiple subarrays.

A PV array can be monopolar or bipolar. A monopolar array or subarray is a typical DC circuit that has two conductors in the output circuit, with positive (+)

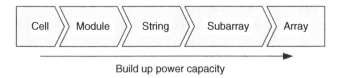

Figure 1.4 PV power capacity built from cell to array.

Figure 1.5 Bipolar PV array formed from two monopolar subarrays.

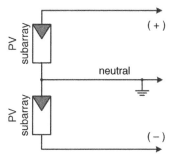

and negative (−) polarities. A bipolar PV array includes a neutral point, as shown in Figure 1.5, and is formed from two monopolar subarrays. Ideally, the two monopolar subarrays should be identical in power and voltage levels. The neutral point is grounded at a central point in the interconnected system. A company called AE Solar Energy used to be the major producer of utility-interaction inverters, which were designed for the bipolar array configuration and large-scale PV power systems. The inverter accommodates the output of the bipolar PV array and is rated at up to ±600 V.

1.2 Blocking Diode

PV components are direct current sources, so the reverse flow of current into the PV source circuit should be prevented. Blocking diodes can be used, installed in series with the PV output string in order to block reverse currents. They are often referred to as "string diodes." To distinguish them from the bypass diode, Figure 1.6 illustrates a typical PV source circuit with the integration of both bypass and blocking diodes, denoted D_{bp} and D_{bl}, respectively. The bypass diodes are standard components that are commonly integrated inside the junction boxes of PV modules, as shown in Figure 1.3. The blocking diodes are parts of the overall PV source circuit, as shown in Figure 1.6, and are optionally implemented when required.

Blocking diodes have been widely used for direct battery-charging applications due to their advantages of effectiveness, safety, reliability, and because they are maintenance-free. However, their disadvantage is a forward voltage drop that results in significant power losses in the PV source circuit. For example, the forward voltage of a typical 600-V/12-A rated diode is about 1 V. Considering that the PV string current is 7 A, each blocking diode introduces about 7 W of conduction loss, which generates heat and creates a hot spot. Furthermore, the failure of a blocking diode will cause a complete loss of the protection function and might lead to the failure of the entire string. The latest PV systems are developed for high efficiency and tend to avoid use of blocking diodes. Since all PV modules show a certain level of tolerance of reverse

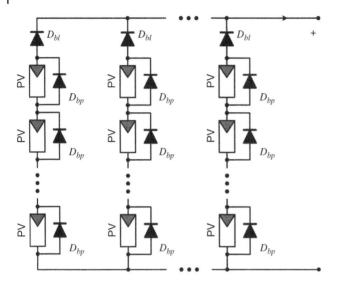

Figure 1.6 PV circuit with bypass diodes and blocking diodes.

current, manufacturers always provide the upper limit of reverse current that causes no damage to the PV product. Fuses and DC circuit breakers have been recently used in series connection with the individual PV string in order to protect the circuit and stop reverse current up to the maximum limit. The power losses are significantly lower than those caused by blocking diodes.

1.3 Photovoltaic Cell Materials and Efficiency

The PV effect can happen in many materials that absorb light and turn a portion of the energy into electricity. Solar cells are made of materials that are designed and formulated to produce strong PV effect. This can be measured by the conversion efficiency of the irradiance to electrical power. If a solar cell is claimed as being 15% efficient, it indicates that the electric power output of a 1 m^2 cell receiving 1000 W/m^2 irradiance at 25°C would be 150 W. Common PV cells are made of mono-crystalline silicon, multi-crystalline silicon, thin films, organic materials, and so on. It should be noted that mono-crystalline and multi-crystalline silicon are also referred to as single-crystalline and poly-crystalline silicon. The crystal growth process during manufacture is behind the formation of the two different types of crystalline-based solar cells. The Czochralski and Siemens processes are commonly used for making PV materials. The process followed includes doping, metallization, and texturing in order to construct solar cells.

The counterparts of crystalline silicon cells are thin-film cells. The common ones are summarized in Table 1.2. One of the most successful companies in the thin-film PV industry is First Solar, which uses cadmium telluride (CdTe) technologies. Even though the efficiency of CdTe-based products is generally lower than that of crystalline silicon cells, the technology has significantly lower material and manufacturing costs.

Organic solar cells are made of thin layers of organic materials. These technologies are under development and are rarely applicable for high power systems.

Table 1.2 Common PV materials.

Composition	Acronym
Mono-crystalline silicon or single-crystalline silicon	Mono-c-Si
Multi-crystalline silicon or poly-crystalline silicon	Poly-c-Si
Cadmium telluride	CdTe
Copper indium gallium selenide	CIGS
Amorphous silicon	a-Si

The efficiency of single junction cells is usually lower than 20% due to physical limits and technical constraints. Multi-junction cells have been invented in order to increase conversion efficiency. These are made up of multiple p-n junctions, which allows absorption of multiple light wavelengths through multiple layers. Efficiencies of over 30% have been reported. Their high price limits their application to aerospace or concentrated PV (CPV) systems, where high power density is particularly desirable. CPV is a technology that focuses sunlight using lenses or mirrors. The implementation minimizes the usage of PV material, which was significantly more expensive 20 years ago. The solar concentration ratio is commonly measured as the number of "suns," where the one-sun condition represents non-concentrated light. With the decreasing cost of PV materials, CPV is no longer as attractive as previously. Due to the specialist nature of multi-junction cells and CPV, the technology will not be further discussed in this book.

1.4 Test Conditions

Photovoltaics are inherently an intermittent energy resource since the electricity production depends on the instantaneous environmental conditions. The output power not only stops at night, but also varies significantly through the day and the season. As an example, Figure 1.7 shows a PV panel's power output, as measured in Vancouver, Canada. Broken clouds caused dramatic variations in the PV power output over the first 2.5 h. The output became significantly low during the last 40 min due to cloud coverage. Therefore, the intermittent nature of PV power production should be always considered when planning either standalone or grid-connected PV power systems.

The irradiance is the density of radiation incident on a given surface. It is usually expressed in units of watts per square meter (W/m^2). The PV cell temperature also plays an important role in determining the output. Defined in IEC 60904, the standard test conditions (STC) correspond to a solar irradiance of $1000\,W/m^2$, a device temperature of $25°C$, with a reference solar spectral irradiance of air mass 1.5 (AM1.5). The standard is commonly applied to evaluate power capacities and conversion efficiencies of PV cells or modules. The rating of PV power systems is usually based on the accumulation of the PV module capacity at STC. The International Electrotechnical Commission (IEC) is the international standards and conformity assessment body for all fields of electrotechnology. The standards relevant to PV products will be discussed in Chapter 3.

According to IEC 61215, PV performance can also be measured at the nominal operating cell temperature (NOCT), which is defined as the equilibrium mean of solar cell

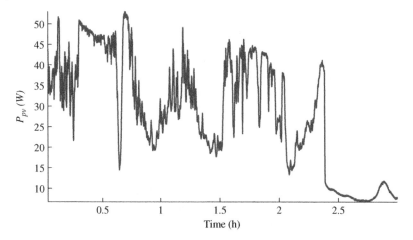

Figure 1.7 PV module 3-h output, as measured on 16 June 2006.

Table 1.3 SRE for measurement of NOCT.

Term	Value
Tilt angle	45° from the horizontal
Total irradiance	800 W/m²
Ambient temperature	45°C
Wind speed	1 ms⁻¹

junction temperature within an open-rack mounted module in the standard reference environment (SRE), as shown in Table 1.3. Measured at STC and/or NOCT, the values of open-circuit voltage, short-circuit current, and power output of PV cells or modules define the product specifications and performance indices.

It is sometimes confusing to distinguish the terms of solar irradiance, insolation, and radiation, since all are used to describe the sunlight strength. Solar radiation is a general term that refers to the electromagnetic nature of sunlight, which is the radiant energy emitted from the sun. The total radiation on a surface includes the direct radiation from the sun, radiation diffused by the atmosphere, and radiation reflected by other objects. Insolation represents the quantity of solar radiation energy received on a surface of a certain size during a certain amount of time. The units can be kWh/m² or Wh/m². The strength of radiation is commonly measured by the level of irradiance, of which the unit is kW/m² or W/m². The term "irradiance" is the instant measure of light density, and will be used in the rest of this book.

1.5 PV Module Test

PV products are usually tested indoors using simulated resources since the outdoor environment is generally hard to control. A fully controlled environment can provide the

Figure 1.8 Laboratory for PV module testing.

standard test conditions, variable irradiance levels, and regulated ambient temperatures. Calibration is also easier indoors than outdoors. A laboratory system for PV module testing is shown in Figure 1.8. It has a dark chamber, a solar simulator, a computer, and a measurement system. The system is located in the Masdar Institute of Science and Technology, Abu Dhabi, UAE.

The dark chamber is designed to mount the light box and PV module for testing. The inside temperature of the dark chamber can be set at a desired level. The solar simulator is a controllable light source that mimics sunlight, with the same or a very similar spectrum, and can be regulated to give different irradiance levels. The laboratory setup can be calibrated for STC with the reference of AM1.5. The light is usually pulsed over a short period to avoid the significant temperature rises that are commonly caused by long-term light exposures. The measurement system includes an electronic load that can be controlled to trace the output (open circuit or short circuit). A high-speed data acquisition system is also included to record data from the PV module output.

1.6 PV Output Characteristics

The output characteristics of PV cells or modules are commonly represented by the current–voltage (I–V) and power–voltage (P–V) curves. In some special cases, the voltage–current (V–I) and power–current (P–I) curves are also used to represent the PV output characteristics. Generally, they are transferable from one to another. Figure 1.9 shows typical I–V and P–V curves for a PV cell output. The normalized curves can also be used to represent the outputs of PV modules, strings, and arrays when all the solar cells are tested under uniform conditions. The curves show the three important points and four important values, as described in Table 1.4. The data are usually presented for STC, which is considered as the nominal rating.

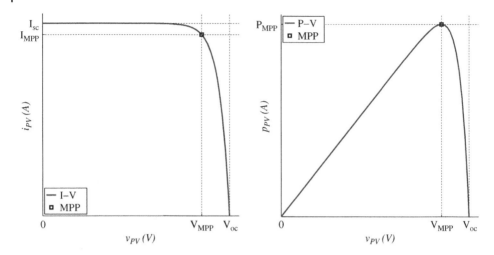

Figure 1.9 PV output characteristics: left, normalized I–V curve; right, normalized P–V curve.

Table 1.4 Four important values representing PV output characteristics.

Symbol	Description
V_{OC}	The open-circuit voltage, measured when the PV output terminal is open-circuit showing zero current.
I_{SC}	The short-circuit current, measured when the PV generator terminal is short-circuited.
I_{MPP}	The current measured at the MPP.
V_{MPP}	The voltage at the MPP.

The P_{MPP} is the highest power level for a certain environmental condition, and is calculated as $P_{MPP} = V_{MPP} \times I_{MPP}$.

The P–V curve clearly shows the maximum power point (MPP), which represents the highest power output (P_{MPP}) that the PV generator can produce under certain environmental conditions. The MPP is located in the "knee" area of the I–V curve, and is represented by the current (I_{MPP}) and voltage (V_{MPP}), as shown in Figure 1.9. The open-circuit voltage, V_{OC} is the highest voltage level of the PV generator under a given test condition. The short-circuit current, I_{SC}, is the highest current level of the PV generator under the test condition. The power output is zero at either open-circuit or short-circuit conditions.

It is usually safe to connect PV generator terminals in short circuit since the output current is always limited by the short-circuit level, which depends on the instantaneous environmental conditions, particularly the irradiance. Short circuits can be used for safety protection when any electrical shock happens. It should be noted that the values of V_{OC}, I_{SC}, I_{MPP}, and V_{MPP} vary with environmental conditions. As a result, maximum

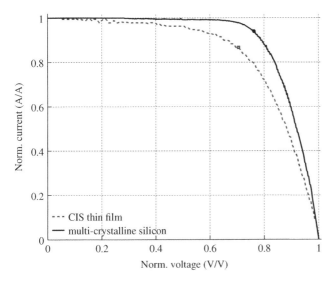

Figure 1.10 Normalized I–V curve to represent the PV generator outputs and the difference in fill factor. CIS, copper indium diselenide.

power point tracking (MPPT) is required to locate the instantaneous MPP depending on the solar irradiance, cell temperature, or other uncertainties.

The shape of the I–V and P–V curves also depends on the cell technology and manufacturing process used. Figure 1.10 shows the measured I–V curves from two different PV modules: models ST10 and BP350, made of copper indium diselenide and multi-crystalline materials, respectively. The I–V curve of the ST10 model looks gentler than that of the BP350. The ratios of MPP voltages are 71.59% and 76.24% of the open-circuit voltage for the ST10 and BP350 respectively. The MPP currents are 86.63% and 93.96% of the short-circuit currents, for the ST10 and BP350, respectively.

The fill factor (FF) is a term that is used to describe the shape of the PV output. Its value is calculated as:

$$FF = \frac{I_{MPP} \times V_{MPP}}{I_{SC} \times V_{OC}} \tag{1.1}$$

The FF has been used as an indicator in PV material research since the ideal PV cell has a rectangular shaped I–V curve, with FF = 1. PV material research has tried to push achievable FF values higher, but it is not a significant measure for practical PV power applications since the performance of PV cells is evaluated on many other measures too, such as cost effectiveness and reliability. The FF values of crystalline-based PV cells are generally higher than those of thin-film devices.

The FF should be considered when optimizing the MPPT parameters since it corresponds to the difference in the PV output curves. The values of FF for the ST10 and BP350 modules are 0.62 and 0.66, respectively. It should be noted that the value of FF depends on the testing conditions: irradiance and temperature and so on. For a fair comparison among various PV materials, the FF values should be evaluated at STC.

1.7 PV Array Simulator

Outdoor evaluation of PV systems enables behavior of real PV arrays to be examined in natural sunlight. However, the outdoor environment is commonly considered difficult because the solar irradiance and ambient temperature are not controllable (Xiao et al. 2013). To perform a fair comparison of PV systems, simulators are commonly used.

Researchers tend to use controllable light and power sources to simulate the sunlight and PV generator outputs, respectively. A PV array simulator is a DC power supply, the output of which mimics PV output characteristics. It should not be confused with a solar or sun simulator, which is the artificial light source that was introduced in Section 1.5. It is common to use solar simulators to test PV outputs at the cell and module levels, but they are impractical at string and array levels.

The PV array simulator can be used for indoor testing of power interfaces developed for PV applications. The output is programmable, to give specific values of the open-circuit voltage, short-circuit current, the MPP, and the corresponding I–V curves. The level of solar irradiance and the cell temperature can also be predefined and programmed to simulate environmental variations.

PV array simulators have been developed in the kilowatt power range for simulating PV strings and arrays. Examples include the products manufactured by Chroma ATE Inc. Others are only at the hundreds-of-watts level, and are used to simulate the output of PV modules. One popular set of models is the E4350 and E4360 series produced by Agilent Technologies. An E4350B model is shown in Figure 1.11. The output ratings are given as:

- maximum output power: 480 W
- maximum output voltage to simulate the open-circuit voltage: 65 V
- maximum output current to simulate the short-circuit current: 8 A
- peak-to-peak voltage ripple: 125 mV.

One key feature of a PV array simulator is the accuracy with which it simulates the I–V curve and represents the open-circuit voltage, short-circuit current, and the MPP. Another important measure, often neglected by users, is the speed of dynamic

Figure 1.11 Agilent 4350B PV array simulator.

response: the time in transition from one steady state to another. A real PV module or array is formed by semiconductors and shows a significantly high dynamic bandwidth. This can be explained by noting that the I–V output curve responds immediately whenever the load condition changes. However, it is impossible for a switching-mode power supply to mimic the same response. They are constrained in their response time or dynamic bandwidth.

It has been reported that a PV array simulator is too slow to test high-speed MPPT performance. For example, according to the product manual, the settling time of the E4350B is 25 ms. It is impossible to test an MPPT algorithm with a tracking speed of more than 40 Hz. However, a lot of MPPT algorithms and corresponding power-conditioning circuits have been developed for significantly higher speeds.

Another drawback results from the well-known disadvantages of switching-mode power supplies: self-resonance, output waveform ripples, and noise. Conventional DC power supplies show the tradeoff between the filter size and dynamic response. The filter can be sized with significant inductance and capacitance to mitigate ripples. However, the approach lowers the speed of the dynamic response. As a result, for the latest technologies, it is recommended that a PV array simulator be used for proof of concept. However, for accurate comparison of PV system performance, including both dynamics and steady-state performance, the suitability of PV array simulators should be carefully considered. It should be kept in mind that the PV array simulator is an imperfect electronic device and that it can interact with electronic power converters during tests.

1.8 Power Interfaces

In some cases, the PV generator is directly coupled to the load without any power interface. The majority of PV power systems are equipped with power-conditioning circuits for the interfaces between the generators and loads. A converter is the equipment that changes electrical voltage levels or waveforms.

Switching-mode power converters have high conversion efficiency and compact size, and are commonly used for PV power interfaces, battery power interfaces, and grid power interfaces. In power electronics, hundreds of converter topologies have been developed. Table 1.5 lists the DC/DC topologies that are commonly used for power interfaces on the PV side and battery side.

Table 1.6 lists the converter topologies that are commonly used for DC-to-AC conversion. They are also known as inverters, and change not only the voltage level but also the waveform. The input is from the DC source – PV generators or batteries – and the output can be connected to the AC grid or an AC load. The analysis, design, simulation, dynamic modeling, and control of the topologies are explained in Chapters 5–7.

1.9 Standalone Systems

Standalone PV systems supply power to local loads, and so are independent of the grid distribution network. Their history can be traced back to the 1950s, when PV technology was widely used for space power supplies. Solar radiation is more intense

Table 1.5 Converters used for DC/DC power interfaces.

Topology	Isolation	Description
Boost converter	No	Ratio of DC output voltage to DC input voltage not less than 1.
Buck converter	No	Ratio of DC output voltage to DC input voltage not larger than 1.
Full-bridge isolated buck converter	Yes	When the winding turn ratio of the transformer is reset to 1:1, it produces the DC output voltage lower in magnitude than the DC input voltage.
Buck–boost converter	No	The voltage conversion ratio is flexible: higher, lower, or equal to 1. However, the input and output port do not share a common ground point.
Flyback converter	Yes	When the winding turn ratio of the transformer is reset to 1:1, the analysis can be based on the principle of the buck–boost converter.
Tapped inductor converter	No	The operation follows the principle used for the boost topology except for the tapping connection of the inductor. Potential for high step-up conversion ratio of voltage.
Dual active bridge	Yes	A bidirectional DC/DC converter that can be used as the battery power interface to support DC grids or DC links.

Table 1.6 Converters for DC/AC power interfaces.

Topoloty	Description
H-bridge inverter	Conversion from DC to single phase AC.
Voltage source inverter (VSI)	Conversion from DC to three-phase AC, controlled by AC voltage regulation.
Current source inverter (CSI)	Based on the same circuit as VSI, and used for DC to three-phase AC conversion, but controlled by AC current regulation.

in space because it is not attenuated by the atmosphere and not blocked by clouds. The outer atmosphere is considered an ideal environment for solar PV power generation. However, their recent application has mostly been on Earth. The major applications of standalone systems include satellites, spacecraft, space stations, remote homes, villages, street lights, communication sites, water pumps, and vehicles. Nowadays, these systems are mainly installed in areas where grid connections are unavailable. Grid-connected PV systems have clear advantages for massive solar power production since the utility grid is a significant energy buffer that can accommodate the intermittency of solar power generation.

Standalone PV systems can supply either DC, AC, or both, depending on the load requirement. In some applications, PV generators can supply loads directly or through power interfaces without significant energy storage. Power-conditioning equipment might be needed for voltage conversion in directly coupled systems. The majority of

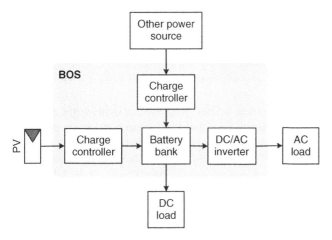

Figure 1.12 Typical standalone system configuration, with PV generator and energy storage. BOS, balance of system.

applications require energy storage, because of the intermittent nature of PV power generation.

Hybrid systems take power from wind turbines, fuel cells or conventional engine-based generators, as well as solar power, as shown in Figure 1.12. Without other resources, a PV-battery system is not a hybrid system, because the energy storage system is an energy buffer, but not another electrical production source. The balance of system (BOS) consists of all equipment between the power sources and the loads. Charge controllers are commonly required to charge the battery bank. Filters, means of disconnection, and protection devices are also important components of the BOS, but they are not illustrated in Figure 1.12.

The DC/AC voltage source inverter (VSI) produces AC waveforms from DC sources. The majority of AC loads are supplied with a sine wave AC source. Some take square waves or modified square waves, allowing simple and low-cost topologies to be used in the VSI. The waveforms are illustrated in Figure 1.13, based on a frequency of 50 Hz. The

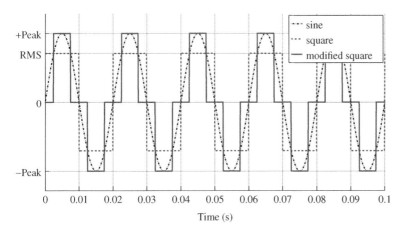

Figure 1.13 Typical waveforms output by voltage source inverter.

total harmonic distortion of the square waveform is 48.3%. It should be noted that the pure sine-wave output requirement for standalone PV systems is no longer as demanding as before, because most modern devices tend to use DC supplies and are more tolerant of differing power supplies. However, in certain environments, noise and electromagnetic interference might be concerns where the AC output does not have a pure sinusoidal waveform.

The most impressive PV standalone system is Solar Impulse 2, a lightweight airplane solely powered by solar energy. Starting from Abu Dhabi on 9 March 2015, it completed the first entirely solar-powered flight around the world. The distance of 40 000 km is considered as the longest solo solar flight ever achieved. The historic voyage took more than 27 months, with the plane touching down again in Abu Dhabi on 26 July 2016. The flight record is summarized in Table 1.7.

A Swiss team initialized the ambitious idea of developing an aircraft purely powered by solar energy with battery support. The mission aimed to promote clean technologies. The plane was powered by 17 000 solar cells built into the body and wings. The longest and the most difficult non-stop flight took 5 days (118 h) and 7212 km from Nagoya, Japan to Honolulu, USA. To date, this is a world record for an uninterrupted flight. It was achieved by the pilot, Andre Borschberg. A photo taken in March 2015 when the Solar Impulse 2 was stationed in Abu Dhabi, is shown in Figure 1.14.

From certain points of view, the aircraft should be classified as a motorized glider. The wingspan is 73 m, which is bigger than that of a Boeing 747 jumbo jet. However,

Table 1.7 Record of Solar Impulse journey.

Leg	From	To	Distance (km)	Start date	Total (h)	Average speed (km/h)
1	Abu Dhabi, UAE	Muscat, Oman	772	3/09/15	13	34
2	Muscat, Oman	Ahmedabad,India	1468	3/10/15	15	97
3	Ahmedabad,India	Varanasi,India	1170	3/18/15	13	89
4	Varanasi,India	Mandalay, Myanmar	1536	3/19/15	13	104
5	Mandalay, Myanmar	Chongqing, China	1450	3/30/15	20	71
6	Chongqing, China	Nanjing, China	1241	4/21/15	17	71
7	Nanjing, China	Nagoya, Japan	2852	5/30/15	44	65
8	Nagoya, Japan	Hawaii, USA	7212	6/28/15	118	61
9	Hawaii, USA	San Francisco, USA	4086	4/21/16	62	65
10	San Francisco, USA	Phoenix, USA	1113	5/02/16	16	70
11	Phoenix, USA	Tulsa, USA	1570	5/12/16	18	86
12	Tulsa, USA	Dayton, USA	1113	5/21/16	16.6	67
13	Dayton, USA	Lehigh valley, USA	1044	5/25/16	16.8	62
14	Lehigh valley, USA	New York, USA	265	6/11/16	4.7	57
15	New York, USA	Seville, Spain	6765	6/20/16	71	95
16	Seville, Spain	Cairo, Egypt	3745	7/11/16	48.8	77
17	Cairo, Egypt	Abu Dhabi, UAE	2694	7/24/16	48.6	55

Source: www.solarimpulse.com.

Figure 1.14 Solar Impulse 2.

the plane can host only one pilot and weights approximately the same as a family car. From Table 1.7, the flight speed varies from 34 km/h to 104 km/h, and the average speed is about 70 km/h, slower than the majority of vehicles traveling on express highways. It usually takes less than 5 h to drive a car from Abu Dhabi to Muscat and less than one hour for commercial flights. However, Table 1.7 shows that Solar Impulse 2 took 13 h. Furthermore, the solar-powered aircraft is very sensitive and vulnerable to weather conditions because of its lightness: it weighs only 2300 kg. The onboard battery bank was seriously damaged during the 7212-km flight from Nagoya to Hawaii. The damage caused significant delays.

There is no doubt that Solar Impulse inspired the world towards the use of clean technologies and renewable energy. People around the world admire the courage and the effort of the founders and pilots in promoting solar energy for aviation. However, the technical data show that Solar Impulse aircraft might be a bad application of PV technology. The current power density of PV materials is generally too low to serve as the sole power source for either ground vehicles or aircraft. Due to size constraints, transport vehicles generally demand high power density by weight and volume, which rules out current PV technologies. However, this cannot stop the PV power generation being used as a secondary power source in hybrid systems for transportation. For example, PV power might be used to reduce fuel consumption in ground vehicles or to enable unmanned aerial vehicles to stay longer in the sky.

Recently, another standalone system that does not integrate with battery storage has become popular. PV power is used as an ancillary source for isolated electric networks based on fossil-fueled generation. The aim is fuel saving in remote places where the fuel cost is high and fossil-fueled generation causes air pollution. It is also a cost-effective solution because no batteries are used.

Without a significant energy buffer, system integration can be a challenge when PV power penetration is more than 25%. However, high penetration can achieve high fuel savings. Since the load fluctuations might produce significant disturbances, control and coordination of wind and PV power interfaces becomes critical to maintaining grid frequency and voltage. Communication is generally required to optimally coordinate the operations of the generators. Utility-scale examples can be found at www.sma.de, in the shape of PV–diesel hybrid systems. Battery storage is optional, and not mandatory in such systems. Current battery technologies are expensive, when the short lifespans and the operating-environment constraints are considered.

1.10   AC Grid-connected Systems

One drawback of a standalone system is that the PV array is usually oversized, so as to accommodate the worst-case scenarios in terms of solar power generation. Solar energy is wasted if the generated power cannot be stored or consumed. Recently, an increasing number of systems have been connected to the AC electric distribution network. The electrical production and distribution network – the grid – is a utility system, which is external to and not controlled by the PV system. Such systems are also called interactive systems or grid-connected systems, and can be operated for full-time maximum power injection. Figure 1.15 shows three examples of grid-connected systems, with and without battery storage. The key component in grid-connected PV systems is the utility-interactive inverter, which performs the DC/AC conversion and the required interconnection functions.

A simple grid-connected system, mainly used for small-scale applications, is illustrated in Figure 1.15a. Local loads can be supplied through the grid interconnection.

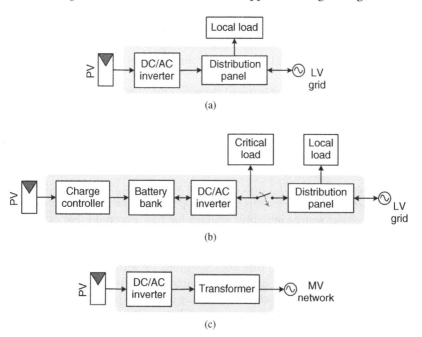

Figure 1.15 Grid-connected PV power systems with PV generators and inverters: (a) simple system; (b) grid-connected system with battery storage; (c) PV power plant.

Any excess electricity produced by the system can be fed into the grid. The block labeled PV can have different power capacities; in other words it could be a PV module, string, or array.

A grid-connected PV system can also be supplemented with battery storage, as shown in Figure 1.15b. Since the battery bank provides a significant energy buffer, the system can be operated in either grid-connected or standalone mode in case of a power outage. According to the definition of Article 705 in the National Electric Code (NEC 2014), such a DC/AC conversion unit is called a multimode inverter, and is capable of both utility interactive operation and standalone power supply. The system can be connected or disconnected from the grid through a switch, as shown in the diagram. The standalone mode is also termed "off-grid" mode. The energy stored in the battery bank can be used to power any critical load and it is charged when solar energy is available. Even in grid-connected mode, the system is very flexible and can be used for grid support functions, such as peak shaving and reactive power, thanks to the significant energy buffering. Peak shaving is the process of reducing the amount of energy demanded from a utility company during peak hours. Energy storage can provide a fast response way to do this.

PV power plants are large-scale systems that include more than one PV array and multiple grid-connected inverters, all grouped together for bulk power generation. The system capacity can reach 1 MW or more. Transformers are generally needed to step up the voltage from low to medium (MV) level, which is efficient for power transmission. Figure 1.15c illustrates the concept of such systems. The PV array can have hundreds or thousands of PV modules in series and parallel connection.

1.11 DC Grid and Microgrid Connections

Even though a grid connection is mainly relevant for AC grids, DC microgrids and DC distribution systems have been drawing significant attention and have shown the potential to compete with conventional AC systems.

High-voltage direct current (HVDC) systems have been installed worldwide and proved to be effective and reliable for bulk and long-distance transmission of electrical power. DC is also attractive for low-voltage applications. DC power supplies are used in most modern electronic devices in the home: phones, computers, printers, monitors, TV screens, LED lights, and so on. Modern data centers have modular and scalable DC power supplies that significantly reduce power consumption. Motors, including induction and permanent magnetic synchronous types, are traditionally driven by three-phase AC supplies. Thanks to the development of modern power electronics and control engineering, DC has become fundamental to supplying high-performance variable-speed motor drives. Various techniques can be used to make the motor's operation smooth, responsive, and efficient, all based on DC power supplies. However, two-stage conversion – AC to DC to AC – is always required for these applications, which causes significant losses due to the existing AC supply system (Patterson 2012). In general, HVDC and low-voltage DC (LVDC) have been proven to be effective for industrial applications. Medium-voltage DC (MVDC) systems at the distribution level are not as widely discussed as HVDC and LVDC ones. Their potential advantages over traditional medium-voltage AC systems in the areas of efficiency and cost-effectiveness are still under investigation.

The common issues in AC systems, such as power factors, current distortion, and synchronization, are no longer a great problem for DC-based systems. Thanks to the nature of the DC output, PV power is considered ideal for DC microgrids. The PV power interconnection for DC grids is straightforward because DC/DC conversion is simpler and more efficient than DC/AC conversion. Furthermore, the system costs can be lowered since multiple-stage conversions between DC and AC can be avoided. Other distributed components, such as fuel cells and batteries, also output DC.

An example of LVDC microgrids is illustrated in Figure 1.16. The system shows two DC distribution panels that are connected through a common DC bus. The system is based on a modular design that can accommodate multiple power sources and loads, sharing the same DC bus. The interconnection becomes easier than in an AC system since the DC voltage is the only variable that needs to be controlled. Each power unit is equipped with a power-conditioning circuit that is independently and optimally controlled. Normally, the DC/DC power interfaces for PV generators are operated for MPPT since the maximum solar energy harvesting is wanted. The battery storage modules balance the difference between generation and load. The charge and discharge of the battery modules are controlled in order to regulate the DC bus voltage. The PV power generator is only allowed to be away from the MPP if the generation would overpower both the load and the battery. The coordination of resource and load can be based on either a centralized or a decentralized approach. Among various sharing strategies, the droop method is one algorithm that is commonly applied for decentralized control and coordination (Huang et al. 2015). There is also an option to include bidirectional DC/AC converters for AC grid interconnection. Therefore, a DC microgrid supplied by PV power is flexible, being configurable as an off-grid (standalone), AC grid-connected, or DC grid-connected system.

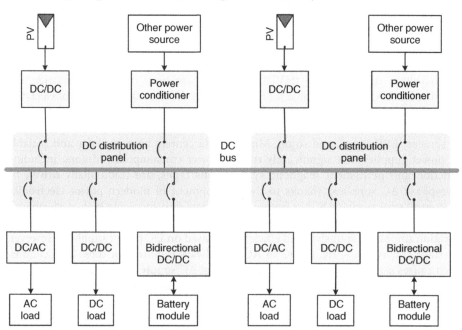

Figure 1.16 DC microgrid with PV power generation.

1.12 Building-integrated Photovoltaics

Special note should be taken of building integrated PVs (BIPVs). According to the Article 690 of the National Electric Code (NEC 2014), BIPV is where PV cells, devices, modules, or modular materials are integrated into the outer surface or structure of a building and serve as its outer protective surface. The term generally refers to PV modules that replace conventional building materials, such as the roof or facade. The term should not be misused and conflated with PV modules that are applied to buildings or roofs, but do not replace conventional building components.

An example of a BIPV system is shown in Figure 1.17. This installation is located in Langley, Canada. Designed for residential applications by MSR Innovations, the 20-kW PV system, known as SolTrak, also serves as the building's roof. The grid interconnection is three-phase, but formed by three single-phase inverters, each is rated as 7 kW. The SolTrak system incorporates two unique innovations: polymer roofing tiles that can have solar cells integrated directly into them, and a proprietary track mounting system that is easily installed, allowing the tiles to be snapped into place with no tools (Metten et al. 2012). The installation of the PV system is therefore simple, rapid, and safe and the installation cost is significantly reduced. Figure 1.18 includes three photos that illustrate the electrical connections and snap-in installation of SolTrak tiles. By reducing the up-front capital cost of solar power, the SolTrak system is designed to lower the biggest barrier to residential adoption of BIPV technology.

The Langley SolTrak installation is the largest of a number of BIPV projects that the author was involved in during his time at MSR. As discussed further below, BIPV systems bring unique engineering challenges. This 20-kW installation allowed the innovative electrical and mechanical designs to be validated, as well as revealing how to adapt to the regulatory requirements.

Figure 1.17 Example of building integrated photovoltaics in Langley, Canada.

Figure 1.18 SolTrak installation technology.

It sounds ideal that the BIPV should act as both building material and PV power generator. However, the technology is not widely used since the application is constrained by the way it is integrated into the building. For the highest annual PV power generation, the installation of PV arrays should have the ideal orientation and tilt angle, which is different from region to region. However, this is usually not possible for BIPV power generation because of the building's shape, location, and style. For example, in most cases, the upright angle of building facades is not ideal for solar power generation. In addition, the BIPV must meet building safety codes. Regional building codes can be very strict in certain areas due to the requirements of rain proofing and resistance to fire and earthquakes and so on. BIPV systems also raise safety concerns. For example, a grid-connected PV system is usually rated as 600 V or more, which can cause electric shocks. For BIPVs the building material must have insulation suitable for such a hazardous voltage rating. The appearance of PV products is another factor that causes resistance to their integration into buildings.

The level of PV power generation is mainly determined by the annual solar irradiances. However, people often ignore another important factor: the temperature. With the same level of solar irradiance, cooler cell temperatures generally yield more solar power. For example, the temperature coefficient of the A-300 solar cell, a Sunpower product, is $-0.38\%/°C$; in other words, every $10°C$ increase of the cell temperature means 3.8% less power. In many countries, the PV cell temperature can rise to more than 60° in summer, which results in significant power reductions. Therefore, a BIPV power system should be designed with full consideration of the thermal effect, especially for hot weather conditions. The thermal constraint is another factor that prevents BIPVs from widespread application. In general, proper ventilation or cooling is important for all PV system installations in order to improve system efficiency and prolong lifespans.

1.13   Other Solar Power Systems

Another category of solar power systems is concentrated solar power (CSP), also known as "concentrating solar power" or "concentrated solar thermal systems." It should be clear that their concept of power generation is completely different from that of PV and CPV systems. CSP uses mirrors or lenses to concentrate a large area of sunlight onto a small area to generate high temperatures. One example is the Shams 1 project located in the desert region of Abu Dhabi, UAE. The power plant occupies 2.5 km^2 and is rated at 100 MW. The system includes 258 000 mirror collectors to concentrate direct solar irradiation and generate heat inside the absorber tube. The system operation is illustrated in Figure 1.19.

The fluid inside the pipes transfers the generated heat to an exchanger in order to produce $380°C$ steam from water. The steam temperature is raised to $540°C$ by the booster heater, which relies on other energy resources. The higher temperature gives better efficiency in the steam turbine. The electric generator converts the mechanical energy into electricity. The voltage is stepped up for medium- or high-voltage transmission and grid connection. Shams 1 can be seen as it was when it was under construction in 2011 in

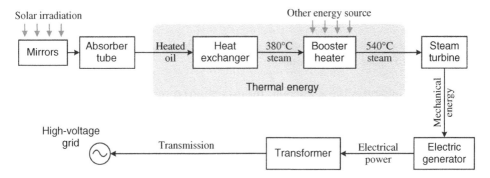

Figure 1.19 System diagram of Shams 1.

Figure 1.20 Shams 1 during construction in 2011.

Figure 1.20. The picture shows the mirror collectors and fluid-filled pipes. More information can be found at http://masdar.ae/en/energy/detail/shams-1.

1.14 Sun Trackers

Sun trackers are devices to adjust the direction and/or tilt angles of solar power generators so that they continue to face the sun over the course of a day. This process is called solar tracking, and aims to enhance levels of energy harvesting. Solar trackers are needed for CPV and CSP systems since direct sunlight is required to make them work. In the example shown in Figure 1.20, the mirrors can be mechanically adjusted to different tilt angles. For a regular PV system rather than a CPV one, the tracker system can improve the system efficiency. However, the costly installation and maintenance prevents widespread adoption. Recently, regular PV systems without sun trackers have become more and more cost-effective.

Solar tracking should not be confused with MPPT, which is a control algorithm rather than a physical device. MPPT is common in all PV power systems since it can boost the PV power output regardless of any variation in environmental conditions.

Problems

1.1 What is the ideal direction and tilt angle if you design a PV system in your local area? Justify your specification.

1.2 A PV cell measures 156×156 mm. The maximum power output is 3.2 W under the test conditions of 800 W/m^2 and 25°C. Calculate the conversion efficiency under the test conditions.

1.3 What is the rating of standard test conditions? What is the normal level of solar irradiance from the atmosphere?

1.4 Besides the PV cell materials introduced in this chapter, what new technologies are currently being investigated for PV power generation?

1.5 Find an example of a multi-junction solar cell and discuss the technology and the achieved efficiency.

References

Huang PH, Liu PC, Xiao W and El Moursi MS 2015 A novel droop-based average voltage sharing control strategy for DC microgrids. *IEEE Transactions on Smart Grids* **6**(3), 1096–1106.

Metten E, Roddick T and Scultety J 2012 Photovoltaic solar roof tile assembly system. US Patent 8,196,360.

NEC 2014 National Electrical Code. NFPA 70, National Fire Protection Association.

Patterson BT 2012 DC, come home: DC microgrids and the birth of the enernet. *Power and Energy Magazine, IEEE* **10**(6), 60–69.

Stark C, Pless J, Logan J, Zhou E and Arent DJ 2015 Renewable electricity: Insights for the coming decade. Technical Report NREL/TP-6A50-63604, JISEA: Golden, Colorado, February.

Xiao W, Zeineldin HH and Zhang P 2013 Statistic and parallel testing procedure for evaluating maximum power point tracking algorithms of photovoltaic power systems. *Photovoltaics, IEEE Journal of* **3**(3), 1062–1069.

2

Classification of Photovoltaic Power Systems

Photovoltaic (PV) systems are playing an increasingly significant role in electricity grids and there have been changes in system configurations in recent years. Classification of PV systems has become important in understanding the latest developments in improving system performance in energy harvesting. This chapter discusses the architecture and configuration of grid-connected PV power systems.

In general, grid-connected PV power systems can be categorized into two main groups: centralized MPPT (CMPPT) and distributed MPPT (DMPPT) (Femia et al. 2008). In contrast to conventional approaches, the subclassification of DMPPT systems is then based on the level at which maximum power point tracking (MPPT) is applied (Xiao et al. 2016): string, module, submodule, and cell where the MPPT function can be applied. This is explained in the following sections.

2.1 Background

Grid-connected PV systems are conventionally classified by their power capacity: small-, intermediate-, and large-scale systems (Ramakumar and Bigger 1993). A system that is less than 50 kW is typically considered as small scale. A PV system that can produce more than 1 MW is commonly considered either large- or utility-scale. System capacities between 50 kW and 1 MW are designated intermediate scale.

Due to the modular nature of PV arrays and their distributed installation, the boundaries of the system categories are often unclear. The accumulated capacity of a PV system can be more than 1 MW, which qualifies as the large scale. However, the installation can be composed or grouped by hundreds or thousands of small-scale or intermediate-scale grid-connected systems. The majority of grid-connected inverters are usually under 1 MW in capacity. In some cases, it is hard to distinguish exactly a single large-scale system from a set of small or intermediate-scale systems.

A grid-connected PV system can integrate with battery storage for operation in either standalone or grid-connected mode. With significant energy storage, battery storage can send active power to the grid during peak hours. They also effective in smoothing the power generation and minimizing the impact of solar energy intermittency. Therefore, one classification of PV systems is whether they are integrated with battery storage or not. However, the majority of grid-connected PV systems do not include battery storage.

Photovoltaic Power System: Modeling, Design, and Control, First Edition. Weidong Xiao.
© 2017 John Wiley & Sons Ltd. Published 2017 by John Wiley & Sons Ltd.
Companion Website: www.wiley.com/go/xiao/pvpower

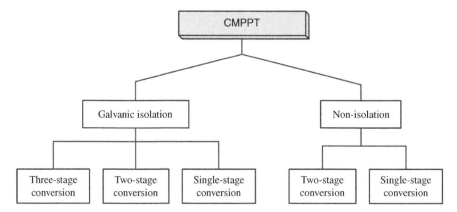

Figure 2.1 Classification of grid-connected PV power systems with centralized maximum power point tracking.

A recent classification focuses on the voltage level of the point of common coupling (PCC), the point at which the PV power production and distribution network couple in the interactive system. Low-voltage connected systems are usually installed close to electricity users, such as residential or commercial buildings. Such systems are mainly controlled by MPPT algorithm, in order to inject active power from PV generators into the grid. A unity power factor is required at the PCC, which is maintained by grid-connected converters. Furthermore, to guard against grid faults, so-called "anti-islanding" protection is normally required, the term referring to a function to detect grid status and cease PV power injection when power outages happen (Basso and DeBlasio 2004).

With the fast increase of PV power generation, these systems are more and more prominent in electric power grids. Utility-scale power plants, which are usually connected to medium-voltage networks or even the high-voltage grid, have a significant impact on grid capacity and stability. Grid operations are planned and regulated to maintain the stability of voltage and frequency. Static grid-support measures include reactive power injection, so that the power factors of such grid-connected converters can be operated at "non-unity" power factors (Xiao et al. 2014). Active power regulation and fault ride-through can be used for dynamic grid support, approaches which are recommended by European grid codes (El Moursi et al. 2013; Kou et al. 2015).

In this chapter, all grid-connected systems are classified by the level at which MPPT becomes active: centralized (CMPPT) and distributed (or decentralized) (DMPPT) systems. The structures of CMPPT and DMPPT systems are illustrated in Figures 2.1 and 2.2, respectively. The classification provides a clear framework for identifying the differences among system architectures and configurations of grid-connected PV systems.

2.2 CMPPT Systems

A PV array comprises several strings connected in parallel to achieve the desired power rating. Each string is formed of several PV modules in series, so as to meet the input

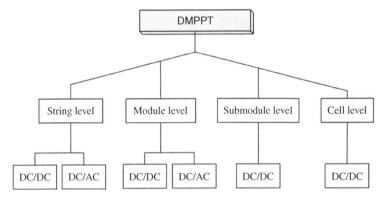

Figure 2.2 Classification of grid-connected PV power systems with distributed maximum power point tracking.

voltage requirements of the grid-connected converters. CMMPT systems track the maximum power point using a centralized inverter at the array level (Romero-Cadaval et al. 2013).

Functional grounding is required by certain grid codes and PV module manufacturers to ensure the safety of grid connections and the reliability of the modules. However, the requirements are inconsistent from region to region and country to country. A grounded PV system is defined as having DC conductors (either positive or negative) bonded to the equipment grounding system, which in turn is connected to earth. Grounding the PV system ensures safety by preventing electrical shock (Bower and Wiles 1994). The grid-connected inverter provides galvanic isolation to support common grounding for both DC and AC. Grid-connected systems typically include transformers in order to accommodate the required galvanic isolation. The DC/AC stage can be connected to single-phase or three-phase networks depending on the configuration at the point of the grid connection.

A three-stage conversion system is illustrated in Figure 2.3. It includes PV, HFAC, DC, and grid links (where HFAC refers to high-frequency AC). The conversion comes about by the sequence from DC to HFAC to DC to LFAC (the latter standing for low-frequency AC). The PV link is the circuitry that links the power converter to the PV output terminals. The grid link is the circuitry connecting the power converter to the AC grid. The topology is generally complex because of the multiple conversion stages. However, the system uses a high-frequency transformer, which is advantageous in terms of its small size, low weight, and low cost. This topology is also referred to as a "mixed-frequency inverter" in the literature (Freitas 2010). One implementation of such a topology is the Xantrex GT series, manufactured by the Schneider Electric Solar Energy Company.

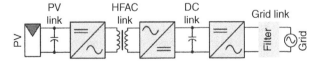

Figure 2.3 Topology of centralized maximum power point tracker with galvanic isolation and three-stage power conversion.

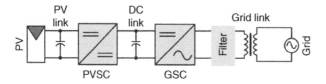

Figure 2.4 Topology of centralized maximum power point tracker with galvanic isolation and two-stage power conversion.

Low-frequency transformers are commonly used to provide galvanic isolation, as shown in Figure 2.4. The two-stage interfacing topology includes PV, DC, and grid links. The DC link is important in this system since it provides energy buffers between the DC/DC and DC/AC converters. The DC/DC converter is referred to as the PV-side converter (PVSC) in this book. The DC/AC converter is referred to as the grid-side converter (GSC). The PV link provides a filtering function between the PV generator and the PVSC. The DC-link circuitry is commonly formed by capacitor banks, which mitigate harmonics caused by the DC/AC conversion and the high-frequency switching operation. High capacitance in the DC link can decouple fast dynamic interactions between the DC/DC and DC/AC stages (Hu et al. 2015). The grid link is designed to provide a filtering function in order to minimize harmonic injection into the grid. Low-frequency transformers, so called "line-frequency transformers," are bulky and heavy, but generally robust and reliable. Their additional advantage is that galvanic isolation is provided exactly at the PCC, thus effectively preventing DC injection.

Figure 2.5 shows a one-stage conversion system, including only DC and grid links. The system converts the PV array output directly to LFAC through the grid-connected PV inverter, and a low-frequency transformer gives galvanic isolation. In such systems, the PV link is the same as the DC link, so only the latter is mentioned.

Certain grid codes allow ungrounded PV systems. Galvanic isolation is no longer mandatory, allowing transformers to be avoided. The grid-connected inverters are commonly referred to as transformer-less. Such systems aim for higher efficiency than their isolated counterparts thanks to the elimination of transformer loss. An ungrounded grid-connected PV system including two-stage conversion is shown in Figure 2.6a. The PV, DC, and grid links are clearly indicated. In the equivalent single-stage conversion system, the PV link is merged with the DC link, which is indicated only as the DC link in Figure 2.6b. The GSC is the key component, converting DC into AC for grid connection and performing the MPPT function.

Even though the multiple-stage conversion system circuits look complicated, the design and evaluation of the PVSC and GSC can be decoupled. A significant capacitance appears at the DC link, which gives separation of the high-frequency dynamics. In the multiple-stage conversion systems, the control of the PVSC is mainly for the MPPT function, so as to give the highest solar energy harvesting. DC-link voltage regulation is

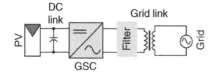

Figure 2.5 Topology of centralized maximum power point tracker with galvanic isolation and single-stage power conversion.

Figure 2.6 Centralized maximum power point tracker without galvanic isolation: (a) with two-stage conversion; (b) with single-stage conversion.

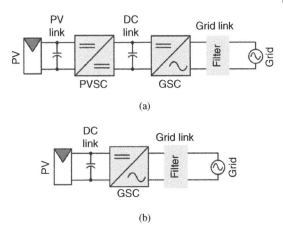

one of the major control functions of the GSC. Other grid-connected functions, such as anti-islanding, power factor regulation, active power throttling, and fault ride-through are also implemented to control the GSC. The DC-link configuration in multiple stage conversion not only divides the control function into two separate tasks but also gives flexibility to implement modular DC/DC MPPT units that will achieve more effective energy harvesting (El Moursi et al. 2013).

The single-stage power interface has shown the advantage of simplicity and high conversion efficiency. However, it should be based on an integrated design since the DC/AC inverter must be able to undertake MPPT and other functions required by the grid interconnection. Furthermore, the MPPT dynamic performance is no longer decoupled from the DC link, which is the same as the PV link. When the system is for a single-phase grid interconnection, a significant capacitance at the DC link is required to mitigate the double-line frequency ripples. This generally lowers the speed of MPPT due to the slow dynamic response at the PV link.

2.2.1 Power Loss due to PV Array Mismatch

Ideally, a solar array should be always constructed of PV modules, all with the same electrical characteristics. It should be installed in a shading-free environment. However, it is impossible to avoid either shading or other mismatch conditions. PV module mismatches result from conditions, such as moving clouds, the shadows of trees and buildings, dust, uneven temperature distributions, aging, or manufacturing imperfections. Perching birds and bird droppings can also cause unavoidable and unpredictable shading and mismatch conditions in CMPPT systems.

PV array mismatches have a disproportionate impact on system performance, because the solar cell with the lowest output limits the current through other elements in the string. Significant power losses have been reported for CMPPT systems due to unbalanced generation among the cells (Xiao et al. 2007).

The mismatch impact can be demonstrated by a simple test. Figure 2.7 shows a test stand where two PV modules are installed on a shared frame with the same tilt angles and directions. Both modules are the same model, BP350, and were manufactured on the same date. The setup ensures that any testing can be conducted at the same time

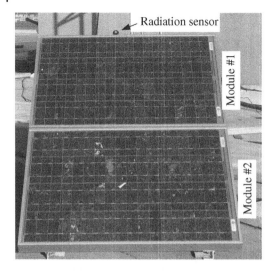

Radiation sensor

Module #1

Module #2

Figure 2.7 Two PV modules installed on the same frame for a partial shading study.

Table 2.1 Specification of photovoltaic module BP350.

Parameter	Specification
Cell material	Multi-crystalline
Number of cells	72
Cell configuration	36×2
MPP voltage$^\varsigma$	17.3 V
MPP current$^\varsigma$	2.89 A
Short-circuit current$^\varsigma$	3.20 A

$^\varsigma$ Data at standard test conditions.

and under the same environment conditions. The specification of the PV module is summarized in Table 2.1.

The initial test shows that the output characteristics of the two modules are slightly different, even though the test conditions are the same. The current–voltage (I–V) and power–voltage (P–V) curves for the two modules are plotted in Figure 2.8. Under the same testing condition, the peak power values are almost identical, at 21.52 W and 21.49 W. The difference is noticeable in the output characteristics in terms of I–V and P–V curves between the two modules. The MPPs are located at the points (16.16 V, 1.33 A) and (16.55 V, 1.30 A) for the first and second module, respectively. The difference is not unexpected, since the PV manufacturer specifies that the tolerance of power output is ±5%.

The initial study has the two PV modules in series. Figure 2.9 shows the output characteristics, of which the output current at the MPP is limited by the lowest value, of 1.30 A. The MPP is measured as 42.88 W, which is lower than the sum of the individual PV modules, $21.52 + 21.49 = 43.01$ W. There is a 0.31% power loss even though the two healthy PV modules only show slight differences in output characteristics.

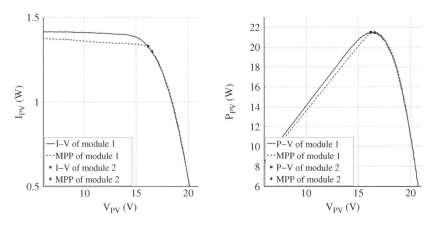

Figure 2.8 Plots of data acquired by I–V tracer without PV partial shading.

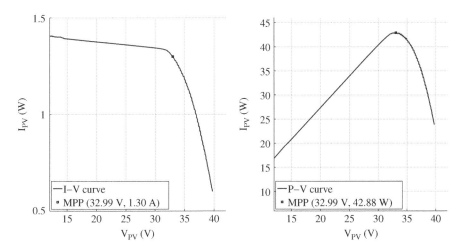

Figure 2.9 Plots of data acquired by I–V tracer without partial shading of PV.

A mismatch condition is then created: one cell in module 2 is intentionally shaded. This represents a 1/72 partial shading condition since there are 72 cells in each PV module. The I–V and P–V curves are plotted in Figure 2.10. The unshaded module has the MPP located at (16.16 V, 1.33 A), giving 21.49 W in power. The shaded module produces 15.44 W, with the MPP location at (17.19 V, 0.90 A). The single-cell shading causes a 28% power loss compared to the unshaded condition. The sum of the available power from both modules becomes 36.93 W.

When the unshaded module and the partially shaded module are connected in series, the I–V and P–V curves of the terminal output are as plotted in Figure 2.11. Two power peaks are seen in the P–V plot. Neither can represent the true available power, which is equal to the sum of the individual maximum powers of the two PV modules. The MPP corresponds to a voltage of 35.10 V and a current of 0.93 A, and the power is measured as 32.65 W. The single-cell shading leads to a 22% power loss when the

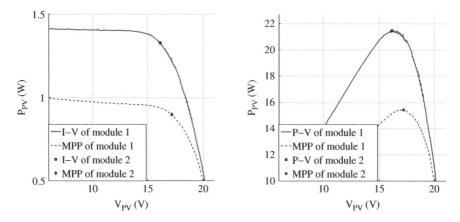

Figure 2.10 Plots of data acquired by I–V tracer with one cell shaded.

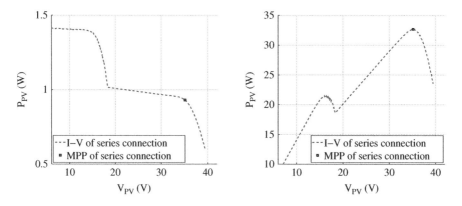

Figure 2.11 Plots of data acquired by I–V tracer with one cell shaded when two modules are connected in series.

shaded and unshaded modules are connected in series. The total loss can be divided into contributions of 14% and 8%, for the loss from the shading alone and that from the hidden MPP. The curves shown in Figure 2.10 can be used to predict the output when more unshaded PV modules are connected with the shaded module in series.

The above discussion is based on the condition of one cell out of 72 shaded. In the real world, the shading pattern will be more complicated and can be caused by many unpredictable factors. The study described in this section demonstrates the drawbacks of the CMPPT system and the importance of developing DMPPT systems.

2.2.2 Communication and Data Acquisition for CMPPT Systems

In CMPPT systems, hundreds and thousands of cells are interconnected to form the PV array. Mismatch effects should be always considered in the design stage, since they might result in significant power losses. Data acquisition and communication systems are commonly utilized in centralized MPPT systems to report on real-time status in order to detect any mismatch conditions. Data acquisition is commonly implemented using grid-connected inverters to sense and record the connection status, real and reactive power output, and the voltage at the interconnection point (Yu et al. 2011). The generation data can be transmitted to any place in the world for monitoring and storage, using the latest communication technologies of the Internet and cloud computing. Figure 2.12 shows a typical grid-connected system with data acquisition and communication functions.

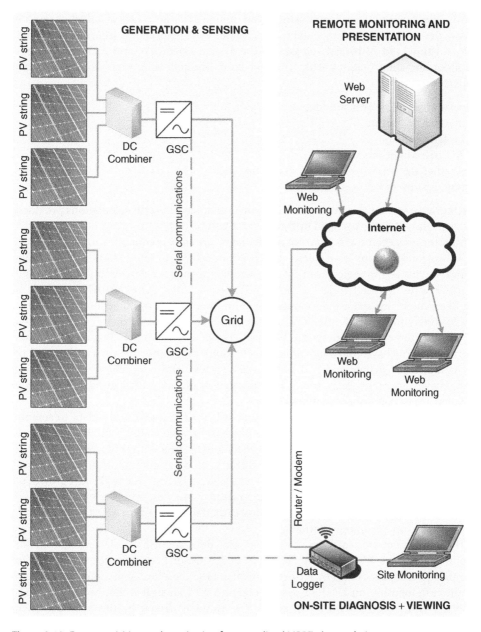

Figure 2.12 Data acquisition and monitoring for centralized MPPT photovoltaic power systems.

The generation data are collected by digital controller units implemented inside the inverters. Serial communication technologies, such as the RS-485 protocol or power line communication (PLC), are commonly used to transmit data from individual inverters to the data logger. RS-485 is a serial communication protocol that is commonly used in industry. It has significant advantages over the previous serial communication standard, RS-232. The PLC technology uses existing electrical wires to transport data and support relatively high bandwidths. Either Ethernet or USB interfaces are used for on-site debugging and monitoring. Even though solar farms are installed in remote areas, the system owner and technical support team can access information about the real-time and historical statuses of the system operation, and receive warnings of about any abnormalities at the inverter level. The following variables are typically collected:

- solar array DC power production
- inverter AC output
- inverter status
- AC grid conditions
- weather station data, including solar irradiance
- temperature of key components.

Accurate data can also be used for the purpose of feed-in tariff calculations provided the metering device is certified by the relevant authorities.

Monitoring systems can incorporate the latest wireless communication technology. It is obviously more convenient to communicate without a significant cabling requirement. Technologies used include wireless local area networks (WLANs) and wireless wide area networks (WWANs) (Yu et al. 2011). One WLAN standard is ZigBee, which was mainly developed for home network communications. The term "WiFi" refers to the IEEE 802.11 WLAN technologies. Either of these can be utilized to monitor the detailed operation of a grid-connected PV system. Figure 2.13 shows a grid-connected system using wireless technology to monitor the operating status of PV modules, PV strings, and grid-connected inverters. The comprehensive monitoring and data acquisition allows for fast identification of any fault caused by PV array mismatches or component malfunctions.

The data logger that collects data through WLAN can transmit the data to remote areas through WWAN. Public cellphone carriers have attracted great interest, since WWAN could be used to transmit the collected data to a data center, from where it would be distributed to individual users through modular communication devices or computers. The cellular network has covered a significant area due to the fast growth of mobile networks. Wireless communication can give low-cost and convenient installation even though the system bandwidth is lower than that of modern wired counterparts, such as fiber-optic communication.

The implementation of communication and data acquisition systems allows for monitoring of the PV system operation, but it can not provide a direct solution to minimize the losses caused by PV array mismatches. Therefore, more and more recent studies are focusing on DMPPT systems. Figure 2.14 illustrates the trend away from CMPPT and towards DMPPT systems. The trend is driven by the level at which MPPT is implemented. Lower-level MPPT gives higher solar power output in the

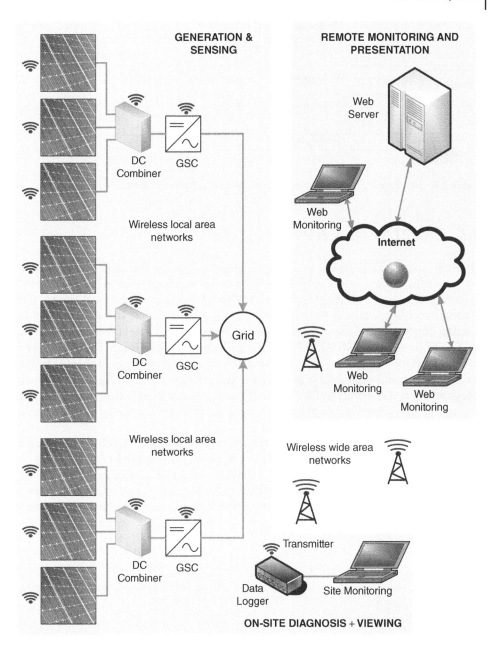

Figure 2.13 Three-level monitoring of photovoltaic power systems using wireless communication technologies.

event of mismatch conditions, because the individual issue can be isolated without affecting other parts of the system. DMPPT systems at the string and module levels are commercially available. DMPPT at the submodule and cell levels is still under research and development.

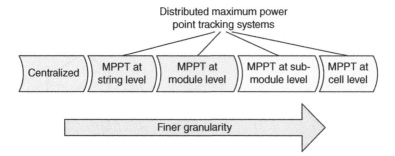

Figure 2.14 Trend towards distributed maximum power point tracking systems.

2.3 DMPPT Systems at PV String Level

DMPPT has attracted significant research attention, aiming to address the issue of PV array mismatches (Femia et al. 2008). It has been widely adopted in commercial PV inverter products at the string level. The DMPPT solution has also been adopted in DC microgrid configurations. A PV array typically comprises multiple strings connected in parallel. The concept of the string inverter was introduced to avoid mismatches among strings. In contrast to CMPPT, the DC/AC grid-connected unit is rated and connected to individual PV strings instead of the whole PV array. Solar energy is collected by the string inverters and supplied to the AC interconnection, as shown in Figure 2.15, which shows multiple PV strings and PV links. The grid link is a series of AC filters that are required to guarantee the injection power quality. Transformers can be integrated into the grid link for galvanic isolation.

Figure 2.16 is another example of a grid-connected system, with three distributed maximum power point trackers at the string level. The output of each string is modulated by an independent DC/DC converter, which is is controlled for MPPT. The common DC bus can be linked to either a DC microgrid or an AC grid through a centralized DC/AC inverter. Therefore, the power degradation caused by any mismatch effect at string level is minimized since the power of each PV string is individually extracted and processed.

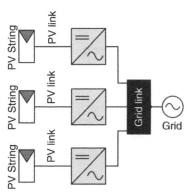

Figure 2.15 Distributed maximum power point trackers at PV string level sharing AC link.

Figure 2.16 Distributed maximum power point trackers at PV string level sharing DC link.

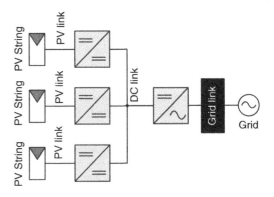

The DC/DC and DC/AC converters can be either independent units or integrated inside one enclosure with the DC/AC inverter. Some commercial systems implement the DC/DC units inside combiner boxes to perform independent string-level MPPT (Romero-Cadaval et al. 2013). One example is the Huawei SUN2000-42KTL grid-connected inverters.[1] The 42-kW inverter unit can connect up to eight PV strings, the operating status of which can be monitored. The unit includes four independent MPPT units in order to minimize the mismatch impacts among the PV strings.

2.4 DMPPT Systems at PV Module Level

Module-integrated converters provide independent MPPT operations within each PV module, which allows for local optimization and reduces power losses resulting from mismatches and partial shading. It should be noted that the diagrams in this section are mainly for conceptual purposes since many neglect the presence of DC and AC filters.

2.4.1 Module-integrated Parallel Inverters

The output terminals of module-integrated parallel inverters (MIPIs) are connected in parallel with the AC network. Each PV module is integrated and connected to one MIPI, as illustrated in Figure 2.17. The MIPI units are usually PV microinverters or AC modules, which directly convert the PV module voltage, typically 22–45 V, to the

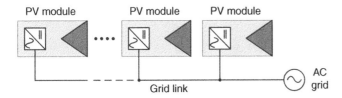

Figure 2.17 Configurations of module-integrated parallel inverters for grid interconnections.

---

1 www.huawei.com/solar.

low-voltage AC grid level (Xiao et al. 2013). The concept of AC modules refers to PV modules having AC output terminals, since the DC/AC conversion stages are integrated inside the junction boxes of the PV panels. However, PV microinverters or MIPIs can be an independent unit installed outside the PV modules. The parallel interconnection eliminates the single point failures that are common in series connections. In addition, the system becomes highly modular and flexible because the parallel structure can be easily expanded or modified. Besides the normal benefits of a parallel structure, the direct DC/AC conversion of MIPIs brings the additional advantages of a highly modular solution for a DC/AC grid connection and simple system wiring; DC wiring is integrated in the module and AC wiring is easy for most electricians.

The drawbacks of MIPIs are also clear: the overall system cost per watt is higher than that of the centralized counterpart. Furthermore, the conversion efficiency is not as good as in a system based on PV string configuration. Due to the voltage limit of a single PV module, the MIPI should be designed for a high voltage boosting ratio in order to reach the grid voltage level. The MPPT effectiveness is influenced by the impact of double-line frequency ripple in single-phase systems. Lastly, harsh outdoor operating environments influence the lifetime and reliability of all electronic components. The manufacturers of such products include Enphase (USA), Petra (USA), Sunpower (USA), AP systems (USA), Enecsys (UK), and Involar (China).

Two-stage conversion is generally required because of the very high voltage-conversion ratio (Edwin et al. 2014). As shown in Figure 2.18, the DC–DC stage steps up the voltage from the PV module to a higher level for grid interconnection. The DC/AC unit converts the DC voltage to the AC line voltage through pulse width modulation. Due to the high capacitance across the DC link, as shown in Figure 2.18a, the voltage can be kept steady to provide an energy buffer for energy transfer from DC to AC form.

The topology in Figure 2.18b does not apply high capacitance at the DC link to maintain the steady DC-link voltage. It allows it to fluctuate with the double-line frequency of the grid voltage. In the second stage, the DC current is unfolded into AC form and injected into the grid. The energy buffer for DC to AC conversion is allocated

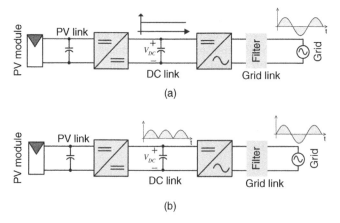

Figure 2.18 Two types of module-integrated parallel inverter: (a) DC link for steady voltage (b) current-unfolding approach.

across the PV link. The unfolding approach aims to minimize the high-frequency switching loss at the DC/AC conversion stage. Since the majority of MIPIs are designed to be connected to a single-phase AC grid, significant capacitance must be present at the PV link in order to mitigate double-line frequency voltage ripples. The drawback of the high capacitance lies in the capacitor lifetime and the slow dynamics across the PV module terminal. However, industry has shown that this topology can achieve more than 96% efficiency, which is significant for the high step-up ratio of the DC to AC voltage conversion (Edwin et al. 2012).

Data acquisition and communication can be implemented in the MIPI to report the status of individual PV modules and the converter in real time. PLC is usually utilized for data communication. Another option is WLAN, which was discussed in Section 2.2.2.

2.4.2 Module-integrated Parallel Converters

In contrast to MIPIs, module-integrated parallel converters (MIPCs) perform only DC/DC conversion, integrating PV modules in parallel on a common DC bus. The DC bus can be linked to either DC microgrids or AC grids through a centralized DC/AC inverter, as shown in Figure 2.19. MIPC-based PV systems take the advantage of the parallel structure of their counterparts, MIPIs. They provide the simplest solution for connection to a DC microgrid. In contrast to the interconnection to a single-phase AC grid, MIPCs show superior performance since the double-line frequency ripple is no longer present in the DC/DC conversion stage. However, for AC grid integration, MIPC systems are not as modular as MIPI structures, but share the same disadvantages in terms of high voltage-conversion ratios, relatively low conversion efficiencies, and harsh outdoor operating conditions.

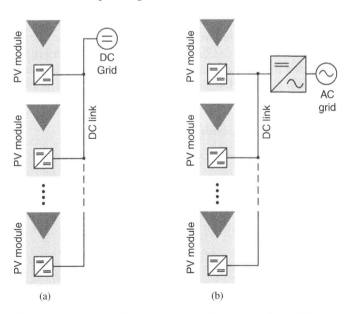

(a)　　　　　　　　　　　　(b)

Figure 2.19 System configurations of module-integrated parallel converters for: (a) DC grid connection; (b) AC grid connection.

The MIPC-based system was commercially developed by eIQ Energy Inc, USA. The tapped-inductor topology was designed for high-step-up voltage gain and conversion efficiency (Krzywinski 2015). The tapped inductor can be described as an autotransformer, which provides no galvanic isolation. Derived from the standard boost topology, the winding ratio of the tapped inductor offers flexibility to step up the DC/DC conversion to a relatively high level. It also utilizes an interleaved structure including two conversion phases, which reduces the filter sizes in both input and output ports. High voltage stress is always presented to the output rectifier, which can be made worse by the leakage inductance of the tapped inductor. The design and analysis of the tapped-inductor topology is further discussed in Section 5.1.5.

2.4.3 Module-integrated Series Converters

Module-integrated series converters (MISCs), also known commercially as DC power optimizers, are integrated with PV modules to perform MPPT and DC/DC conversion. The idea was originally proposed in the 2004 Annual Conference of the IEEE Industrial Electronics Society and subsequently published in *IEEE Transactions on Industrial Electronics* (Roman et al. 2006). However, the reported efficiency was not high enough to be used in practice. More recently, thanks to efficiency enhancements, commercial examples of MISCs have become available from National Semiconductor, SolarEdge, Tigo, and Xandex.

In contrast to MIPCs, the outputs of MISCs are serially connected to form a DC string, as shown in Figure 2.20. The DC string can be connected in parallel with other DC strings in order to create a DC link, which can be used for a DC microgrid or an AC grid through a centralized DC/AC inverter. The structure provides an ideal solution for a DC microgrid. The stack structure can build up the voltage of the DC string through multiple MISCs, each of which can be operated at a low conversion ratio which achieves a high conversion efficiency. The drawbacks are that the structure is not as modular as parallel configurations such as MIPIs and MIPCs. Reliability can also be a concern since a series connection can be affected by a single point failure. Extra protection should be considered and implemented to ensure reliability. Installations are mostly located outdoors, similar to other DMPPT units at the module level.

2.4.4 Module-integrated Differential Power Processors

Module-integrated differential power processors (MIDPPs) were introduced to balance operations of individual PV modules and to eliminate mismatches

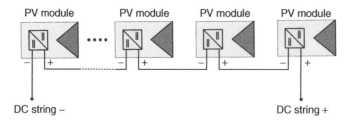

Figure 2.20 Configurations of module-integrated series converters to form a DC link.

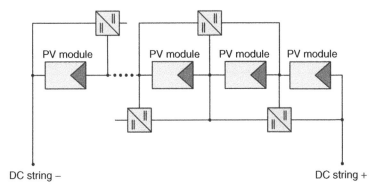

Figure 2.21 Configurations of module-integrated differential power processors to form a DC link.

(Blumenfeld et al. 2014). MIDPPs are multi-port DC/DC converters, which provide a bypass route in parallel with the PV strings, forming a DC string, as shown in Figure 2.21. The DC string can be connected in parallel with other DC strings in order to create a DC link of the desired capacity. The system takes advantage of a standard string configuration since the MIDPPs carry only the mismatch current. The concept is similar to the active balancing methods used for mitigating battery mismatches.

Switched capacitor topologies are used to achieve the goal of differential power processing. The DC bus voltage is formed by series-connected PV modules, which can be coupled to either DC microgrids or AC grids through a centralized DC/AC inverter. The system concept sounds ideal because the power interface is active only in the case of mismatch conditions. However, many constraints prevent practical implementation. Complex wiring is one of the drawbacks in comparison with module-level DMPPT units. The reliability is also not as good as a parallel structure because the series connection can be subject to short-circuit faults. Communication among the MIDPPs is generally required to achieve MPPT at the module level, which adds costs to the system.

2.4.5 Module-integrated Series Inverters

A new configuration utilizing module-integrated series inverters (MISIs) has recently been proposed (Jafarian et al. 2015). The output terminals of the MISIs are series connected, to form a stacked AC string, as shown in Figure 2.22. This arrangement is also referred to as "cascaded AC modules."

The system has advantages over conventional parallel MIPI structures. High conversion efficiency can be expected, and low-voltage components can be used because a MISI system avoids high conversion ratios thanks to the structure of the AC voltage stack. However, it exhibits the disadvantages of series connections, such as single point failures and complication in coordination of all the MISIs. Furthermore, the series connection of the AC output terminals is more difficult to control and coordinate than the DC stack used for MISCs. A central unit, shown as the grid link in Figure 2.22, is essential to coordinate the operation of the MISIs. Grid monitoring, protection, and filtering functions are commonly implemented in the central unit, ensuring safety and power quality.

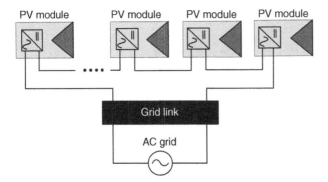

Figure 2.22 Series configurations of module-integrated series inverters for AC grid interconnection.

2.5 DMPPT Systems at PV Submodule Level

PV submodules are not independent units, but parts of a laminated PV panel. Laminated crystalline PV modules are commonly made up of 60 or 72 solar cells arranged in three or four submodules. Figure 1.3 illustrated an example in which each submodule consisted of 24 PV cells in series connection with parallel connected bypass diodes. Without losing generality, the following discussion and illustration is based on the three-submodule-per-panel configuration, which is common in commercial PV products. In reducing mismatches, DMPPT at the submodule level provides better output than module-level approaches.

The output voltage of a submodule is usually less than 15 V. The voltage level matches the requirements of laptop computers and other mass-produced portable devices, so the components to construct the submodule converters are widely available, and are compact, high performance, and low cost.

Due to the low voltage of the submodule output, parallel structures are uncommon for grid-connected applications. Three architectures based on series connections are introduced in the following subsections.

2.5.1 Submodule-integrated Series Converters

Submodule-integrated series converters (subMISCs) are integrated with PV submodules to perform MPPT and DC/DC conversion functions, as shown in Figure 2.23.

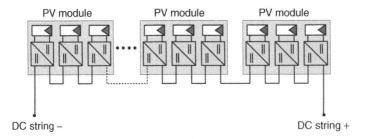

Figure 2.23 Series configurations of submodule-integrated series converters to form a DC link.

The output terminals of subMISCs are connected in series to form a string, which provides a voltage stack for grid integration. The DC bus formed by subMISCs can be directly linked to a DC microgrid. For AC grid interconnections, subMISC-based systems process power in two stages: DC/DC and DC/AC conversion. To balance the output currents among subMISCs, the preferred topology uses synchronous buck converters thanks to their high efficiency and wide conversion ranges. Due to the cascaded connection to form the voltage stack, any open-circuit failure of subMISCs breaks the complete string, as shown in Figure 2.23. However, protection can be added to avoid a whole-system malfunction, bypassing or short-circuiting the faulty subMISC.

One issue that prevents the integration of subMISCs is that the submodules inside commercial PV panels are internally connected in series prior to PV lamination. Without breaking the interconnection, subMISC systems can not be implemented. Therefore, subMISC systems require that PV panel manufacturers revise the electrical layout of their PV panels prior to lamination. Systems based on subMISCs have not been commercialized up until now, but they might one day achieve high conversion efficiency, integrated inside the junction box of PV modules.

2.5.2 Submodule-integrated Differential Power Processors

The concept of differential power processors, discussed in Section 2.4.4, can be extended to the submodule level, an arrangement referred to as as a submodule-integrated differential power processor (subIDPP). The power in each submodule directly passes through the string of submodules, while the subIDPPs in the parallel path process only the mismatch power, as shown in Figure 2.24. It appears an ideal solution since the losses are lower than in subMISC-based systems, which convert PV power full time. Unlike subMISC-based systems, subIDPP-based systems do not need the electrical connections inside the PV lamination to be modified. The system can be sized, designed, and constructed just as for conventional CMPPT systems. In theory, the subIDPP activates power conversion for energy harvesting only in the case of unbalanced generation. From a reliability point of view, the subIDPP system is advantageous since the connection of converters provides a parallel path, which can protect the system in case of short circuits or open-circuit failures of

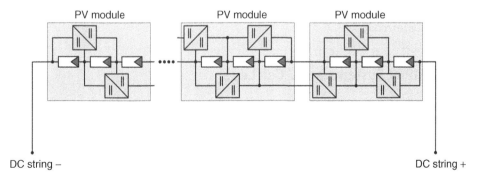

Figure 2.24 Series configurations of submodule-integrated differential power processors to form a DC link.

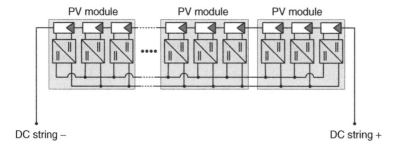

Figure 2.25 Series configurations of submodule-integrated isolated-port differential power processors to form a DC link.

submodules. The drawbacks of the subIDPP solution lie in the complexity of the topologies and the wiring requirements, which result in low conversion efficiency compared to subMISC systems. Communication is generally required to coordinate distributed MPPT by the subIDPPs. In outdoor installations, the additional wiring requirement for bypassed current and communication will significantly increase the system cost.

2.5.3 Isolated-port Differential Power Processors

Another differential power processing solution is known as an isolated-port differential power processor (subIPDPP), as shown in Figure 2.25. Since the outputs of subIPDPPs share a common ground, galvanic isolation is required in the converter topology. Each submodule is integrated in parallel with a dedicated converter, which makes it similar to a subMISC in terms of connectivity. A simple control strategy of voltage equalization of submodules can be used for MPPT. The converters can be turned off when no mismatch is detected. When there is unbalanced generation in the submodules, the mismatch power is processed through converters to adjust the string current and eliminate the impact. From the connectivity point of view, the wiring is complicated, which makes practical implementations difficult outdoors. Furthermore, transformers are required to provide galvanic isolation, and these are costly and difficult to accommodate in the junction boxes of commercial PV modules. The need for additional weather-resistant enclosures for each PV module will also significantly increase the implementation cost.

2.6 DMPPT Systems at PV Cell Level

An on-chip integrated power management architecture has been proposed to achieve MPPT at PV cell level; the fully integrated circuit is claimed to eliminate partial shading issues completely (Shawky et al. 2014). Figure 2.26 illustrates the implementation of DMPPT at the cell level. The system adopts the topology of synchronous DC/DC boost converters. To limit the size, the switching frequency is set to 500 kHz. The solution sounds ideal to push the MPPT operation down to the finest granular level, but there are drawbacks in the system's complexity and high cost, which arise because of the significant numbers of DC/DC power units required for a typical

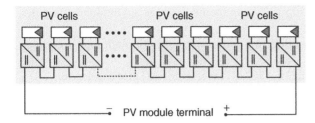

Figure 2.26 Integrated power management architecture for DMPPT at PV cell level.

grid-connected PV system. There is also a concern to match the DC/DC converter lifetime with that of the PV cells under the same harsh environment of direct sunlight. Furthermore, it is difficult to design a high-efficiency DC/DC converter for the low input voltages and high output currents of typical PV cells; the maximum power output of a six-inch crystalline-based cell is about 4 W, at 0.5 V and a current of 8 A. Few studies these days are focused on cell-level MPPT because of this difficulty and complexity.

2.7 Summary

This chapter has given a classification of grid-connected PV systems. The majority of grid-connected PV power systems can be categorized as centralized MPPT (CMPPT) or distributed MPPT (DMPPT). In contrast to conventional approaches, the subclassification of DMPPT systems focuses on the level at which MPPT is applied: string, module, submodule, or cell level. Each application has been discussed in terms of its advantages and disadvantages. In summary, the cost of DMPPT systems is generally higher than CMPPT systems because of the complex circuitry required.

The CMPPT solution was the initial approach for large-scale PV power systems. More and more DMPPT systems have been developed to deal with the significant power losses caused by PV mismatch conditions, which cannot be easily avoided in real-world environments. The DMPPT approaches have been commercialized at the string and module levels. Significant research is pushing into even finer levels, such as the submodule. Submodule-level DMPPT is a promising approach since it can adopt the low-voltage power converters developed for computer and communication devices. The voltage stack can be built to match the voltage required by the grid, while avoiding high voltage-conversion ratios. However, the integration of subMISC requires the modification of the electrical wiring inside conventional PV panels.

The above discussion provides a clear framework for readers to understand the architectures that are widely used for grid-connected PV systems. To avoid confusion, some important definitions in this chapter are summarized:

- *PV submodule*: one section of the laminated PV module. A common crystalline-based PV module includes three or four sections that can be utilized as the submodules. Each submodule usually consists of 15–24 solar cells in series connection.

- *PV link*: the circuitry connecting PV components, such as submodules, modules, strings, or arrays.
- *HFAC link*: the circuitry connecting two conversion units through high-frequency AC (see Figure 2.3).
- *DC link*: the circuitry connecting the DC/AC inverter for AC grid connection.
- *DC string*: the circuitry that is formed by series-connected DC/DC converters in parallel with or without PV components.
- *Grid link*: the circuitry connecting the DC/AC stage to the AC grid.
- *PV-side converter (PVSC)*: the DC/DC converter that is coupled to the PV link.
- *Grid-side converter (GSC)*: the DC/AC converter that is coupled to the grid link.

Problems

2.1 Identify an industrial product that is used as an inverter for a grid-connected PV system. Identify at which level the MPPT function is applied.

2.2 Find another criterion to classify and distinguish different PV power systems.

2.3 List some factors in your locale that might cause partial shading or other PV mismatch conditions.

References

Basso T and DeBlasio R 2004 IEEE 1547 series of standards: interconnection issues. *Power Electronics, IEEE Transactions on* **19**(5), 1159–1162.

Blumenfeld A, Cervera A and Peretz MM 2014 Enhanced differential power processor for PV systems: Resonant switched-capacitor gyrator converter with local MPPT. *Emerging and Selected Topics in Power Electronics, IEEE Journal of* **2**(4), 883–892.

Bower W and Wiles JC 1994 Analysis of grounded and ungrounded photovoltaic systems *Photovoltaic Energy Conversion, 1994, Conference Record of the Twenty Fourth. IEEE Photovoltaic Specialists Conference – 1994, 1994 IEEE First World Conference on*, vol. 1, pp. 809–812.

Edwin F, Xiao W and Khadkikar V 2012 Topology review of single phase grid-connected module integrated converters for PV applications *IECON 2012 – 38th Annual Conference on IEEE Industrial Electronics Society*, pp. 821–827.

Edwin F, Xiao W and Khadkikar V 2014 Dynamic modeling and control of interleaved flyback module integrated converter for PV power applications. *Industrial Electronics, IEEE Transactions on* **61**(3), 1377–1388.

El Moursi MS, Xiao W and Kirtley Jr JL 2013 Fault ride through capability for grid interfacing large scale PV power plants. *IET Generation, Transmission & Distribution* **7**(9), 1027–1036.

Femia N, Lisi G, Petrone G, Spagnuolo G and Vitelli M 2008 Distributed maximum power point tracking of photovoltaic arrays: Novel approach and system analysis. *Industrial Electronics, IEEE Transactions on* **55**(7), 2610–2621.

Freitas C 2010 How inverters work. *Home Power Magazine*. http://www.homepower.com/
    articles/solar-electricity/equipment-products/how-inverters-work.

Hu Y, Du Y, Xiao W, Finney S and Cao W 2015 DC-link voltage control strategy for
    reducing capacitance and total harmonic distortion in single-phase grid-connected
    photovoltaic inverters. *IET Power Electronics* **8**(8), 1386–1393.

Jafarian H, Mazhari I, Parkhideh B, Trivedi S, Somayajula D, Cox R and Bhowmik S 2015
    Design and implementation of distributed control architecture of an AC-stacked PV
    inverter *Energy Conversion Congress and Exposition (ECCE), 2015 IEEE*, pp. 1130–1135.

Kou W, Wei D, Zhang P and Xiao W 2015 A direct phase-coordinates approach to fault
    ride through of unbalanced faults in large-scale photovoltaic power systems. *Electric
    Power Components and Systems* **43**(8–10), 902–913.

Krzywinski G 2015 Integrating storage and renewable energy sources into a DC microgrid
    using high gain DC DC boost converters *DC Microgrids (ICDCM), 2015 IEEE First
    International Conference on*, pp. 251–256.

Ramakumar R and Bigger J 1993 Photovoltaic systems. *Proceedings of the IEEE* **81**(3),
    365–377.

Roman E, Alonso R, Ibañez P, Elorduizapatarietxe S and Goitia D 2006 Intelligent PV
    module for grid-connected PV systems. *Industrial Electronics, IEEE Transactions on*
    53(4), 1066–1073.

Romero-Cadaval E, Spagnuolo G, Franquelo L, Ramos-Paja C, Suntio T and Xiao W 2013
    Grid-connected photovoltaic generation plants: components and operation. *IEEE
    Industrial Electronics Magazine* **7**(3), 6–20.

Shawky A, Helmy F, Orabi M, Qahouq J, Dang Z *et al*. 2014 On-chip integrated cell-level
    power management architecture with MPPT for PV solar system *Applied Power
    Electronics Conference and Exposition (APEC), 2014 Twenty-Ninth Annual IEEE*,
    pp. 572–579.

Xiao W, Edwin FF, Spagnuolo G and Jatskevich J 2013 Efficient approaches for modeling
    and simulating photovoltaic power systems. *Photovoltaics, IEEE Journal of* **3**(1),
    500–508.

Xiao W, Elmoursi M, Khan O and Infield D 2016 A review of grid-tied converter topologies
    used in photovoltaic systems. *IET Renewable Power Generation*. In press.

Xiao W, Ozog N and Dunford WG 2007 Topology study of photovoltaic interface for
    maximum power point tracking. *Industrial Electronics, IEEE Transactions on* **54**(3),
    1696–1704.

Xiao W, Torchyan K, Moursi E, Shawky M and Kirtley JL 2014 Online supervisory voltage
    control for grid interface of utility-level PV plants. *Sustainable Energy, IEEE Transactions
    on* **5**(3), 843–853.

Yu FR, Zhang P, Xiao W and Choudhury P 2011 Communication systems for grid
    integration of renewable energy resources. *IEEE Network* **25**(5), 22–29.

3

Safety Standards, Guidance and Regulation

All PV and related products must be manufactured in accordance with safety and performance regulations. Manufacturers and end users treat product certification as the reference of trust. A number of codes and standards have been developed for safe and reliable development of PV power systems. It should be noted that safety standards are different from country to country. The following introduction is based on several well-established standards and provides a fundamental understanding of the safety practice of PV power engineering.

3.1 Certification of PV Modules

The common safety standards of PV module products include IEC-61215, IEC-61646, IEC-61730, and UL-1703, details of which are given in Table 3.1. As well as the PV cells, other components used for PV module construction should be evaluated to ensure their safety, reliability, quality, and performance. These include polymeric materials, junction boxes, connectors, and connecting cables.

The International Electrotechnical Commission (IEC) is a not-for-profit, non-governmental organization, which publishes consensus-based standards and manages conformity assessment systems for electrical and electronic products. In Europe, the leading organizations for certification are the Germany-based TÜV Rheinland and VDE, the Association for Electrical, Electronic, and Information Technologies. In North America, the UL-1703 standard is commonly applied and PV modules are validated by either the Underwriters Laboratories (UL) or the Electrical Testing Laboratory (ETL), a division of the Intertek Group.

For certification, PV module manufacturers must submit the product specification, full material lists, product samples, and related documents to the certification agencies. After an initial evaluation, the certification body starts lab tests to assess construction and performance. For example, the performance evaluation in UL-1703 includes:

- temperature tests
- voltage/current/power measurements
- leakage current tests
- strain relief test
- push test
- cut test

Photovoltaic Power System: Modeling, Design, and Control, First Edition. Weidong Xiao.
© 2017 John Wiley & Sons Ltd. Published 2017 by John Wiley & Sons Ltd.
Companion Website: www.wiley.com/go/xiao/pvpower

Table 3.1 Standards related to PV products.

Standard	Title
IEC 61215	Crystalline silicon terrestrial photovoltaic (PV) modules – Design qualification and type approval
IEC 61646	Thin-film terrestrial photovoltaic (PV) modules – Design qualification and type approval
IEC 61730	Photovoltaic (PV) module safety qualification:
	Part 1 – Requirements for construction
	Part 2 – Requirements for testing
UL 1703	Standard for flat-plate photovoltaic modules and panels
UL 2703	Standard for mounting systems, mounting devices, clamping/retention devices, and ground lugs for use with flat-plate photovoltaic modules and panels
UL 3703	Standard for solar trackers for use with distributed energy resources

IEC, International Electrotechnical Commission; UL, Underwriters Laboratories.

- bonding path resistance test
- dielectric voltage withstanding test
- wet insulation resistance test
- reverse current overload test
- terminal torque test
- impact test
- water spray test
- accelerated aging test
- temperature cycling test
- humidity test
- corrosive atmosphere test
- salt spray test
- metallic coating thickness test
- hot-spot endurance test
- arcing test
- mechanical loading test
- wiring compartment secure test
- fire tests.

This comprehensive evaluation guarantees the product's performance, quality, and safety. When all tests have been passed, the certification can be issued. The end users can search for certified products through the online databases provided by the certification agencies.

UL 2703 provides the safety guidance not only for the construction requirements for the PV mounting rack but also the bonding and grounding requirements that ensure electrical continuity and the prevention of shocks. When solar trackers are used, UL 3703 defines the safety requirement. A solar tracker is a device that orients any solar-related device toward the sun. In PV systems, the solar tracker can be used to minimize the angle of incidence between the incoming sunlight and a PV panel in order to maintain maximum power generation. This is generally essential in CPV and CSP systems. For non-concentrating PV systems, fixed mounting structures are common because of their mechanical simplicity, low cost, and reliability. A solar tracker should not be confused with maximum power point tracking (MPPT), which is an algorithm for controlling power-conditioning systems.

Clarification should be given to the labels "UL Listed" and "UL Recognized".

- *UL Listed* means the PV product has achieved full certification
- *UL Recognized* is used for certified components for PV products.

A product can be assembled only from UL recognized components, but is not necessarily UL listed. For certified PV products, the end users should always look for the UL Listed mark, which shows the certification of the end product, and not just individual components.

3.2   Interconnection Standards

Table 3.2 lists the safety standards of UL 1741 and IEC 62109, which must be met by PV converters for grid interconnection. UL 1741 was first introduced in the late 1990s and revised in 2001. It is the product safety standard that defines the testing and certification requirements for inverters, converters, controllers, and other interconnection system equipment. In most jurisdictions of the USA and Canada, compliance with UL 1741 is mandatory for any PV power system that will be connected to the utility grid. The performance testing of UL 1741 is comprehensive, including

- maximum-voltage measurements
- temperature evaluation
- dielectric voltage-withstand test
- output power characteristics
- abnormal tests
- rounding impedance test
- overcurrent protection calibration test
- strain relief test
- reduced spacings on printed wiring boards tests
- bonding conductor test
- voltage surge test
- calibration test

Table 3.2 Inverter-related standards.

Standard	Title
UL 1741	Standard for Inverters, Converters, Controllers and Interconnection System Equipment for Use With Distributed Energy Resources
IEC 62109-1:2010	Safety of power converters for use in photovoltaic power systems–Part 1: General requirements
IEC 62109-2:2011	Safety of power converters for use in photovoltaic power systems–Part 2: Particular requirements for inverters

IEC, International Electrotechnical Commission; UL, Underwriters Laboratories.

- over-voltage test
- current withstand test
- capacitor voltage determination test
- stability evaluation
- static load
- compression test
- rain and sprinkler tests.

In grid-connected PV systems, the term of islanding refers to the condition in which the distributed resource (DR) continues to supply electric power even though the electric utility is malfunctioning. Islanding involves a potential of safety risk to utility workers. Therefore, the function of anti-islanding or islanding protection is mandatory in UL 1741 against the continuous power supply from the inverter, which is isolated from the remainder of the electric utility system. To meet the requirement, research has been conducted to detect the grid failure accurately and efficiently regardless of the presence of harmonics and noise. Anti-islanding includes both passive and active methods. The active anti-islanding algorithm is more effective than passive counterparts to avoid a non-detection zone, but the conventional algorithm introduces disturbance to the utility grid.

IEC 62109 was developed by IEC and published in 2010. It is a newer safety standard than UL 1741 and applies to power conversion equipment for use in PV systems. The standard has also been used by UL, who issued it as UL 62109-1 to harmonize US and international regulation. According to UL, the requirement is to ensure that the design and system interconnection provide adequate protection for the operator and the surrounding area. Similar to UL 1741, IEC 62109-2 covers covers important testing requirements for safety purposes.

Grid-connected PV systems are having to become more grid-friendly and actively participate in grid stabilization when required. To accommodate this change, the Institute of Electrical and Electronics Engineers (IEEE) developed standard 1547 and released it in 2003 (Basso and DeBlasio 2004). In industry, the combination of UL 1741 and IEEE 1547 was recently applied to grid-connected PV products to regulate both safety and utility support functions. The IEEE 1547 series provides standards for and guidance on the interconnection of distributed generation resources into power grids. The series includes six standards and one draft standard, as listed in Table 3.3.

The terms defined by IEEE 1547 are widely used in this book. They are illustrated in Figure 3.1. The electric power system (EPS) is separated into the area EPS and the local EPS. The dashed lines in Figure 3.1 shows the boundaries of local EPSs. The point of common coupling (PCC) is the connection point between an area EPS and a local EPS. The distributed resource and the point of distributed resource connections are shown as DR and PDRC, respectively.

DRs should cease to power an area EPS if abnormal voltages and/or frequencies are detected. This requirement is commonly referred to as "anti-islanding protection." According to IEEE 1547, the abnormal condition and corresponding clearing time are as defined in Tables 3.4 and 3.5. It should be noted that the frequency definition assumes 60 Hz as the nominal frequency, so the criterion is not applicable to any grid based on 50 Hz, but provides a reference to define abnormal frequencies in term of percentages. Moreover, the clearing time is adjustable from 0.16 to 300 s when a low

Table 3.3 IEEE 1547 series of standards.

Standard	Published	Amended	Title
IEEE 1547	2003	2014	Standard for interconnection and interoperability of distributed energy resources with associated electric power systems interfaces
IEEE 1547.1	2005	2015	Standard for conformance test procedures for equipment interconnecting distributed resources with electric power systems
IEEE 1547.2	2008	n/a	Application guide for IEEE 1547 standard for interconnecting distributed resources with electric power systems
IEEE 1547.3	2007	n/a	Guide for monitoring, information exchange, and control of distributed resources interconnected with electric power systems
IEEE 1547.4	2011	n/a	Guide for design, operation, and integration of distributed resource island systems with electric power systems
IEEE 1547.6	2011	n/a	Recommended practice for interconnecting distributed resources with electric power systems distribution secondary networks
IEEE 1547.7	2013	n/a	Guide to conducting distribution impact studies for distributed resource interconnection
IEEE 1547.8	draft	n/a	Recommended practice for establishing methods and procedures that provide supplemental support for implementation strategies for expanded use of IEEE Standard 1547

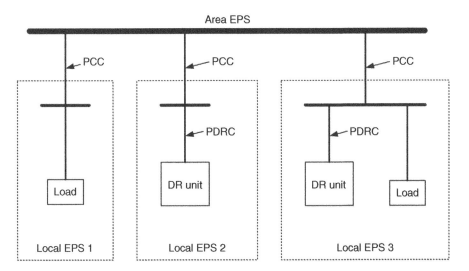

Figure 3.1 Interconnection terms defined by IEEE 1547.

Table 3.4 Clearing time in response to abnormal voltages.

Voltage range	Maximum clearing time if DR ≤ 30 kW (s)	Default clearing time if DR > 30 kW (s)
$V < 50\%\ V_{nom}$	0.16	0.16
$50\%\ V_{nom} \leq V \leq 88\%\ V_{nom}$	2.00	2.00
$110\%\ V_{nom} < V < 120\%\ V_{nom}$	1.00	1.00
$V > 120\%\ V_{nom}$	0.16	0.16

V_{nom}, nominal system voltage; DR, distributed resource capacity.

Table 3.5 Clearing time in response to abnormal frequency.

DR size (kW)	Frequency range (Hz)	Maximum clearing time (s)	Default clearing time (s)
≤ 30	> 60.5	0.16	n/a
	< 59.3	0.16	n/a
> 30	> 60.5	n/a	0.16
	< 57.0	n/a	0.16

system frequency is detected, but is not lower than 57.0 Hz. The definition of clearing time is different for a system rated at less than 30 kW and more than 30 kW. The default clearing time is recommended for large systems, and the standard defines the maximum clearing time for small-scale systems.

Power quality is another important specification in IEEE 1547. DC current injection should be minimized so that it is not greater than 0.5% of the full rated output current at the PDRC. Harmonic current injection should be lower than the maximum limits. The odd harmonic limits defined by IEEE 1547 are shown in Table 3.6. The even harmonics should be limited to 25% of the odd harmonic limits. It should be noted that the total demand distortion (TDD) is different from the total harmonic distortion (THD):

- TDD is the ratio of harmonic current distortion to the maximum demand load current
- THD is the ratio of the sum of the powers of all harmonic components to the power of the fundamental frequency.

Under full load conditions, the value of THD is equivalent to that of the TDD. When the injected current is lower than the rated value, the TDD is generally lower than the THD. The definition of TDD indicates only the harmonic impact of the applied DR on the overall system. The limits of individual odd harmonics that are defined in IEEE 1547 are slightly different from the 2001 version of UL 1741.

Table 3.6 Limits of maximum harmonic current distortion in odd order.

Harmonic order (h)	$h < 11$	$11 \leq h < 17$	$17 \leq h < 23$	$23 \leq h < 35$	$35 \leq h$	TDD
Percentage (%)	4.0	2.0	1.5	0.6	0.3	5.0

3.3   System Integration to Low-voltage Networks

In Germany, the VDE 4105 code of practice is generally applied for PV interconnections with low-voltage networks. In the USA, the requirements for inverter installations and grid interconnections are governed by the National Electric Code, Articles 690 and 705 (NEC 2014). The articles provide an important and comprehensive guideline to circuit sizing and safe and reliable current protection. The articles identify key components of solar PV systems, which are defined as PV source circuits, PV output circuits, inverter input circuits, inverter, inverter output circuits, and grid connections. A typical grid-connected PV system is shown in Figure 3.2. The important components are listed and explained in Table 3.7.

3.3.1   Grounded Systems

Equipment grounding is required not only for PV power systems but for any electrical system with a hazard voltage rating. For safety, all exposed metal parts that carry no current in PV power systems should be grounded to earth. This is common practice in electrical engineering. A grounded PV system is defined as one in which the DC conductors (either positive or negative) are bonded to the equipment grounding system, which in turn is connected to the earth. Grounding the PV system addresses safety protection, preventing electrical shocks, because one of the DC terminals can be defined as "live" instead of having both floating, as in ungrounded systems. System grounding is sometimes referred to as "functional grounding" to distinguish it from the equipment grounding. Functional grounding is required by a lot of local jurisdictions

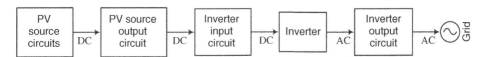

Figure 3.2 A grid-connected PV system as defined by Article 690 of the National Electric Code.

Table 3.7 Key components in solar PV systems.

Terms	Description
Photovoltaic source circuits	The circuits between modules and from modules to the common connection point(s) of the DC system. Usually refers to the circuitry to form the PV array.
Photovoltaic output circuit	The circuit conductors between the PV source circuits and the inverter input circuit. The common components include DC combiner boxes, DC overcurrent protection devices, and cables to form the PV output circuit.
Inverter input circuit	Conductors between the inverter and the PV output circuits. These usually include DC disconnects and cables to form the inverter input circuit.
Inverter output circuit	Conductors between the inverter and the utility service equipment for electrical production and distribution network. The common components for the inverter output circuit include AC disconnect, AC overcurrent protection devices, and cables.

in the USA and by several PV module manufacturers, such as First Solar and Sunpower, in order to ensure the safety of grid connections and the reliability of PV products. However, the requirements are inconsistent from region to region and country to country. The latest version of Article 690 allows ungrounded PV sources and output circuits.

In AC distribution systems, the neutral point is often connected to earth. In a grounded PV system, one of the DC terminals is connected to the neutral point through the same grounding. Therefore, galvanic isolation through the power conversion is required for functional grounding in grid-connected PV systems. In general, an ungrounded system has better conversion efficiency and lower cost than a grounded system thanks to its simplicity. However, the decision to use a grounded or ungrounded system is difficult because many aspects must be considered: safety, reliability, installation environment, power quality, and so on.

3.3.2 DC Ground Fault Protection

DC ground fault protection is specified in Article 690 as a way to reduce fire hazards. The system should be able to detect a ground fault in DC circuits, including conductors and DC components. When a fault is detected, the flow of fault current will be interrupted and the fault should be flagged for maintenance and correction. However, it cannot generally protect personnel from electric shocks.

A current flow without any DC ground fault is shown in Figure 3.3. The DC conductor carrying negative current is earthed. Since the value of I_{DC+} is equal to that of I_{DC-}, the ground connection conducts no current, indicated as $I_{GND} = 0$.

A DC ground fault is illustrated in Figure 3.4. When a human receives an electric shock from the positive terminal, the I_{PV+} current becomes different from I_{DC+} since part of the current is deviated to the earth through the human body. The ground current is no longer zero, which can be detected by the inverter, as shown in Figure 3.4a. The ground fault protection will open the switch (SW) to disconnect from the PV output circuit. However, the function can not protect personnel from electric shocks since there is no dedicated switch to separate the human from the solar generators, as shown in Figure 3.4b.

3.3.3 Voltage Specification

In grid-connected PV systems, the AC voltage rating is the grid voltage at the PDRC. The design follows conventional grid codes that were established for AC systems. The

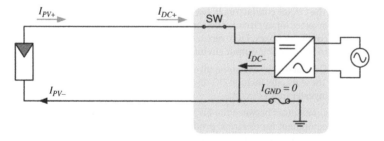

Figure 3.3 Normal operation without DC ground fault.

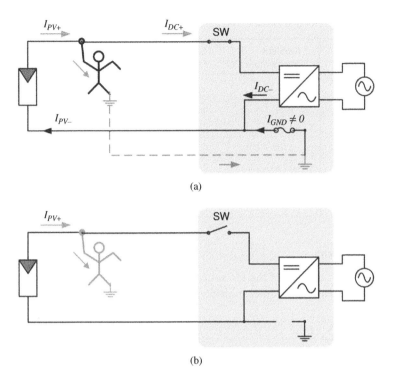

(a)

(b)

Figure 3.4 DC ground fault and protection: (a) DC ground fault detected; (b) protection action responding to DC ground fault.

maximum DC voltage is first calculated as the sum of the rated open-circuit voltages of the series connected PV modules. Further, the voltage should be corrected for the lowest expected ambient temperature as follows:

$$V_{pvdc}(max) = K_T \times \sum_{1}^{N_{series}} V_{oc} \tag{3.1}$$

where V_{oc} and N_{series} refer to the open-circuit voltage of the PV modules at STC and the number of series-connected modules. The value of the correction factor, K_T, can be derived from the PV product datasheet or the criteria in Article 690. Based on the product datasheet, the correction factor can be determined as follows:

$$K_T = 1 + \beta_T(T_{cold} - 25) \tag{3.2}$$

where β_T and T_{cold} represent the temperature coefficient provided by the manufacturer and the expected coldest temperature at the installation location.

The maximum DC voltage should be used to determine the voltage rating for cables, disconnects, overcurrent protection devices, and other devices in the DC circuits. Article 690 provides a reference for the correction factors for Si crystalline modules, and the details are summarized in Table 3.8. These can be used for either the reference or the correction calculation when the temperature coefficient is unavailable in the product datasheet. It should be noted that the expected coldest temperature is just an estimate since it is hard to accurately predict.

Table 3.8 Voltage correction factors for Si crystalline-based PV circuits.

Ambient temperature (°C)	Correction factor (K_T)	Ambient temperature (°C)	Correction factor (K_T)
24 to 20	1.02	19 to 15	1.04
14 to 10	1.06	9 to 5	1.08
4 to 0	1.10	−1 to −5	1.12
−6 to −10	1.14	−11 to −15	1.16
−16 to −20	1.18	−21 to −25	1.20
−26 to −30	1.21	−31 to −35	1.23
−36 to −40	1.25		

3.3.4 Circuit Sizing and Current

In Article 690, the maximum current of the PV source circuit (PVsc) is the sum of the short-circuit current of the parallel modules multiplied by 125%:

$$I_{PVsc}(max) = 1.25 \times \sum_{1}^{N_{parallel}} I_{sc} \tag{3.3}$$

where I_{sc} and $N_{parallel}$ refer to the short-circuit current of PV modules at STC and the number of parallel connected modules.

The maximum current of the output circuit (PVOC) is the sum of the parallel source circuit maximum currents:

$$I_{PVOC}(max) = \sum_{1}^{N_{PVsc}} I_{PVsc}(max) \tag{3.4}$$

where $I_{PVsc}(max)$ and N_{PVsc} refer to maximum current of the PVsc and the number of them in parallel connection.

The circuit conductors and overcurrent devices should be sized to carry not less than 125% of the maximum currents as calculated in (3.3) and (3.4) for the PV source circuits and PV output circuit, respectively. This is casually referred to as "the Rule of 1.56" since it is equal to 1.25×1.25, as described above. Cables used for outdoor installations should be sunlight- and moisture-resistant. When ambient temperatures are higher than 30°C, the ampacity of cables should be derated by the factors found in the electrical codes. It is usually recommended to use cables with a temperature rating of 90°C or more. In any DC portion of the PV power system, it is important to apply overcurrent protection (OCP) devices, such as circuit breakers, which are certified for DC circuits. Fuses are commonly used as the OCP devices. Meanwhile, the device should meet the requirement in terms of voltage, current, and interrupt ratings, as calculated in (3.1), (3.3), and (3.4), with consideration of the corresponding correction factors.

3.3.5 Cable Selection

Since the PVsc connects the modules to each other and to the common connection point(s) of the DC system, cable selection is important for long-term reliability and

Table 3.9 Cable types.

Type	Temperature rating (°C)	Description
THHN	90	Dry and damp locations, flame retardant
THWN-2	90	Dry and wet locations, flame retardant
RHH	90	Dry and damp locations, flame retardant
RHW-2	90	Dry and wet locations, flame retardant
USE-2	90	Underground service entrances, sunlight resistant
XHHW-2	90	Dry and wet locations, flame retardant

safety. Due to the need for outdoor installation, the cable should withstand moisture, high temperatures, and intense UV. The cable should be rated at 90°C or even higher. Table 3.9 lists common wires that can be used for PV source circuits and their constraints. The flame-retardant feature allows cables to be safely installed inside conduits. The sunlight-resistant feature allows cables to be installed outdoors without any further protection against UV damage. From the list, a USE-2 cable can be used directly for the PVsc. However, a regular USE-2 cable cannot be installed in a conduit since it is not flame retardant. A "solar cable" is usually USE-2 for sunlight resistance, but other types can be used too. This makes on-site installation easy and flexible. Cable ampacity should be derated by appropriate factors with the consideration of high ambient temperatures. The cables of the PVsc and PVOC should not share the same raceway.

3.3.6 Connectors and Disconnects

According to Article 690, if the installation is readily accessible and the voltage is higher than 30 V, connectors should be of the latching or locking type, which require tools to open them. For reasons of safety and reliability, this requirement prevents the interconnection from being accidentally disconnected. For DC connections, the PV connectors should be distinguishable with respect to polarity to avoid incorrect connections. Another important requirement in the article is the installation location of disconnects. For safety purposes, they should be at a readily accessible location.

3.3.7 Grid Interconnections through Power Distribution Panels

The power distribution panel is also called the load center, since electricity is conventionally delivered to the local distribution company through the service entrance at the main switch. AC circuit breakers are the safety protection devices in the panel, which are designed to prevent overcurrents and potential fire hazards. These devices protect against overcurrents or overloads by stopping the flow of electricity when it exceeds safe levels.

The point of distributed resource connection of a small-scale PV system can use the same LV power distribution panel for grid interconnections, as shown in Figure 3.5. PV power generation mainly supports local loads, and surplus power is transferred to the grid through the power distribution panel, which is available in buildings. The system allows customers to offset electric power drawn from the utility. This is called net metering. The system is low cost and highly efficient since the PV power is directly used

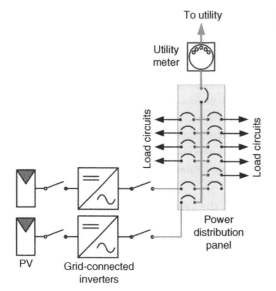

Figure 3.5 Grid interconnection to LV network through power distribution panel.

by the loads, avoiding transmission losses. Furthermore, maintenance and cleaning are easy because the system is accessible and close to end users. Such systems have been installed in Australia, the USA, and Germany. However, warning signs are critical for such installation for safety reasons.

3.3.8 Marking

The article requires labeling of PV power sources and system component rating. For example, a warning sign at the coupling device must indicate the dual power supply, especially for residential applications. Otherwise, the PV power integration would be a potential risk to maintenance workers since the electric power is not only from the grid but also from the PV source circuit.

3.4 System Integration to Medium-voltage Network

With the spread of PV generated power, utility-scale PV systems have drawn significant attention to the question of ensuring grid stability. The intermittent nature of PV power generation has a negative impact on transient system response in utility grids.

A technical guideline on generating plant connections to and parallel operation with medium-voltage networks was published in 2008 by BDEW, the German Association of Energy and Water Industries.[1] The European Network of Transmission System Operators (ENTSO-E), which represents 42 electricity transmission system operators from 35 countries across Europe,[2] published guidelines entitled the "Network Code on Requirements for Grid Connection Applicable to all Generators" in the European Union Official Journal in 2016.

---

1 Bundesverband der Energie- und Wasserwirtschaft e.V.
2 https://www.entsoe.eu/.

Table 3.10 Voltage levels for PV power integration.

Voltage definition	Description
Low voltage (LV)	The LV grid handles the distribution of electricity to end consumers. It commonly refers to 400 V, three-phase or 230 V, single phase.
Medium voltage (MV)	Electricity is distributed to large consumers at levels from 6 to 30 kV.
High voltage (HV)	Electricity is transmitted regionally at the voltage levels from 60 to 110 kV.

The general definitions of three voltage levels are summarized in Table 3.10. The "extra-high" voltage level is not included in the table since voltages of 220–600 kV are mainly used for long-distance transmission. It should be noted that the voltage levels in Table 3.10 should be used for reference only since they might be different from country to country.

The BDEW guideline mainly focuses on stability considerations when large-scale PV power systems are integrated into medium-voltage (MV) networks or high-voltage (HV) grids. Such systems were introduced in Section 1.10 and illustrated in Figure 1.15c. Several important aspects were discussed in the guideline, including active power throttling, reactive power support, and fault ride-through (FRT) (El Moursi et al. 2013). In North America, FRT is commonly referred to as "low-voltage ride-through."

3.4.1 Active Power Throttling

The normal operation of a PV system is MPPT for the most effective solar energy harvesting. However, when the grid is overpowered, the frequency deviates from the nominal value. Active power throttling (APT) takes effect if the frequency deviation reaches a certain level. It should be noted that active power throttling should be coordinated among all distributed generators and monitored by the network operator. According to the BDEW guideline, APT reduces power injection when the grid frequency is detected as being above 50.2 Hz. This means that the PV power system no longer operates at the maximum power point. The reduction is expressed as:

$$\Delta P = 20 \times P_{50.2} \times \frac{50.2 - f}{50} \tag{3.5}$$

where $P_{50.2}$ refers to the recorded power level when the frequency is 50.2 Hz, which is just before the throttling action is activated. It is also illustrated in Figure 3.6.

APT is disabled when the grid frequency is back to its normal value of 50.05 Hz. The difference between 50.05 Hz and 50.2 Hz provides a hysteresis band to prevent any absurd control actions. When the normal frequency is detected and confirmed, the system should be able to terminate the APT operation and resume MPPT. As shown in Figure 3.6, the distributed generation (DG) system should be disconnected from the grid if the frequency deviates severely from the nominal value: either less than 47.5 Hz or higher than 51.5 Hz.

3.4.2 Fault Ride-through

A large-scale PV power system that is connected to an MV network should stay connected and contribute to the grid even if there is a severe grid voltage disturbance.

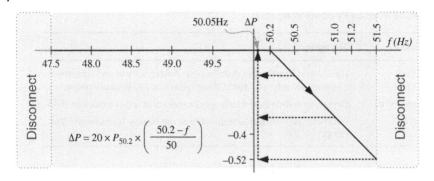

Figure 3.6 Active power throttling for medium voltage integration.

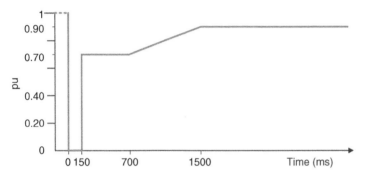

Figure 3.7 Fault ride-through for medium-voltage integration.

This is because any disconnection may further degrade the voltage, during and after the disturbance. Figure 3.7 shows how FRT works. The limit line indicates that the system interconnection should not be disconnected during the first 150 ms of any voltage dip. If this is an instantaneous fault, the grid voltage recovers gradually. The DG system is requested to remain connected to the MV network so long as the grid voltage is above these limit lines versus the time period, as shown in Figure 3.7.

Various solutions have been presented in the literature for the dynamic grid support in terms of FRT, for both solar and wind systems (El Moursi et al. 2013; Huang et al. 2014, 2013; Kou et al. 2015). In general, FRT for PV power systems is relatively straightforward in comparison with wind power systems. The reason is that PV power generation is fully controllable and stoppable without any voltage or mechanical stress. The deviation from the maximum power point of the open-circuit voltage can effectively limit the PV power output from the peak to zero.

3.4.3 Reactive Power Support

Voltage control in electrical power systems is important to maintain the network voltage profile. Generally, increasing reactive power causes the network voltage to rise, and decreasing it results in a voltage drop. Therefore, the BDEW guideline recommends that a large-scale PV system should be able to produce reactive power to support the stability of grid voltage, especially during grid faults, to assist fast grid-voltage recovery (Xiao et al. 2014). The value of the power factor at the PCC can deviate from unity, and range from 0.95 under-excited to 0.95 over-excited. The network operator can schedule and assign a target value of the reactive power supply to individual PV power plants through a predefined scheme or real-time communication. Energy buffers such as capacitor banks, which are available in grid-connected PV inverters, are used to support reactive power production from grid-connected PV systems.

3.5 Summary

Standards and electrical codes are important guidelines for developing PV power systems with full consideration of safety, quality, and reliability. This chapter has briefly introduced several important standards that are widely used in Europe and North America. The regulations set out in standards and codes also provide a useful reference for future research and development. The regulation of PV modules and grid-connected products is referred to the standards and test procedures developed by UL and the IEC. LV interconnections follow the requirements defined in Article 690 of the NEC. MV grid connections refer to the technical guideline of BDEW. Large-scale PV power systems that are connected to MV networks should stay connected and contribute to the grid in the event of severe grid voltage disturbances since any disconnection may further degrade voltages.

The product safety certification for inverters, converters, controllers, and other interconnection system equipment is guided by the UL 1741 and IEC 62109 standards. Equipment grounding should not be confused with the PV grounding system. According to the definition of grounded PV systems, the DC conductors (either positive or negative) should be connected to earth at a central point. The term "functional grounding" is sometimes used to distinguish this from ordinary equipment grounding.

Each code or standard for grid interconnection is only applied in a certain region. For example, the grid frequency defined by Article 690 refers to 60 Hz, which is different from Europe and most Asian countries. However, the deviation of voltage and frequency in percentage terms can be applied to 50-Hz systems, so that other countries can define their local electric code for PV power systems.

Regulation also provides a guideline for academic research. For example, significant research has been undertaken into making a seamless transition between standalone mode and the grid-connected mode for PV power systems. However, electrical codes might require a certain waiting time before PV power systems reconnect, when the grid is recovering from a blackout.

Problems

3.1 Find your local electrical code for PV system grid interconnections.

3.2 Find your local electrical code for the specification of LV, MV, and HV.

3.3 Evaluate the local climate and find the temperature correction required for voltage when a certain PV product is used.

3.4 Find a solution to protect humans from electric shocks from PV outputs.

3.5 Visit a local PV power system, evaluate if all components have proper certification marks, find if the installation follows the relevant code and guidance, and identify any potential hazards in the system based on the knowledge gained from this chapter.

3.6 Study the Network Code on Requirements for Grid Connection Applicable to all Generators, issued by ENTSO-E. Summarize the important functions that are required for static and dynamic grid support.

References

Basso T and DeBlasio R 2004 IEEE 1547 series of standards: interconnection issues. *Power Electronics, IEEE Transactions on* **19**(5), 1159–1162.

El Moursi MS, Xiao W and Kirtley Jr JL 2013 Fault ride through capability for grid interfacing large scale PV power plants. *IET Generation, Transmission & Distribution* **7**(9), 1027–1036.

Huang PH, El Moursi MS, Xiao W and Kirtley J 2014 Fault ride-through configuration and transient management scheme for self-excited induction generator-based wind turbine. *Sustainable Energy, IEEE Transactions on* **5**(1), 148–159.

Huang PH, El Moursi MS, Xiao W and Kirtley Jr JL 2013 Novel fault ride-through configuration and transient management scheme for doubly fed induction generator. *IEEE Transactions on Energy Conversion* **28**(1), 148–159.

Kou W, Wei D, Zhang P and Xiao W 2015 A direct phase-coordinates approach to fault ride through of unbalanced faults in large-scale photovoltaic power systems. *Electric Power Components and Systems* **43**(8–10), 902–913.

NEC 2014 National Electrical Code. NFPA 70, National Fire Protection Association.

Xiao W, Torchyan K, Moursi E, Shawky M and Kirtley JL 2014 Online supervisory voltage control for grid interface of utility-level PV plants. *Sustainable Energy, IEEE Transactions on* **5**(3), 843–853.

4

PV Output Characteristics and Mathematical Models

I–V and P–V curves are commonly used to illustrate the outputs of PV cells, modules, strings, or arrays, as discussed in Section 1.6. Computer simulation is an important tool to reproduce the behavior of PV power systems in response to various environmental conditions and load disturbances. A computational model can be developed to recreate the PV generator output under variations in irradiance and temperature. There are various modeling approaches that have been presented in literature to represent PV output characteristics.

Doping creates an interface between two types of semiconductor material – p-type and n-type – inside a single crystal of a semiconductor. These p-n junctions are the elementary units of most semiconductor electronic devices: diodes, transistors, and integrated circuits. A crystalline-based PV cell is also constructed from a large area of silicon p-n junction, so a diode model is naturally used to represent the output characteristics of crystalline-based solar cells.

A model to represent crystalline-based PV cells is usually formed from the equivalent circuits. These models are usually categorized into two main types: single-diode models (SDMs) and double-diode models (DDMs), and their equivalent circuits are shown in Figures 4.1 and 4.2, respectively.

An ideal model for PV cells should have a current source in parallel with a diode. However, due to various non-ideal factors, the equivalent circuit of the standard SDM also includes one shunt resistor and one series resistor, as illustrated in Figure 4.1. In this book, the model is defined as the complete single diode model (CSDM) in order to distinguish it from other simplified single-diode models (SSDMs). The current–voltage characteristics according to the equivalent circuit are expressed as

$$i_{pv} = i_{ph} - i_d - \frac{v_d}{R_h} \tag{4.1}$$

The I–V characteristics of the p-n junction diode is nonlinear, and it can be represented in exponential form:

$$i_d = i_s \left[e^{\left(\frac{q v_d}{k T_c A_n} \right)} - 1 \right] \tag{4.2}$$

where v_d is equal to

$$v_d = v_{pv} + i_{pv} R_s \tag{4.3}$$

This is based on the theory of Shockley (1949). The symbol T_c is the absolute temperature of the p-n junction. The ideality factor of the diode, A_n, is a measure of how closely

Photovoltaic Power System: Modeling, Design, and Control, First Edition. Weidong Xiao.
© 2017 John Wiley & Sons Ltd. Published 2017 by John Wiley & Sons Ltd.
Companion Website: www.wiley.com/go/xiao/pvpower

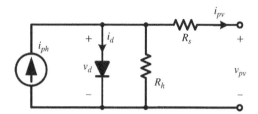

Figure 4.1 Equivalent circuit of single-diode model.

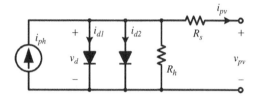

Figure 4.2 Equivalent circuit of double-diode model.

the diode follows the ideal diode equation, which is defined as $A_n = 1$. The value of A_n typically ranges from 1 to 2. Values of the p-n junction constants, model parameters, and variables are listed in Tables 4.1–4.3, because they are frequently used in the chapter. To represent the relation of v_{pv} and i_{pv}, or the I–V curve, five unknown parameters should be identified to give the CSDM: i_{ph}, i_s, A_n, R_s, and R_h. Thus, the model is sometimes referred to as the five-parameter model.

The DDM includes one current source, two diodes, and shunt and series resistances, as shown in Figure 4.2. The output current of the DDM is expressed as

$$i_{pv} = i_{ph} - i_{d1} - i_{d2} - \frac{v_d}{R_h} \tag{4.4}$$

where the currents in the two diodes can be expressed as:

$$i_{d1} = i_{s1}\left[e^{\left(\frac{qv_d}{kT_cA_{n1}}\right)} - 1\right] \tag{4.5}$$

$$i_{d2} = i_{s2}\left[e^{\left(\frac{qv_d}{kT_cA_{n2}}\right)} - 1\right] \tag{4.6}$$

and where the diode voltage is

$$v_d = v_{pv} + i_{pv}R_s \tag{4.7}$$

Table 4.1 PV model coefficients.

Symbols	Term definition	Value
E_{STC}	Irradiance at STC*	1000 W/m²
k	Boltzmann constant	1.38×10^{-23} J/K
q	Charge	1.6×10^{-19} C
T_{CS}	PV cell temperature at STC*	298°K
V_{TCS}	Thermal voltage of p-n junction at STC*	25.7 mV

*Standard test conditions (STC): 1000 W/m², AM1.5 standard reference spectrum, and 298 K (or 25°C).

Table 4.2 PV model parameters and constants.

Symbols	Definition	Unit
α_T	Temperature coefficient on PV current	(A/A)/K
β_T	Temperature coefficient on PV voltage	(V/V)/K
γ_T	Irradiance coefficient on PV power	(W/W)·m^2/W
v_T	Irradiance coefficient on PV voltage	(V/V)·m^2/W
A_n	Diode ideality factor in the SDM	n/a
A_{n1}	Diode ideality factor 1 in the double-diode model	n/a
A_{n2}	Diode ideality factor 2 in the double-diode model	n/a
I_{MS}	PV current at the maximum power point (MPP) at the STC	A
I_{ph}	PV photon current at the STC	A
I_{SCS}	PV short-circuit current at the STC	A
I_{SS}	PV short-circuit current at the STC	A
P_{MPP}	PV power at the maximum power point (MPP) at the STC	W
R_s	Series resistance	Ω
R_h	Shunt resistance	Ω
V_{MS}	PV voltage at the maximum power point at the STC	V
V_{OCS}	PV open-circuit voltage at the STC	V

Table 4.3 PV model variables.

Symbols	Definition	Unit
E_a	Solar irradiance	W/m^2
I_M	Instant MPP current	A
i_{ph}	PV photon current	A
i_{pv}	PV cell output current	A
i_d	Diode current	A
i_s	Diode reverse-bias saturation current	A
T_C	PV cell temperature	K
v_d	Diode voltage	V
V_M	Instant MPP voltage	V
v_{OC}	PV open-circuit voltage	V
v_{pv}	PV cell terminal voltage	V
v_t	Thermal voltage of p–n junction	V

Seven unknown parameters must be identified in the DDM: i_{ph}, i_{s1}, i_{s2}, A_1, A_2, R_S, and R_h. The model coefficients and variables are as set out in Tables 4.1–4.3. The DDM is commonly considered to be a more comprehensive model of PV cell output characteristics than the SDM. Two more tuning parameters appear in the model than in the SDM, in order to accurately reproduce the output characteristics of PV cells. However, the model is not commonly used due to its complexity (Huang et al. 2016;

Mahmoud et al. 2013). The model in (4.4) includes two independent terms for diode current, which increases the complexity for parameter identification and computer simulation. Improper tuning of the parameters can prevent the claimed advantage of accuracy being achieved. Parameter identification for the DDM is also very sensitive to initial conditions (Romero-Cadaval et al. 2013).

Due to the complexity of the DDM, the SDM is normally used for simulating PV power systems, because it offers a reasonable trade-off between simplicity and accuracy. For this reason, the DDM will not be further discussed.

4.1 Ideal Single-diode Model

Based on the p-n junction structure for both PV cell and diode, the ideal single-diode model (ISDM) is a current source in parallel with a diode, as shown in Figure 4.3. It can be considered the simplest SDM. In comparison with the CSDM shown in Figure 4.1, the series resistance and shunt resistance are removed. The mathematical expressions are therefore:

$$i_{pv} = i_{ph} - \underbrace{i_s \left[e^{\left(\frac{q v_{pv}}{k T_c A_n} \right)} - 1 \right]}_{i_d} \tag{4.8}$$

Without the resistances, three model parameters must be identified in the modeling process: the photon current (i_{ph}), the diode reverse bias saturation current (i_s), and the diode ideality factor (A_n). Three independent constraints are required to identify the three unknown parameters in (4.8).

4.1.1 Product Specification

A common source for identifying values for the PV model parameters is the product datasheets provided by PV cell or module manufacturers. The modeling process is to match the PV output characteristics with these data, which are prepared at STC. Table 4.4 illustrates one such specification. All data refer to the multi-crystalline solar cell, IM156B3, which is manufactured by MOTECH Industrial Inc. To avoid any confusion, it should be noted that the datasheet was downloaded in 2014 showing the version of October 2012. It might not be consistent with the latest data since the manufacturer often releases new product versions. The data that are useful for PV cell modeling and simulation include the STC values of the short-circuit current I_{SCS}, open-circuit voltage V_{OCS}, the operating voltage and current at the maximum power point (MPP) (V_{MS}, I_{MS}), the temperature coefficients (α_T, β_T, and γ_T), and the correction factors for irradiation on the electrical outputs. The symbols α_T, β_T, and γ_T are the

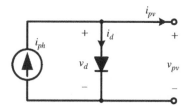

Figure 4.3 Equivalent circuits of the ideal single-diode model.

Table 4.4 Sample solar cell data.

	Basic information		
Manufacturer	Model	Cell material	Dimensions
MOTECH	IM156B3-164	Multi-crystalline	$156 \times 156 \pm 0.5$ mm

	Electrical performance at STC			
Efficiency	P_{MPP}	I_{MS}	V_{MS}	I_{SCS}	V_{OCS}
16.4 %	3.99 W	7.85 A	0.509 V	8.38 A	0.614 V

Temperature coefficients		
α_T (%/°C)	β_T (%/°C)	γ_T (%/°C)
0.06	−0.33	−0.40

Correction factors for irradiance		
E_a (W/m²)	Voltage correction (V/V)	Current correction (A/A)
1000	1.000	1.000
800	0.989	0.798
600	0.972	0.597
200	0.911	0.192

All symbols shown above refer to Tables 4.2 and 4.3.

temperature coefficients for the correction of the PV cell output current, voltage, and power, respectively. Parameter identification is performed at the PV cell level since the cell is the fundamental unit for construction of modules and arrays.

4.1.2 Parameter Identification at Standard Test Conditions

ISDM parameters are initially determined at STC since the data are available from the product datasheet. For the case study, the modeling uses the data in Table 4.4.

When the terminal of the equivalent circuit, as shown in Figure 4.3, is shorted, the diode current, i_d, is equal to zero. The value of the photon current, i_{ph}, is equal to the short-circuit current, I_{SCS}, which is available from the product datasheet. Thus the photon current, i_{ph} becomes known at STC. Two unknowns remain in the ISDM to represent the diode.

When the terminal of the equivalent circuit, as shown in Figure 4.3, is opened, the output current of the PV cell, i_{pv}, is equal to zero. The value of the diode current, i_d, becomes equal to the photon current, i_{ph}, which is the I_{SCS} at STC. This can be expressed as in (4.9), which includes two unknown parameters of I_{SS} and A_n:

$$I_{SCS} = I_{SS} \left[e^{\left(\frac{V_{OCS}}{V_{TCS} A_n} \right)} - 1 \right] \tag{4.9}$$

where V_{TCS} is the thermal voltage at STC, which is a constant and expressed as

$$V_{TCS} = \frac{kT_{CS}}{q} \tag{4.10}$$

At the MPP for STC, the I–V characteristics in (4.8) can be rewritten as (4.11), which includes two unknown parameters:

$$I_{MS} = I_{SCS} - I_{SS}\left[e^{\left(\frac{V_{MS}}{V_{TCS}A_n}\right)} - 1\right] \tag{4.11}$$

The unknown parameters I_{SS} and A_n can be determined by solving the two nonlinear equations, (4.9) and (4.11). They can also be combined to form one equation:

$$\frac{e^{\left(\frac{V_{MS}}{V_{TCS}}\frac{1}{A_n}\right)} - 1}{e^{\left(\frac{V_{OCS}}{V_{TCS}}\frac{1}{A_n}\right)} - 1} = 1 - \frac{I_{MS}}{I_{SCS}} \tag{4.12}$$

A new variable A_{inv} is defined as the reciprocal of A_n:

$$A_{inv} = \frac{1}{A_n} \tag{4.13}$$

Thus (4.12) can be rearranged as

$$f(A_{inv}) = e^{C_1 A_{inv}} - C_3 e^{C_2 A_{inv}} - 1 + C_3 = 0 \tag{4.14}$$

where the constants are calculated as:

$$C_1 = \frac{V_{MS}}{V_{TCS}} \tag{4.15a}$$

$$C_2 = \frac{V_{OCS}}{V_{TCS}} \tag{4.15b}$$

$$C_3 = 1 - \frac{I_{MS}}{I_{SCS}} \tag{4.15c}$$

The Newton–Raphson (NR) method, or simply the Newton method, is a numerical technique to solve nonlinear equations (Xiao 2007). The derivation of $f(A_{inv})$, as expressed in (4.16), is required for the NR iteration.

$$f'(A_{inv}) = C_1 A_{inv} e^{C_1 A_{inv}} - C_3 C_2 A_{inv} e^{C_2 A_{inv}} \tag{4.16}$$

The solver requires an initial estimate of the root of (4.14), which is defined as the value of $A_{inv}(0)$. Back to the fundamental of the p-n junction, the ideality factor (A_n) usually ranges from 1 to 2. Therefore, the value of $A_{inv}(0)$ can be assigned a value of 0.7 as the initial point.

The software flowchart for parameter identification for the ISDM is illustrated in Figure 4.4. The numerical iteration starts with an initial estimate and continues by the updating procedure:

$$A_{inv}(n + 1) = A_{inv}(n) - \frac{f[A_{inv}(n)]}{f'[A_{inv}(n)]} \tag{4.17}$$

The value of A_{inv} is continuously updated until the output of $f(A_{inv})$ is close to zero. The tolerance is defined as the maximum error ($Err = 10^{-6}$). If the constraints are satisfied, the iteration stops and the value of (A_n) is finalized. This represents the ideality factor

Figure 4.4 Flowchart of the NR method for parameter identification of an ideal single-diode model.

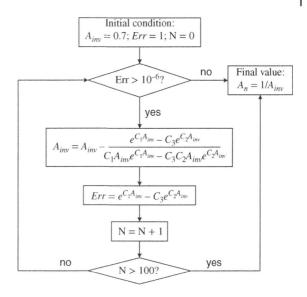

of the ISDM. The identification loop is also assigned the maximum iterative number ($N = 100$). The identification process fails if the maximum iterative number is reached.

The ideality factor (A_n) can also be found using other numerical approaches: the Matlab "solve" function.

With the parameters given in Table 4.4 and the ISDM expression in (4.8), the value of A_n is found to be 1.4798 with both methods. Based on the same software platform, it took 9 µs and 0.54 s to reach the final value of A_n by the NR method and the "solve" function of Matlab, respectively. This shows that the NR method is an efficient numerical solver when a good estimate is applied at initialization.

With the known value of A_n, the another parameter, I_{SS}, can be determined from

$$I_{SS} = \frac{I_{SCS}}{e^{C_2 A_{inv}} - 1} \tag{4.18}$$

This equation is derived from (4.9) at STC. The reverse saturation current, I_{SS}, is determined as 8.17×10^{-7} for the PV cell, MOTECH IM156B3. Thus, based on the values in Table 4.4 and the identified parameters, the current and voltage characteristics at STC can be represented by

$$i_{pv} = 8.38 - 8.17 \times 10^{-7} \times (e^{26.29 V_{pv}} - 1) \tag{4.19}$$

The modeling performance can be visualized by plotting the output next to the product data. The I–V and P–V curves are shown in Figure 4.5. The modeled MPP is the peak power point derived from the modeled P–V curve. The datasheet MPP refers to the value in Table 4.4.

4.1.3 Variation with Irradiance and Temperature

The above discussion refers to standard test conditions (STC) of solar irradiance, air mass, and cell temperature. In reality, conditions can be significantly different from STC and can also vary from region to region and from season to season. The

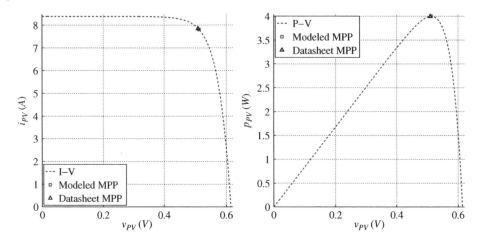

Figure 4.5 I–V and P–V curves for the IM156B3 PV cell.

I–V characteristics of simulation models should respond to variations in both solar irradiance and cell temperature. The PV cell manufacturer usually provides correction coefficients derived from experimental evaluations. The temperature coefficients for PV output current, voltage, and power are denoted α_T, β_T, and γ_T, respectively. Example data are set out in Table 4.4. The following discussion is based on the assumption that the ideality factor, A_n, is constant regardless of environmental variations.

When the operating conditions are different to STC, the expression for i_{ph} should be

$$i_{ph}(E_a, \Delta T) = \frac{E_a}{E_{STC}} I_{SCS}(1 + \alpha_T \Delta T) \tag{4.20}$$

where E_a represents the instant irradiance value and ΔT is the difference between the cell temperature T_C and the temperature at STC, T_{CS}, which is 25°C or 298 K. The constants I_{SCS} and E_{STC} refer to the short-circuit current at STC and the irradiance level at STC of 1000 W/m². The short-circuit current is mostly proportional to the irradiance value, as expressed in (4.20). Since α_T is a positive parameter, the photon current increases or decreases in the same direction as the cell temperature.

The output voltage also varies with changes of cell temperature and solar irradiance. Regarding to the cell temperature variation, the correction factor is denoted by β_T. The irradiance coefficient on voltage, v_T, is nonlinear, and is typically given in a table, as in Table 4.4. A function can be derived to estimate the correction factors for the radiation level. A polynomial function can be used to correct the voltage for irradiance levels:

$$v_T(E_a) = C_{E1} E_a^2 + C_{E2} E_a + C_{E3} \tag{4.21}$$

For this case, the order of the polynomial is 2. Higher orders give better accuracy, but increase the computational complexity.

Based on the data in the product datasheet, the coefficients, $C_{E(1-3)}$, can be identified using a software tool, such as the "fit" function in Matlab. Using the data in Table 4.4, the function for the voltage correction is

$$v_T(E_a) = -1 \times 10^{-7} E_a^2 + 2.315 \times 10^{-4} E_a + 0.8688 \tag{4.22}$$

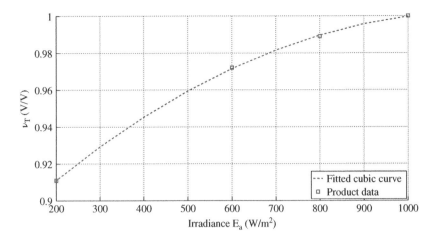

Figure 4.6 Voltage correction factors for modeling and simulation.

The curve fitting match can be evaluated by plotting the function's output and the data given by the manufacturer, as shown in Figure 4.6. The polynomial curve is generally a good match with the product data.

The open-circuit voltage correction for variation of irradiance and cell temperature is

$$v_{oc}(E_a, \Delta T) = V_{OCS}(1 + \beta_T \Delta T)v_T \tag{4.23}$$

The open-circuit voltage is updated by the correction factors β_T and v_T, which are always updated by (4.21).

The values of $i_{ph}(E_a, \Delta T)$ and $v_{oc}(E_a, \Delta T)$ are estimated from (4.20) and (4.23). Their values vary with changes in the solar radiation and cell temperature. The value of the diode saturation current $i_s(E_a, \Delta T)$ can be determined for open-circuit conditions since the photon current is equal to the diode current, as shown in Figure 4.3. The value can be updated by (4.24) for the instantaneous values of $v_{oc}(E_a, \Delta T)$ and $i_{ph}(E_a, \Delta T)$.

$$i_s(E_a, \Delta T) = \frac{i_{ph}(E_a, \Delta T)}{e^{\left[\frac{qv_{oc}(E_a, \Delta T)}{kT_c A_n}\right]} - 1} \tag{4.24}$$

The I–V characteristic equation of a PV cell with variations of the solar irradiance and cell temperature can be written as

$$i_{pv} = i_{ph}(E_a, \Delta T) - \underbrace{i_s(E_a, \Delta T)\left[e^{\left(\frac{qv_{pv}}{kT_c A_n}\right)} - 1\right]}_{i_d(E_a, \Delta T)} \tag{4.25}$$

Figure 4.7 is the flowchart for the real-time simulation based on the ISDM. The inputs include the instantaneous values of the irradiance, E_a, and the cell temperature, T_C. The PV cell current, i_{pv}, is updated every cycle as the output, based on the variation of the cell output voltage, v_{pv}, and the environmental conditions. The variables in (4.25), $i_{ph}(E_a, \Delta T)$, $v_{oc}(E_a, \Delta T)$, and $i_s(E_a, \Delta T)$ can be estimated from (4.20)–(4.24).

Based on the product data in Table 4.4, the I–V curves are simulated and plotted in Figures 4.8 and 4.9 to illustrate the impact of changes in solar irradiance and cell

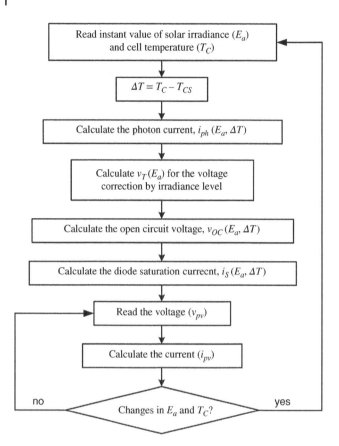

Figure 4.7 Flowchart for simulating PV cell output.

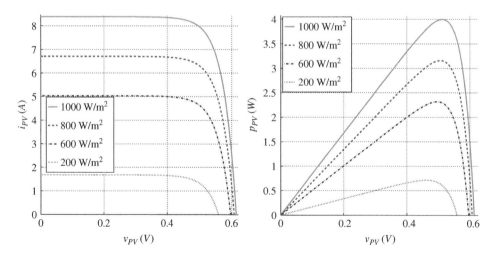

Figure 4.8 Modeled I–V curve of the PV cell IM156B3 with constant temperature (25°C) and variable irradiance.

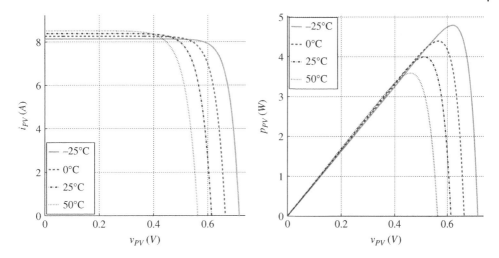

Figure 4.9 Modeled I–V curve of the PV cell IM156B3 with constant irradiance (1000 W/m²) and variable temperature.

temperatures, respectively. It shows that the highest PV generation power is at high solar irradiance and low cell temperature. The power output is almost proportional to the irradiance when the temperature is constant. It should be noted that a lower cell temperature yields higher power output. Therefore, ventilation should be considered for PV installations.

The advantages of the ISDM lie in the straightforward parameterization process and the simple structure for model implementation, thanks to which potentially fast simulation speeds can be expected. The modeling process guarantees that the output I–V curve at STC passes through the three critical points, $(V_{OCS}, 0)$, $(0, I_{SCS})$, and (V_{MS}, I_{MS}), which are provided by the product datasheet.

4.2 Model Accuracy and Performance Indices

Model accuracy can be measured at STC, where the manufacturer's data are available for comparison and verification. The open-circuit voltage of PV generators at STC is the reference for DC voltage rating in PV power systems. The difference in the open-circuit voltage between the model output and the product data can be quantified by the performance index:

$$D_{OC} = \sqrt{\left(\frac{\tilde{V}_{OCS}}{V_{OCS}} - 1\right)^2} \tag{4.26}$$

where $\tilde{V}_{OCS}$ and V_{OCS} represent the values of the open-circuit voltage that are generated from the simulation model and the product data, respectively.

The short-circuit current provides an upper current limit for rating the overcurrent protection devices, cables, and DC disconnects. D_{SC} is the performance index, which indicates how accurately the simulation model output represents the short-circuit

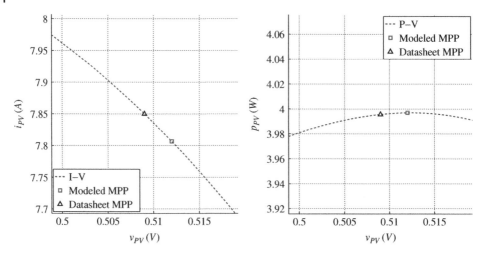

Figure 4.10 Zoom-in plots of I–V and P–V for the IM156B3 PV cell.

current from the product data. It is calculated by

$$D_{SC} = \sqrt{\left(\frac{\tilde{I}_{SCS}}{I_{SCS}} - 1\right)^2} \tag{4.27}$$

where $\tilde{I}_{SCS}$ and I_{SCS} represent the values of the short-circuit current from the simulation model and the product data, respectively.

Following the modeling process using the ISDM, there is no error at the short-circuit and open-circuit points since both constraints are included for parameter identification. Based on the model example, as shown in (4.19), the current is 8.38 A when the voltage is 0 V, and the current is zero when the voltage is 0.614 V. Therefore, the performance indices of D_{OC} and D_{SC} are not relevant for the ISDM. Following the P–V curve, as shown in Figure 4.5, the modeled MPP can be identified at the point (0.512 V, 7.81 A). The point is different from the true MPP point of (0.509 V, 7.85 A) that is provided by the product datasheet. The deviation is noticeable in Figure 4.10, which zoomed in around the MPP.

The constraint for the MPP is the condition in (4.11). This can only guarantee that the model-produced I–V curve passes through the MPP given by the manufacturer's datasheet. In fact, the ISDM modeling process does mean that the MPP given in the product datasheet is the maximum value in the modeled P–V function. Thus, it is normal that there is a difference between the model and datasheet MPPs.

The modeling accuracy of the MPP is therefore important since it represents the product power rating. The concern is the value and location of the modeled MPP in comparison with the actual product outputs. The distance from the true MPP in the P–V plot can be used as a measure of the modeling accuracy. The true MPP location is normalized to values of 1 and 1, which represent the maximum power (P_{MPP}) and the corresponding PV voltage (V_{MS}), respectively. The distance from the true MPP is normalized and calculated as

$$D_{MPP} = \sqrt{\left(\frac{\tilde{P}_{MPP}}{P_{MPP}} - 1\right)^2 + \left(\frac{\tilde{V}_{MPP}}{V_{MS}} - 1\right)^2} \tag{4.28}$$

where $\tilde{P}_{MPP}$ and $\tilde{V}_{MPP}$ represent the maximum power and the corresponding voltage, respectively, which are generated from the PV model (Huang et al. 2016).

Therefore, the D_{MPP} is considered as a measure of the accuracy of the simulation model output in representing the true MPP. In the modeling example, D_{MPP} is calculated as 0.0059. For a small-scale PV power system of 1000 IM156B3 PV cells in series connection, the ISDM model indicates that the MPP is located at a voltage of 512 V, which is 3 V different from the output of the real array (509 V). The deviation of the MPP is 11.63 W for this specific modeling example. If the error is acceptable, the ISDM is a suitable model, exhibiting the advantages of modeling simplicity and simulation efficiency.

The performance indices D_{OC}, D_{SC}, and D_{MPP} assume that values for the open-circuit, short-circuit, and maximum power points are available from product datasheets. However, a complete dataset can be obtained either from experimental measurements or from the manufacturer. This enables the full I–V curve of a specific PV product to be plotted. In this case, an additional performance index can be calculated to define the complete curve-fitting accuracy. This modeling performance index was defined by Mahmoud et al. (2013). Using the PV cell voltage as the reference, the difference of the PV current between the model and experimental measurements can be calculated by way of the root-mean-square deviation (RMSD). The RMSD is computed as the square root of the mean-square error and is mathematically expressed as

$$RMSD(I) = \sqrt{\frac{\sum_{j=1}^{N} (\tilde{I}_j - I_j)^2}{N}} \qquad (4.29)$$

Based on the reference voltage sample, $\tilde{I}_j$ represents the current sample of the model output and I_j represents the current sample of the actual value. N is the number of samples. Similarly, the RMSD for the power deviation can be calculated as

$$RMSD(P) = \sqrt{\frac{\sum_{j=1}^{N} (\tilde{P}_j - P_j)^2}{N}} \qquad (4.30)$$

where $\tilde{P}_j$ represents the power sample of the model output and P_j represents the power sample of the actual value.

The RMSD is not a uniform performance index for PV cells. The output current rating of solar cells varies greatly from one to another due to the differences in size and materials used. The normalized root-mean-square deviation (NRMSD) is also recommended by Mahmoud et al. (2013) to quantify the modeling accuracy. A relative measurement allows different solar cells to be compared on the same platform even if their power capacities are very different. The short-circuit current at STC can be used as the normalizing base. NRMSD is calculated as

$$NRMSD(I) = \frac{RMSD(I)}{I_{SCS}} \qquad (4.31)$$

where I_{SCS} is the short-circuit current at STC, which represents the reference base of the PV current. It is a uniform performance index for comparing modeling accuracy of different solar cells and different modeling techniques. Similarly, the NRMSD can be determined for the power output as

$$NRMSD(P) = \frac{RMSD(P)}{P_{MPP}} \qquad (4.32)$$

where P_{MPP} is the maximum power at STC, which is the reference base for normalization of the PV output power.

The performance index discussed above is based on STC, which is normal for evaluating PV cell products and simulation model performance. Furthermore, the data at STC are readily available from product manufacturers for quantitative comparison. Performance indices, such as D_{MPP} and *NRMSD*, can be used to evaluate model accuracy based on conditions different from STC. Reliable data are essential for advanced and fair performance analysis when variations of irradiance and temperature are considered.

4.3 Simplified Single-diode Models

Besides the ISDM, the CSDM is often simplified to another two versions by neglecting either the series resistor or the shunt resistor, as shown in Figures 4.11 and 4.12.

One type of SSDM makes the series resistor, R_s equal to zero, but includes the shunt resistor R_h in the equivalent circuit. The model is referred to as SSDM1 in this book, and the circuit is shown in Figure 4.11. Including one additional unknown parameter, R_h, in contrast with the ISDM, the I–V expression becomes

$$i_{pv} = i_{ph} - i_s \left[e^{\left(\frac{q v_{pv}}{k T_c A_n} \right)} - 1 \right] - \frac{v_{pv}}{R_h} \tag{4.33}$$

Another type of SSDM makes the shunt resistor R_h equal to ∞, but includes the series resistor R_s in the model circuit. In this chapter, the model is referred to as SSDM2, and this is shown in Figure 4.12. In comparison with the ISDM, the I–V expression includes an additional unknown parameter, R_s:

$$i_{pv} = i_{ph} - i_s \left[e^{\left(\frac{q v_d}{k T_c A_n} \right)} - 1 \right] \tag{4.34}$$

where

$$v_d = v_{pv} + i_{pv} R_s \tag{4.35}$$

The models of SSDM1 and SSDM2 are sometimes referred to as four-parameter models because of the fourth unknown parameter: either R_h or R_s. The fourth constraint for

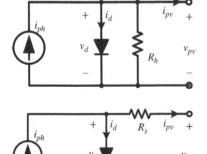

Figure 4.11 Equivalent circuit of simplified single-diode model with shunt resistor.

Figure 4.12 Equivalent circuit of simplified single-diode model with series resistor.

parameter identification can be derived from

$$\left.\frac{dp_{pv}}{dv_{pv}}\right|_{(V_{MS}, I_{MS})} = 0 \qquad (4.36)$$

This is based on the extremum value theorem, which implies that the maximum of the P–V function can be identified at the critical point. The critical point is the MPP (V_{MS}, I_{MS}), and the derivative there must be equal to zero.

4.3.1 Parameter Identification: Part One

To identify four unknown parameters, the fourth constraint is based on (4.36). For SSDM1 at STC, the derivative operation results in

$$I_{SCS} - I_{SS}\left[\left(1 + \frac{V_{MS}}{V_{T_{CS}} A_n}\right) e^{\left(\frac{V_{MS}}{V_{T_{CS}} A_n}\right)} - 1\right] - \frac{2V_{MS}}{R_h} = 0 \qquad (4.37)$$

Applying the constants and variables that are defined in (4.15) and (4.13), this can be simplified to

$$I_{SCS} - I_{SS}[(1 + C_1 A_{inv})e^{(C_1 A_{inv})} - 1] - \frac{2V_{MS}}{R_h} = 0 \qquad (4.38)$$

Referring to Figure 4.11, the photon current i_{ph} is equal to the short-circuit current. The photon current i_{ph} is also equivalent to the sum of the diode current and the current through the shunt resistor at open-circuit conditions. Therefore, the equivalence at STC is expressed by

$$I_{SCS} = I_{SS}[e^{(C_2 A_{inv})} - 1] + \frac{V_{OCS}}{R_h} \qquad (4.39)$$

At the MPP and STC, the I–V characteristics can be written as

$$I_{SCS} - I_{MS} = I_{SS}[e^{(C_1 A_{inv})} - 1] + \frac{V_{MS}}{R_h} \qquad (4.40)$$

In principle, the three unknown parameters for SSDM1, A_n, R_h, and I_{SS}, can be determined using a system solver from (4.38)–(4.40). At STC, the model output is constrained to match the open-circuit voltage (V_{OCS}), the short-circuit current (I_{SCS}), and the MPP location (V_{MS}, I_{MS}) from the manufacturer's data.

In SSDM1, the unknown parameters can be defined as a vector:

$$x_k = \begin{bmatrix} I_{SS}(k) \\ A_{inv}(k) \\ G_h(k) \end{bmatrix} \qquad (4.41)$$

and the NR method can then be applied to identify the unknown parameters. An unknown G_h is introduced, which is $1/R_h$. and the index, k, represents the NR operation in discrete time. The NR iteration is

$$x_{k+1} = x_k - J(x_k)^{-1} F(x_k) \qquad (4.42)$$

where $F(x_k)$ is conformed by (4.43), which is derived from the three constraints in (4.38)–(4.40).

$$F_k = \begin{bmatrix} I_{SCS} - I_{SS}(k)[e^{C_2 A_{inv}(k)} - 1] - V_{OCS}G_h(k) \\ I_{SCS} - I_{MS} - I_{SS}(k)[e^{C_1 A_{inv}(k)} - 1] - V_{MS}G_h(k) \\ I_{SCS} - I_{SS}(k)\{[1 + C_1 A_{inv}(k)]e^{C_1 A_{inv}(k)} - 1\} - 2V_{MS}G_h(k) \end{bmatrix} \tag{4.43}$$

The partial derivatives (4.38)–(4.40) are shown as J_k, which is a 3×3 matrix expressed in the discrete time by the index, k:

$$J_k = \begin{bmatrix} J_{11}(k) & J_{12}(k) & J_{13}(k) \\ J_{21}(k) & J_{22}(k) & J_{23}(k) \\ J_{31}(k) & J_{32}(k) & J_{33}(k) \end{bmatrix} \tag{4.44}$$

where

$$J_{11}(k) = -e^{C_2 inv((k)} + 1 \tag{4.45a}$$

$$J_{21}(k) = -e^{C_1 inv((k)} + 1 \tag{4.45b}$$

$$J_{31}(k) = -[1 + C_1 inv(k)]e^{C_1 inv((k)} + 1 \tag{4.45c}$$

$$J_{12}(k) = -I_{SS}(k)C_2 e^{C_2 A_{inv}(k)} \tag{4.46a}$$

$$J_{22}(k) = -I_{SS}(k)C_1 e^{C_1 inv(k)} \tag{4.46b}$$

$$J_{32}(k) = -I_{SS}(k)[C_1 e^{C_1 inv(k) + C_1^2 A_n(k)e^{C_1 inv(k)}}] \tag{4.46c}$$

$$J_{13}(k) = -V_{OCS} \tag{4.47a}$$

$$J_{23}(k) = -V_{MS} \tag{4.47b}$$

$$J_{33}(k) = -2V_{MS} \tag{4.47c}$$

The solver can start with the estimation of the values of A_{inv}, G_h, and I_{SS}, which form the initial vector, x_k. If the ISDM parameters have been determined, they can be directly used for initializing A_{inv} and I_{SS}. The value of G_h can start with a zero value. The software flowchart for parameterization is shown in Figure 4.13. When the norm of the incremental vector, Δx_k, is close to zero, the iteration stops and outputs the final parameter vector, x_k. Otherwise, the iteration stops when it reaches 10 000 cycles. The parameters are output and converted to parameters in terms of A_n, I_{SS}, and R_h, which are used to form the output function of the SSDM1.

The parameterization and evaluation is for the PV cell, the parameters of which were given in Table 4.4. For the ISDM, the two important parameters are $A_n = 1.4798$ and $I_{SS} = 8.17 \times 10^{-7}$. Based on the NR iteration, as shown in Figure 4.13, the values of A_n, R_h, and I_{SS} can be identified for SSDM1. It took 27 iteration cycles to identify the parameters. Thus, the current and voltage characteristics at STC can be constructed by following the model function in (4.33). The modeling performance can be visualized by plotting the model output and the product data. The I–V and P–V curves are plotted in Figure 4.14. For comparison, these are plotted together with the output of the ISDM developed in

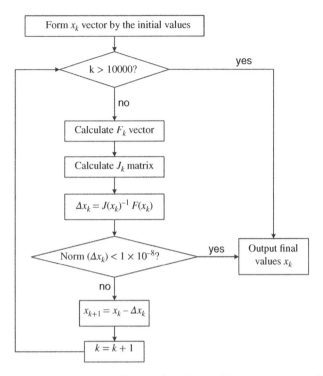

Figure 4.13 Flowchart for identifying PV model parameters using Newton–Raphson iteration.

Section 4.1. The modeled MPP refers to the peak power points that were derived from the SSDM1 and ISDM models.

The model accuracy can be measured at STC, since the manufacturer's data are available for comparison. Following the modeling process using the SSDM1, there is no error at the short-circuit and open-circuit points since both constraints are included for parameter identification. Following the P–V curve, as shown in Figure 4.14, the modeled MPP can be identified at (0.509 V, 7.85 A), with no deviation from the MPP based on the product specification. The advantage of the SSDM lies in the introduction of the fourth constraint to represent the maximum value in PV cell output. The MPP match is clearly shown in Figure 4.15, which zooms in around the MPP. For reference, the MPP modeled by the ISDM is also indicated in the plot.

4.3.2 Parameter Identification: Part Two

All components in the equivalent circuit that is used for the PV cell model should be physically realistic. The values of A_n, R_h, and I_{SS} that were identified for the SSDM1 are 1.5708, $-5.069\,\Omega$, and 2.1×10^{-6} A, respectively. Although a deviation in the MPP value has been avoided, a negative value is identified for R_h. The shunt resistance must obviously be positive. As a result, SSDM1 is not an appropriate model of the IM156B3 cell. The following parameterization will focus on the SSDM2 model shown in Figure 4.12.

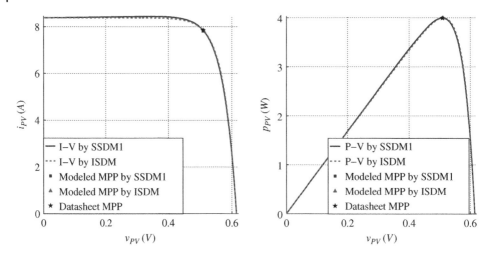

Figure 4.14 I–V and P–V curves of the IM156B3 PV cell modeled using ISDM and SSDM1.

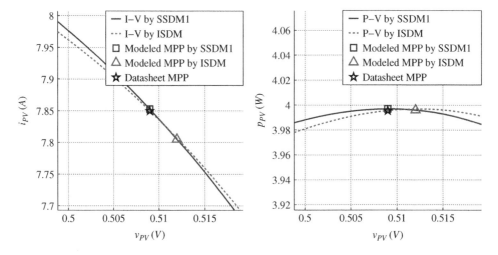

Figure 4.15 Zoom-in plots of I–V and P–V of model outputs for the IM156B3 PV cell.

Using SSDM2 is more complicated than ISDM or SSDM1. One reason is that the series resistance and the current appear inside an exponential term, as shown in (4.34) and (4.35). The I–V characteristics are represented by a coupled equation, $i_{pv} = g(v_{pv}, i_{pv})$, which is different from the direct format, $i_{pv} = g(v_{pv})$ used for ISDM and SSDM1. Furthermore, the photon current, i_{ph}, cannot be simply replaced by the short-circuit current since the series resistance causes a residual voltage at the short-circuit condition.

To remove the coupling and form a direct expression for i_{pv} and v_{pv}, the I–V representation in (4.34) is turned into the format of $v_{pv} = h(i_{pv})$, as shown in (4.48), at STC.

$$v_{pv} = \underbrace{V_{TCS} A_n \ln\left(\frac{i_{ph} + I_{SS} - i_{pv}}{I_{SS}}\right) - i_{pv} R_s}_{h(i_{pv})} \tag{4.48}$$

The power output is therefore expressed as

$$P_{pv} = i_{pv} V_{TCS} A_n \ln \left(\frac{i_{ph} + I_{SS} - i_{pv}}{I_{SS}} \right) - i_{pv}^2 R_s \qquad (4.49)$$

The first constraint for the numerical solver applies the short-circuit condition to (4.34):

$$I_{SCS} = I_{ph} - I_{SS} \left[e^{\left(\frac{I_{SCS} R_s A_{inv}}{V_{TCS}} \right)} - 1 \right] \qquad (4.50)$$

Representing (4.34) for open-circuit conditions, the second constraint can be established as

$$I_{ph} - I_{SS} \left[e^{\left(\frac{V_{OCS} A_{inv}}{V_{TCS}} \right)} - 1 \right] = 0 \qquad (4.51)$$

and from (4.34) at the MPP, the third constraint can be established as

$$I_{MS} = I_{ph} - I_{SS} \left[e^{\left(\frac{(V_{MS} + I_{MS} R_s) A_{inv}}{V_{TCS}} \right)} - 1 \right] \qquad (4.52)$$

From (4.36) and (4.49), the fourth constraint can be defined as

$$\frac{V_{TCS}}{A_{inv}} \ln \left(\frac{I_{ph} + I_{SS} - I_{MS}}{I_{SS}} \right) - \frac{V_{TCS} I_{MS}}{A_{inv}} \ln \left(\frac{1}{I_{ph} + I_{SS} - I_{MS}} \right) - 2 I_{MS} R_s = 0 \quad (4.53)$$

From (4.50)–(4.53), the function, $F(x_k)$, can be built as

$$F(x_k) = \begin{bmatrix} I_{ph} - I_{SS}(k) \left[e^{\left(\frac{I_{SCS} R_s(k) A_{inv}(k)}{V_{TCS}} \right)} - 1 \right] - I_{SCS} \\ \\ I_{ph} - I_{SS}(k) \left[e^{\left(\frac{V_{OCS} A_{inv}(k)}{V_{TCS}} \right)} - 1 \right] \\ \\ I_{ph} - I_{SS}(k) \left[e^{\left(\frac{(V_{MS} + I_{MS} R_s(k)) A_{inv}(k)}{V_{TCS}} \right)} - 1 \right] - I_{MS} \\ \\ \frac{V_{TCS}}{A_{inv}(k)} \ln \left(\frac{\lambda_i(k)}{I_{SS}} \right) - \frac{V_{TCS} I_{MS}}{A_{inv}(k)} \ln \left(\frac{1}{\lambda_i(k)} \right) - 2 I_{MS} R_s(k) \end{bmatrix} \qquad (4.54)$$

where $\lambda_i(k) = I_{ph}(k) + I_{SS}(k) - I_{MS}$. This has four unknowns, which form the vector, x_k, in discrete time:

$$x_k = \begin{bmatrix} I_{SS}(k) \\ A_{inv}(k) \\ G_s(k) \\ I_{ph}(k) \end{bmatrix} \qquad (4.55)$$

From the four unknowns, partial differentiation can construct J, which becomes a 4×4 matrix in discrete time:

$$J(x_k) = \begin{bmatrix} J_{11}(k) & J_{12}(k) & J_{13}(k) & J_{14}(k) \\ J_{21}(k) & J_{22}(k) & J_{23}(k) & J_{24}(k) \\ J_{31}(k) & J_{32}(k) & J_{33}(k) & J_{34}(k) \\ J_{41}(k) & J_{42}(k) & J_{43}(k) & J_{44}(k) \end{bmatrix} \qquad (4.56)$$

where

$$J_{11}(k) = -e^{\left[\frac{I_{SCS}R_s(k)A_{inv}(k)}{V_{TCS}}\right]} + 1 \tag{4.57a}$$

$$J_{21}(k) = -e^{\left[\frac{V_{OCS}A_{inv}(k)}{V_{TCS}}\right]} + 1 \tag{4.57b}$$

$$J_{31}(k) = -e^{\left\{\frac{[V_{MS}+I_{MS}R_s(k)]A_{inv}}{V_{TCS}}\right\}} + 1 \tag{4.57c}$$

$$J_{41}(k) = \frac{V_{TCS}}{A_{inv}}\left[\frac{1}{\lambda_i} - \frac{1}{I_{SS}(k)}\right] + \frac{I_{MS}V_{TCS}}{A_{inv}(\lambda_i)^2} \tag{4.57d}$$

$$J_{12}(k) = -\frac{I_{SS}(k)I_{SCS}R_s(k)}{V_{TCS}}e^{\left[\frac{I_{SS}(k)I_{SCS}R_s(k)A_{inv}(k)}{V_{TCS}}\right]} \tag{4.58a}$$

$$J_{22}(k) = -\frac{I_{SS}(k)V_{OCS}}{V_{TCS}}e^{\left[\frac{V_{OCS}A_{inv}(k)}{V_{TCS}}\right]} \tag{4.58b}$$

$$J_{32}(k) = -\frac{I_{SS}(k)[V_{MS}+I_{MS}R_s(k)]}{V_{TCS}}e^{\left[\frac{[V_{MS}+I_{MS}R_s(k)]A_{inv}(k)}{V_{TCS}}\right]} \tag{4.58c}$$

$$J_{42}(k) = -\frac{1}{A_{inv}^2}\left[V_{TCS}\ln\left(\frac{\lambda_i}{I_{SS}(k)}\right) - \frac{I_{MS}V_{TCS}}{\lambda_i}\right] \tag{4.58d}$$

$$J_{13}(k) = -\frac{I_{SS}(k)I_{SCS}R_s(k)A_{inv}(k)}{V_{TCS}}e^{\left[\frac{I_{SCS}R_s(k)A_{inv}(k)}{V_{TCS}}\right]} \tag{4.59a}$$

$$J_{23}(k) = 0 \tag{4.59b}$$

$$J_{33}(k) = -\frac{I_{SS}(k)I_{MS}A_{inv}}{V_{TCS}}e^{\left\{\frac{[V_{MS}+I_{MS}R_s(k)]A_{inv}}{V_{TCS}}\right\}} \tag{4.59c}$$

$$J_{43}(k) = -2I_{MS} \tag{4.59d}$$

$$J_{14}(k) = 1 \tag{4.60a}$$

$$J_{24}(k) = 1 \tag{4.60b}$$

$$J_{34}(k) = 1 \tag{4.60c}$$

$$J_{44}(k) = \frac{V_{TCS}}{A_{inv}\lambda_i} + \frac{I_{MS}V_{TCS}}{A_{inv}\lambda_i^2} \tag{4.60d}$$

From Table 4.4, the values of A_n, R_s, I_{ph}, and I_{SS} for SSDM2 are identified as 1.3044, 1.6 mΩ, 8.38 A, and 9.3186×10^{-8} A, respectively. In this case, the photon current is the same as the short-circuit current since R_s is low in value. All parameters have a physical meaning for the specific PV cell, IM156B3. The numerical operation follows the NR iteration, as shown in Figure 4.13.

In this way, the current and voltage characteristics at STC can be constructed by following the model function in (4.34). The modeling performance can be visualized by plotting the simulation model output against the product data. The I–V and P–V curves are plotted in Figure 4.16. together with the output of the ISDM model developed in Section 4.1. The modeled MPPs are the peak power points derived from the SSDM2 and ISDM models.

The model accuracy can be measured at STC, where the manufacturer's data are available for comparison. Using SSDM2, there is no error at the short-circuit or open-circuit points since both constraints are included in the parameter identification. Following

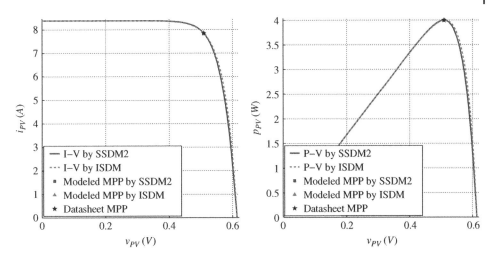

Figure 4.16 I–V and P–V curves of the IM156B3 PV cell modeled using ISDM and SSDM2.

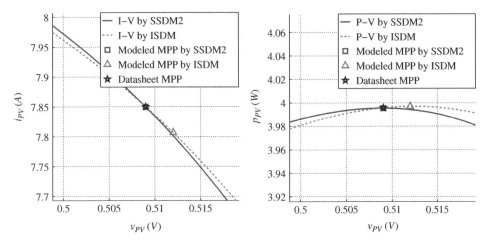

Figure 4.17 Zoom-in plots of I–V and P–V of model output for the IM156B3 PV cell.

the P–V curve, as shown in Figure 4.16, the modeled MPP can be identified at (0.509 V, 7.85 A), with no deviation from the true MPP from the product specification. The MPP match is clearly shown in Figure 4.17, which zooms in around the MPP. For reference, the MPP modeled by the ISDM is also indicated. It should be noted that the modeling process for SSDM2 is a little more complicated than for SSDM1 or ISDM. The significant advantage of the simplified single-diode models, SSDM1 or SSDM2, is that they can match the three important points at STC.

Due to the complexity of parameterization for the SSDM2, it might have convergence issues if the initial values are not properly assigned. Since R_s is low in value, the photon current I_{ph} can be estimated in the same way as the short-circuit current, I_{SCS} at STC. Therefore, the SSDM2 is simplified for identification with three unknowns in x_k, as shown in (4.61).

$$x_k = \begin{bmatrix} I_{SS}(k) \\ A_{inv}(k) \\ G_s(k) \end{bmatrix} \tag{4.61}$$

The functions in F_k can be built as a 3×1 vector:

$$F_k = \begin{bmatrix} I_{SCS} - I_{SS}(k) \left[e^{\left(\frac{V_{OCS} A_{inv}(k)}{V_{TCS}} \right)} - 1 \right] \\ I_{SCS} - I_{SS}(k) \left[e^{\left(\frac{(V_{MS} + I_{MS} R_s(k)) A_{inv}(k)}{V_{TCS}} \right)} - 1 \right] - I_{MS} \\ \frac{V_{TCS}}{A_{inv}(k)} \ln \left(\frac{\lambda_i(k)}{I_{SS}} \right) - \frac{V_{TCS} I_{MS}}{A_{inv}(k)} \ln \left(\frac{1}{\lambda_i(k)} \right) - 2 I_{MS} R_s(k) \end{bmatrix} \tag{4.62}$$

where $\lambda_i(k) = I_{SCS} + I_{SS}(k) - I_{MS}$.

Based on the three unknowns, partial differentiation is applied to construct the J_k, which becomes a 3×3 matrix in discrete time:

$$J_k = \begin{bmatrix} J_{11}(k) & J_{12}(k) & J_{13}(k) \\ J_{21}(k) & J_{22}(k) & J_{23}(k) \\ J_{31}(k) & J_{32}(k) & J_{33}(k) \end{bmatrix} \tag{4.63}$$

where

$$J_{11}(k) = -e^{\left[\frac{V_{OCS} A_{inv}(k)}{V_{TCS}} \right]} + 1 \tag{4.64a}$$

$$J_{21}(k) = -e^{\left\{ \frac{[V_{MS} + I_{MS} R_s(k)] A_{inv}}{V_{TCS}} \right\}} + 1 \tag{4.64b}$$

$$J_{31}(k) = \frac{V_{TCS}}{A_{inv}} \left[\frac{1}{\lambda_i} - \frac{1}{I_{SS}(k)} \right] + \frac{I_{MS} V_{TCS}}{A_{inv}(\lambda_i)^2} \tag{4.64c}$$

$$J_{12}(k) = -\frac{I_{SS}(k) V_{OCS}}{V_{TCS}} e^{\left[\frac{V_{OCS} A_{inv}(k)}{V_{TCS}} \right]} \tag{4.65a}$$

$$J_{22}(k) = -\frac{I_{SS}(k)[V_{MS} + I_{MS} R_s(k)]}{V_{TCS}} e^{\left[\frac{V_{MS} + I_{MS} R_s(k)] A_{inv}(k)}{V_{TCS}} \right]} \tag{4.65b}$$

$$J_{32}(k) = -\frac{1}{A_{inv}^2} \left[V_{TCS} \ln \left(\frac{\lambda_i}{I_{SS}(k)} \right) - \frac{I_{MS} V_{TCS}}{\lambda_i} \right] \tag{4.65c}$$

$$J_{13}(k) = 0 \tag{4.66a}$$

$$J_{23}(k) = -\frac{I_{SS}(k) I_{MS} A_{inv}}{V_{TCS}} e^{\left\{ \frac{[V_{MS} + I_{MS} R_s(k)] A_{inv}}{V_{TCS}} \right\}} \tag{4.66b}$$

$$J_{33}(k) = -2 I_{MS} \tag{4.66c}$$

With the simplified process with three unknowns, the parameterization gives values of A_n, R_s, and I_{SS} for the SSDM2 of 1.3044, 1.6 mΩ, and 9.3186×10^{-6} A, respectively. They are the same as the parameters that are identified using the complex modeling process and 4×4 matrix in J_k.

4.3.3 Variation with Irradiance and Temperature

The I–V characteristics of the simulation models should respond to differences in solar irradiance and cell temperature. The correlation for the ISDM was discussed in Section 4.1.3. The open-circuit voltage for both SSDM1 and SSDM2, $v_{oc}(E_a, \Delta T)$, can be updated by (4.23) in response to variations in irradiance and temperature.

For SSDM1, the correction for the photon current, $i_{ph}(E_a, \Delta T)$, is the same as (4.20). For SSDM2, it can be updated using

$$i_{ph}(E_a, \Delta T) = \frac{E_a}{E_{STC}} I_{ph}(1 + \alpha_T \Delta T) \tag{4.67}$$

since the value of I_{ph} identified is not exactly equal to the short-circuit current because of the series resistance. For SSDM2, the diode saturation current $i_s(E_a, \Delta T)$ can be determined at the open-circuit condition since the photon current is equal to the diode current, as shown in Figure 4.12. The value can be updated by (4.24) for the instantaneous values of $v_{oc}(E_a, \Delta T)$) and $i_{ph}(E_a, \Delta T)$.

For SSDM1, the diode saturation current $i_s(E_a, \Delta T)$ can be determined at the open-circuit condition in the presence of the shunt resistance, R_h. The value can be updated for the instantaneous values of $v_{oc}(E_a, \Delta T)$) and $i_{ph}(E_a, \Delta T)$ using

$$i_s(E_a, \Delta T) = \frac{i_{ph}(E_a, \Delta T) - \frac{v_{oc}(E_a, \Delta T)}{G_h}}{e^{\left[\frac{v_{oc}(E_a, \Delta T)}{v_T A_n}\right]} - 1} \tag{4.68}$$

where $v_T = kT_c/q$. Thus, the current can be updated every simulation cycle for the instantaneous values of the photon current and the diode saturation current representing the variations in solar irradiance and cell temperature using

$$i_{pv} = i_{ph}(E_a, \Delta T) - i_s(E_a, \Delta T) \left[e^{\left(\frac{v_{pv}}{v_T A_n}\right)} - 1\right] - \frac{v_{pv}}{R_h} \tag{4.69}$$

When the SSDM2 is selected as the simulation model, the I–V characteristics and their response to environmental variation are

$$i_{pv} = i_{ph}(E_a, \Delta T) - i_s(E_a, \Delta T) \left[e^{\left(\frac{v_d}{v_T A_n}\right)} - 1\right] \tag{4.70}$$

where, $v_d = v_{pv} + i_{pv} R_s$. This turns into an implicit equation, which requires an iteration loop for simulation. Another way to represent the output of SSDM2 is to use the V–I characteristics instead of I–V. The expression is shown in (4.71) is an explicit function:

$$v_{pv} = v_T A_n \ln\left(\frac{i_{ph}(E_a, \Delta T) + i_s(E_a, \Delta T) - i_{pv}}{i_s(E_a, \Delta T)}\right) - i_{pv} R_s \tag{4.71}$$

Based on the SSDM2 for IM156B3, the I–V curves are simulated and plotted in Figures 4.18 and 4.19 to illustrate the impact of changes in solar irradiance and cell temperature, respectively. The advantages of SSDM2 lie in the elimination of any deviation at the MPP and the matching of the three important points at STC. However, the presence of the series resistance, R_s, causes complexity in the modeling and implementation, as discussed in Section 4.3.2 for ISDM and SSDM1.

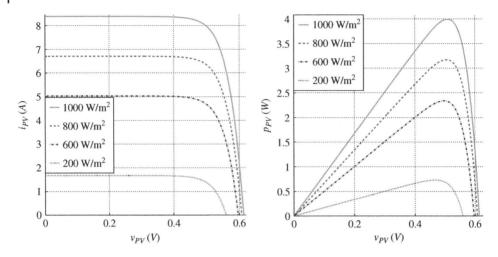

Figure 4.18 I–V curve for IM156B3 modeled with constant temperature (25°C) and variable irradiance.

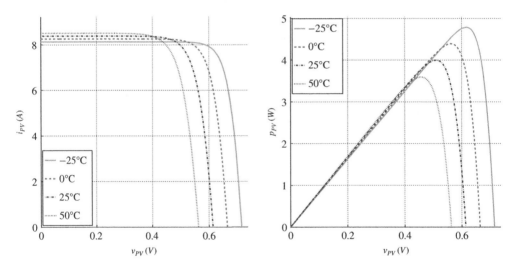

Figure 4.19 Modeled I–V curve of the PV cell IM156B3 with constant irradiance (1000 W/m²) and variable temperature.

4.4 Model Selection from the Simplified Single-diode Models

It should be mentioned that the IM156B3 case study does not mean that SSDM2 is a more suitable equivalent circuit than SSDM1. Selection of either SSDM1 or SSDM2 depends on the PV cell specification. For example, JAC-M6SR-3 is a PV product of JA Solar and is made of mono-crystalline silicon. Its specification is shown in Table 4.5.

The data are from the product datasheet from April 2014. It should be noted that the latest datasheet might show different information since the manufacturer often upgrades products. Following the process described in Sections 4.1 and 4.3, the model parameters can be determined for ISDM, SSDM1, and SSDM2. The comparison of

Table 4.5 Sample data.

	Basic information		
Manufacturer	Model	Cell material	Dimensions (mm)
JA Solar	JAC-M6SR-3	Mono-crystalline	$156 \times 156 \pm 0.5$

		Electrical performance at STC		
Efficiency	P_{MPP}	I_{MS}	V_{MS}	I_{SCS}	V_{OCS}
19.4–19.5%	4.72 W	8.661 A	0.545 V	9.272 A	0.0644 V

	Temperature coefficients	
α_T (%/°C)	β_T (%/°C)	γ_T (%/°C)
0.06	−0.36	−0.36

	Correction factor based on irradiance	
E_a (W/m²)	V_{OC} (V/V)	I_{SC} (A/A)
1000	1.00	1.00
900	1.00	0.90
800	0.99	0.80
500	0.97	0.50
300	0.95	0.30
200	0.93	0.20

All symbols shown above refer to Tables 4.2 and 4.3.

parameters and the model accuracy is shown in Table 4.6. There is no deviation at the short-circuit or open-circuit points for all three. The value of D_{MPP} is presented as the only performance index.

ISDM always has a deviation from the true MPP location. The significance varies from case to case. Both SSDM1 and SSDM2 eliminate the deviation at the MPP for both PV cell examples. However, SSDM1 is more suitable for the JACP6RF cell because of the positive value of R_h. Meanwhile, SSDM2 is more suitable for the IM156B3 cell because it gives a physically realistic value of R_s.

If the product data for the three critical points at STC are the only available information for modeling, it is recommended to choose a simulation model from among ISDM, SSDM1, and SSDM2 (Huang et al. 2016). The model accuracy can be sufficiently evaluated to support the model selection. There is always a tradeoff between the model simplicity and the model accuracy. A more complicated model generally requires more data and computational power for parameterization and simulation. Based on the differences in the model characteristics, a modeling process is recommended and illustrated in Figure 4.20.

The process always starts with the ISDM, thanks to its simplicity of both parameter identification and simulation (Mahmoud et al. 2012). The modeling accuracy in terms of the MPP deviation, D_{MPP}, should be always evaluated. Even though the performance

Table 4.6 Modeling comparison.

Cell	Model	Parameters	D_{MPP}
JACP6RF-3	ISDM	$I_{SS} = 1.9224 \times 10^{-7}$ A, $A_n = 1.4168$, $I_{ph} = 9.272$ A	0.0041
	SSDM1	$I_{SS} = 1.1004 \times 10^{-7}$ A, $A_n = 1.3740$, $I_{ph} = 9.272$ A, $R_h = 10.196\Omega$	$\simeq 0$
	SSDM2	$I_{SS} = 6.0741 \times 10^{-7}$ A, $A_n = 1.5153$, $I_{ph} = 9.272$ A $R_s = -7 \times 10^{-4}\Omega$	$\simeq 0$
IM156B3-164	ISDM	$I_{SS} = 8.1684 \times 10^{-7}$ A, $A_n = 1.4803$, $I_{ph} = 8.38$ A	0.0059
	SSDM1	$I_{SS} = 2.1014 \times 10^{-6}$ A, $A_n = 1.5708$, $I_{ph} = 8.38$ A, $R_h = -5.069\Omega$	$\simeq 0$
	SSDM2	$I_{SS} = 9.3186 \times 10^{-8}$ A, $A_n = 1.3048$, $I_{ph} = 8.38$ A, $R_s = 1.6m\Omega$	$\simeq 0$

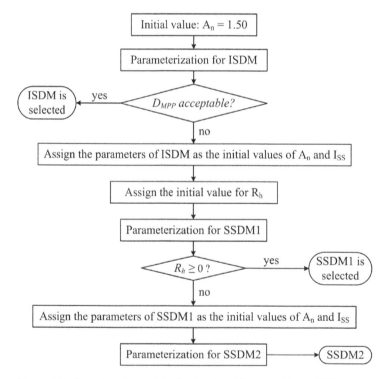

Figure 4.20 Flowchart for model selection from ISDM, SSDM1, and SSDM2.

of the ISDM will not meet the required accuracy, the identified values of A_n and I_{SS} can be used as the initial conditions for parameter identification with a more complicated model: SSDM1 or SSDM2.

SSDM1 is considered the second choice because of its relative simplicity and zero deviation at the MPP. SSDM2 is the last resort because of its complexity and the iteration required for the I–V characteristics. In the case of the IM156B3 cell, the

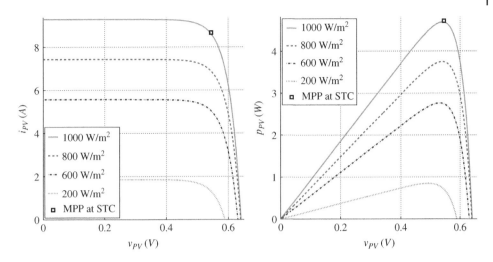

Figure 4.21 I–V curve of the JAC-M6SR-3 cell modeled with constant temperature (25°C) and variable irradiance.

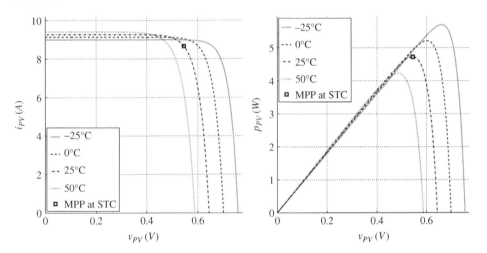

Figure 4.22 I–V curve of JAC-M6SR-3 cell modeled with constant irradiance (1000 W/m²) and variable temperature.

modeling process ultimately selected SSDM2 to achieve zero deviation at the MPP. For the JAC-M6SR-3 cell, the process stops at SSDM1 since the model parameters are physically meaningful. The SSDM1 output for JAC-M6SR-3 with changes in solar irradiance and temperature are shown in Figures 4.21 and 4.22, respectively. If the values of D_{MPP} are insignificant, the ISDM can be adopted for both cell examples.

4.5 Complete Single-diode Model

Most product datasheets provide the I–V and even the P–V curve in low-resolution figures, which serve as a basic reference for the output characteristics. The low resolution and data inaccuracy of the curves might not be appropriate where accurate modeling and simulation are desirable. Sometimes the dataset that is recovered from the datasheet I–V curve does not match the short-circuit, open-circuit, and maximum power points. Such information might be misleading for modeling accuracy.

Due to the imperfections in the manufacturing process, variations in PV cells might be significant. For this reason, the product datasheet usually provides sets of data showing the results at STC for the same PV cell model. For example, the product datasheets of the JAC-M6SR-3 and IM156B3 provide 10 and 16 different samplings of data at STC.

The data in the product specification are generally insufficient to recover the five unknown parameters required for the CSDM. It is recommended to adopt the model only when the full I–V dataset is available for modeling, and the curve-fitting performance is critically important for simulation. However, in reality, the measured data might not accurately reflect the CSDM expression and solving the equations leads to a singular solution. Following the CSDM shown in Figure 4.1 and the expression in (4.1), the I–V characteristics are expressed in a non-explicit form, $i_{pv} = f(i_{ph}, i_s, v_{pv}, i_{pv})$. This implicit equation requires an iterative loop for numerical simulation, which increases the computational load. This generally brings into question whether the CSDM is necessary for PV modeling since the various SSDMs can perform well in matching the data from PV manufacturers.

A parameterization approach to identifying four coefficients instead of five in a CSDM has been proposed by Mahmoud et al. (2013). It results in a reduced-order final model by eliminating either R_s or R_h, which indicates the effectiveness of the SSDMs, SSDM1 and SSDM2. The model output is evaluated by the index for curve-fitting performance in (4.28) and (4.31).

Another approach to identifying the five unknowns in the CSDM adds a further constraint derived from the I–V dataset (Huang et al. 2016). One specific PV module, MSX-83, is selected for the modeling process since its measured I–V data at STC are available in spreadsheet form. The solar module is constructed from 36 multi-crystalline PV cells, and it is a product of Solarex/BP Solar. At STC, the cell specification is derived as $V_{OCS} = 0.59$ V, $I_{SCS} = 5.27$ A, $I_{MS} = 4.85$ A, and $V_{MS} = 0.48$ V, with the assumption that all 36 cells are identical. Based on the I–V dataset, one additional point is selected for the parameter identification. The NR method is applied to identify the parameters. The results show that the CSDM can not only match the important points, but also improve the curve-fitting performance. The evaluation results are summarized in Table 4.7, showing the advantage of CSDM in terms of the low value of RMSD(I). Therefore, if the curve fitting is important, CSDM can be used to improve the modeling accuracy. The pre-condition is that the accurate data is available to represent the full I–V or P–V curves. Otherwise, the significant effort involved in using either CSDM or DDM might be wasted.

4.6 Model Aggregation and Terminal Output Configuration

Solar cells are usually physically encapsulated for weather protection and electrically connected to form modules for high power capacity. For practical power generation, the PV modules are mounted on supporting structures and electrically connected in series or series/parallel to form PV strings or arrays. Assuming the solar radiation

Table 4.7 Modeling comparison for MSX-83 cell.

Model	Parameters	D_{MPP}	RMSD(I)
ISDM	$I_{SS} = 7.3697 \times 10^{-6}$ A, $A_n = 1.7565$, $I_{ph} = 5.27$ A	0.0084	0.2 A
SSDM1	$I_{SS} = 2.8347 \times 10^{-6}$ A, $A_n = 1.5885$, $I_{ph} \simeq 5.27$ A, $R_s = 15$ mΩ	$\simeq 0$	0.1339 A
CSDM	$I_{SS} = 2.1560 \times 10^{-8}$ A, $A_n = 1.1886$, $I_{ph} \simeq 5.27$A, $R_s = 5.3$ mΩ, $R_h = 4.80$ Ω	$\simeq 0$	0.0289 A

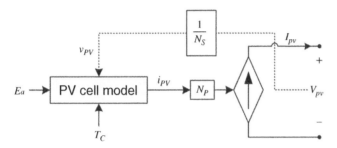

Figure 4.23 Block diagram for PV array aggregation and the terminal connection interface, representing the PV output as a current source.

and cell temperatures are identical for each solar cell, the single-cell model can be aggregated to any size PV array using the number of cells connected in series and parallel (Xiao et al. 2013).

Figure 4.23 illustrates the block diagram for PV array aggregation for simulation purposes, where N_S and N_P represent the numbers of cells that are connected in series and parallel, respectively. The PV cell model is the core for calculating the cell output current, i_{PV}, which is determined by the solar irradiance, E_a, the cell temperature, T_C, and the cell PV voltage, v_{PV}. The simulation model is suitable for the implementation of ISDM and SSDM1 since both have explicit functions for the I–V characteristics: (4.25) for the ISDM and (4.69) for SSDM1. The model implementation is simple and allows fast calculation and simulation. The cell PV voltage, v_{PV}, varies with the power equilibrium between the PV generation and loading condition. The output terminal is illustrated as a controllable current source for circuit-based simulation.

For the CSDM, an iterative solver is needed in the cell model due to the coupled terms between the PV cell voltage (v_{PV}) and current (i_{PV}). For simulation, the implicit method with iteration requires extra computation and a small time step to derive accurate output. In discrete time, the expression is

$$i_{pv}(k) = f[i_{pv}(k-1), v_{pv}(k), E_a(k), T_c(k)] \tag{4.72}$$

where the function f refers to the expression in (4.1).

The simplified model, SSDM2, also has an implicit equation for the I–V characteristics. However, the expression can be mathematically transformed into the PV cell voltage (v_{PV}) in terms of the current (i_{PV}), as expressed in (4.71). This can avoid the iterative term, which causes slow simulation. In this case, the PV array aggregation can be constructed as in Figure 4.24, which adopts (4.71) and a controllable voltage source for circuit-based simulation.

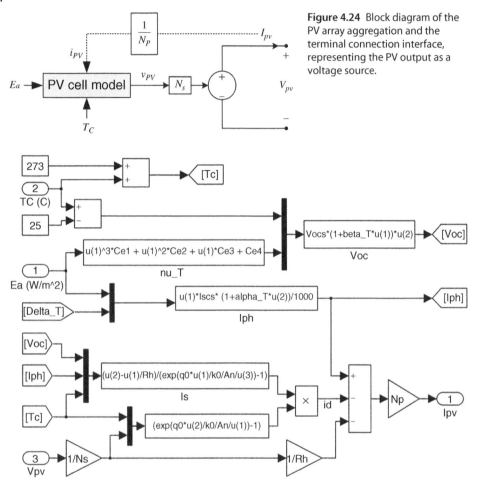

Figure 4.24 Block diagram of the PV array aggregation and the terminal connection interface, representing the PV output as a voltage source.

Figure 4.25 Simulink model of PV array for simulation.

When $N_P = 1$ and $N_S > 1$, the model generally stands for a string or module configuration that can be viewed as a specific case of a PV array. Thus the block diagram in Figures 4.23 and 4.24 illustrates a general form to represent the configuration of a PV module, string, and array, when the solar radiation and cell temperature are uniformly distributed to each solar cell. Using the coefficients N_P and N_S, and the PV cell model, the aggregated model is adjustable and flexible to simulate PV power outputs ranging from watts to megawatts. The ranges of the current and voltage can also be specified according to the parameter configuration of N_P and N_S when the PV cell model is determined. The mathematical representation can be integrated with controllable current or voltage sources, which gives terminal outputs as shown in Figures 4.23 and 4.24, respectively. The PV array model becomes the interconnection interface, which can be used for various terminal-based simulation software, such as Simulink, PSIM, PSPICE, or PSCAD.

A simulation model is built in Simulink using SSDM1, as shown in Figure 4.25. The model includes three inputs: irradiance, cell temperature, and the voltage. These are indicated in the integrated simulation block, as illustrated in Figure 4.26. The output is

Figure 4.26 Integrated Simulink model for simulation.

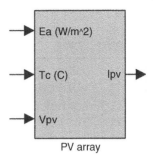

Ea (W/m^2)

Tc (C)   Ipv

Vpv

PV array

the current, which corresponds to the three input variables. The photon current, the open-circuit voltage, and the diode saturation current are calculated in different blocks. One block is for the correction factor of the open-circuit voltage in response to irradiance variations. When the shunt resistance is set to be infinite, the simulation model for the SSDM1 becomes the ISDM. The model will be widely used for simulations in the following chapters. The model can be integrated in one block to represent a PV array, as illustrated in Figure 4.26.

4.7 Polynomial Curve Fitting

Aside from the equivalent circuits of the diode-based models, a polynomial equation represents the electrical characteristics of PV generators (Xiao et al. 2006). The approach requires use of the measured data across the operating range of the PV output from the open-circuit to the short-circuit condition.

It is recommended to use a sixth-order polynomial to present the PV output characteristics of crystalline-based solar modules, which have relatively sharp "knees" in their I–V curves. The I–V and P–V characteristics can be expressed as in (4.73) and (4.74), respectively.

$$i_{pv} = b_{i0} + b_{i1}v_{pv} + b_{i2}v_{pv}^2 + b_{i3}v_{pv}^3 + b_i4v_{pv}^4 + b_{i5}v_{pv}^5 + b_{i6}v_{pv}^6 \tag{4.73}$$

$$p_{pv} = b_{p0} + b_{p1}v_{pv} + b_{p2}v_{pv}^2 + b_{p3}v_{pv}^3 + b_{p4}v_{pv}^4 + b_{p5}v_{pv}^5 + b_{p6}v_{pv}^6 \tag{4.74}$$

An example based on the MSX-83 PV module is now considered. The module is constructed from 36 poly-crystalline PV cells manufactured by the Solarex Corporation. Based on the measured data, the unknown coefficients can be determined by fitting the polynomial to the data. In Matlab, the function, "polyfit," can be used to find the coefficients of a polynomial in order to fit the data in the least-squares sense. For MSX-83, the parameters are identified and shown in Tables 4.8 and 4.9 for (4.73) and (4.74), respectively. The parameter, b_{i0}, represents the short-circuit current of the PV module. The model output is shown in Figure 4.27. The I–V and P–V curves are compared with the measured data points.

Thin-film PV material does not share the same physical composition as the p-n junctions of diodes. Therefore, the diode-based model might not be suitable for simulating their output. A fourth-order polynomial is recommended for modeling the output of thin-film solar modules due to their "soft" knee shape (Xiao et al. 2006). Therefore, the model output in terms of the I–V and P–V characteristics can be expressed as in (4.75)

Table 4.8 Polynomial parameters for the I–V curve of MSX-83.

Parameter	Value	Parameter	Value	Parameter	Value
b_{i0}	5.2734	b_{i1}	0.005609	b_{i2}	−0.012412
b_{i3}	0.0050318	b_{i4}	-7.7579×10^{-4}	b_{i5}	5.0727×10^{-5}
b_{i6}	-1.1933×10^{-6}				

Table 4.9 Polynomial parameters for the P–V curve of MSX-83.

Parameter	b_{p0}	b_{p1}	b_{p2}	b_{p3}	b_{p4}	b_{p5}	b_{p6}
Value	− 0.025625	6.5361	− 0.99799	0.27316	− 0.033358	0.00186	-3.884×10^{-5}

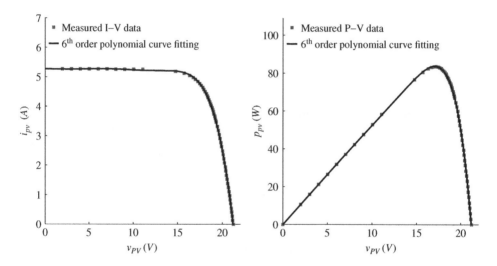

Figure 4.27 Sixth-order polynomial output in comparison with measured data of MSX-83.

and (4.76), respectively. One advantage is that the polynomial order is less than that for crystalline-based modules.

$$i_{pv} = b_{i0} + b_{i1}v_{pv} + b_{i2}v_{pv}^2 + b_{i3}v_{pv}^3 + b_{i4}v_{pv}^4 \tag{4.75}$$

$$p_{pv} = b_{p0} + b_{p1}v_{pv} + b_{p2}v_{pv}^2 + b_{p3}v_{pv}^3 + b_{p4}v_{pv}^4 \tag{4.76}$$

The unknown parameters of the polynomial model can be determined by the least-squares method when the I–V curve is acquired in a discrete-time format. Equation 4.75 is expressed in discrete time as:

$$I_k = b_{i0} + b_{i1}V_k + b_{i2}V_k^2 + b_{i3}V_k^3 + b_{i4}V_k^4 \tag{4.77}$$

corresponding to the sampling index, k. For parameter identification, a number of samples should be acquired. The data can be organized in a vector Y:

$$Y = \begin{bmatrix} I_1 \\ I_2 \\ I_3 \\ \vdots \\ I_M \end{bmatrix} \tag{4.78}$$

The recorded voltage and its polynomial form are assembled to form a matrix:

$$X = \begin{bmatrix} 1 & 1 & 1 & \cdots & 1 \\ V_1 & V_2 & V_3 & \cdots & V_M \\ V_1^2 & V_2^2 & V_3^2 & \cdots & V_M^2 \\ V_1^3 & V_2^3 & V_3^3 & \cdots & V_M^3 \\ V_1^4 & V_2^4 & V_3^4 & \cdots & V_M^4 \end{bmatrix} \tag{4.79}$$

The unknown parameters, b_{i0-4} can also be formed in a vector:

$$\Theta = \begin{bmatrix} b_{i0} \\ b_{i1} \\ b_{i2} \\ b_{i3} \\ b_{i4} \end{bmatrix} \tag{4.80}$$

The regression model can be expressed as

$$Y = X^T \Theta \tag{4.81}$$

where the vector Y includes the recorded value of the current and the matrix X contains the regression variables of the voltage. According to the formula of the least-squares method, the vector of the model parameters can be identified for the best curve fitting as:

$$\Theta = (X^T X)^{-1} X^T Y \tag{4.82}$$

As an example, we use the ST-10 module, which is made of cells based on copper indium diselenide (CIS). The PV module was a product of Siemens Solar GmbH. The unknown parameters of b_{i0-4} can be determined by the least-squares method, as shown in Table 4.10. The I–V characteristics are plotted in Figure 4.28, showing the model output and the measured data points.

The polynomial can also be formed from the P–V output characteristics, as expressed in 4.76. Using the least-squares method, the coefficients can be identified to represent

Table 4.10 Polynomial coefficients for I–V output characteristics of ST-10.

b_{i0}	b_{i1}	b_{i2}	b_{i3}	b_{i4}
0.77423	-4.3718×10^{-3}	6.4798×10^{-4}	-4.893×10^{-6}	-3.5209×10^{-6}

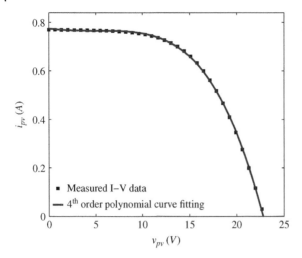

Figure 4.28 I–V characteristics of the fourth-order polynomial output and measured data of ST-10 module.

Table 4.11 Coefficients for P–V characteristics of ST-10.

b_{p0}	b_{p1}	b_{p2}	b_{p3}	b_{p4}
−0.0001945	0.0042023	−0.03151	0.84818	−0.040404

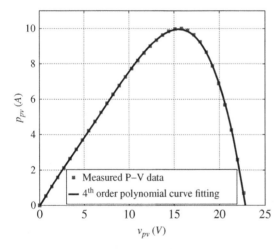

Figure 4.29 P–V characteristics of fourth-order polynomial output and measured data of ST-10 module.

the P–V curve, as shown in Table 4.11. The model output is plotted in Figure 4.29, showing the P–V curve in comparison with the measured data points.

Comparing the polynomial models, the P–V representation is superior to the I–V function since the coefficients are consistent in representing the P–V characteristics. There are extremely low values in the model coefficients for the I–V model, for example, -3.5209×10^{-6}, which is difficult to implement for simulation and practical applications. The P–V based model is also useful for directly identifying the location of the MPP.

Research has shown that the polynomial representation is useful for real-time identification of the optimal operating point of PV power systems in order to achieve MPPT

(Xiao et al. 2006). This approach is valuable for determining the MPP instead of conventional search methods based on the trial and error. The recursive least-squares method can be used to find the model parameters in real-time to accommodate variations of solar irradiance and cell temperature. Furthermore, the polynomial equation does not have the limitation of the diode-based model in representing the output of thin-film PV cells. Due to their softer knee shape in comparison with crystalline-based modules, a lower-order polynomial can be used for modeling thin-film PV modules, as demonstrated by the case study.

4.8 Summary

The diode-based model is well established and suitable for representing PV cells that are made of crystalline silicon since both diodes and PV cells share the same composition and manufacturing process. It is considered a static model since only the I–V characteristics are represented in the mathematical equations. The dynamics regarding response times is not included.

This chapter focused on the SDM to represent PV output characteristics. From the product datasheet, four equations or constraints can be formulated, corresponding to the short-circuit current I_{SCS}, open-circuit voltage V_{OCS}, operating voltage and current at MPP (V_{MS}, I_{MS}) under standard test conditions. The information is insufficient to derive the five unknowns corresponding to the CSDM. As a result, SSDMs are mostly utilized for parameterization.

The modeling starts with the ideal single-diode model (ISDM), which has only three unknowns. It has the significant advantages of simplicity in both parameterization and implementation, and it is very efficient for simulation. The drawback of the ISDM lies in the model deviation from the true MPP. The MPP mismatch happens because the MPP point is given as an ordinary input-output value, but not identified as the extremum value in the power output. The modeling process ensures that the output of the mathematical model passes through the MPP, but is not necessary for this to be the peak power point.

The SSDMs including the SSDM1 and SSDM2 have the advantage of accuracy at three important points and a straightforward modeling process. Four unknown parameters can be identified and can take advantage of the data provided by the PV cell manufacturer for those points. The implicit information used is that the peak of the P–V curve occurs at the voltage point (V_{MS}) which represents the MPP at STC. The models have no mismatch at the modeled MPP, which is an improvement over the ISDM. However, the initial values are critical for the success of the parameter identification when the Newton–Raphson iteration method is used.

The correction factors that are commonly provided by product datasheets are used to construct the simulation model in response to variations in solar irradiance and cell temperature. If the solar radiation and cell temperature are uniformly distributed, the single-cell model can be aggregated to any size of PV array using a number of cells connected in series and parallel.

Equivalent circuits for diodes are mainly used for PV cells made of crystalline silicon. They might not be suitable models for thin-film PV materials since the physics of a diode model does not apply. The proper model to represent thin-film-based PV cells is under investigation because of the diversity of materials used. However, for simulation purposes, the SDM can be used to output I–V characteristics that match thin-film PV cells (Mahmoud et al. 2013). Polynomial equations can also be used to simulate the output of thin-film PV cells or modules.

The model selection for the CSDM should be based on the accurate data that are available to represent the complete I–V characteristics. The modeling performance can be illustrated to show the accuracy of the curve fitting. However, if the data are unreliable, the difficulties in both parameterization and simulation are increased.

It should be noted that the temperature effect on the resistances and ideality factors in diode-based models are not considered in the modeling and simulation process. In reality, the temperature affects such parameters. However, accurate data are required for modeling and performance evaluation. Without accurate data for reliable evaluation, any additional effort or complicated modeling becomes meaningless.

Problems

4.1 Follow the modeling procedure and use the same examples to identify the simulation model parameters for the ISDM and one of the SSDMs. Plot the I–V and P–V curves and compare to the product data for the short-circuit, open-circuit, and maximum power points. Use the predefined performance index to quantify the model accuracy.

4.2 Search online to download the data of a crystalline-based PV cell. Try to find the three critical points at STC and the correction coefficient.

4.3 Ideal single-diode modeling:
  a) Based on the data at STC, try to identify all the required parameters to formulate the ideal single-diode model. Output the I–V and P–V curves to compare with the given data.
  b) Following the performance indices introduced in Section 4.2, calculate the value of the deviations at the three critical points.
  c) Following the introduction in Section 4.1.3, reconstruct the PV ISDM simulation model with correction for solar irradiance and cell temperature. Based on the cell temperature of 25°C and solar irradiance levels of 1000, 800, 600, and 200 W/m², simulate and plot the corresponding I–V and P–V output curves. For a solar irradiance of 1000 W/m², simulate and plot the corresponding I–V and P–V output curves.
  d) Construct the simulation model using Simulink, as described in Section 4.6. This model should be capable of being aggregated to any size with an assumption of identical PV cells.

4.4 Simplified single-diode modeling:
  a) If the D_{MPP} of the ISDM is calculated and higher than 0.1%, the simplified single diode model, SSDM1 or SSDM2, can be used. Following the recommended modeling process, as shown in Figure 4.20, identify the coefficients for either SSDM1 or SSDM2. It should be noted that the suitable model, either SSDM1

or SSDM2, for the selected PV cell can be determined by the value of either the shunt resistor or the series resistor.

b) Plot the I–V and P–V curves of the simplified model and compare with the model output of the ISDM.

4.5 Complete single-diode modeling:
a) If complete and accurate I–V data or P–V data are available, the CSDM can be used to illustrate the curve-fitting performance, which is supposed to be better than using simplified models. The parameter estimation can follow the procedure discussed in (Huang et al. 2016).
b) Demonstrate the modeling performance using the curve-fitting performance indices, as expressed in (4.29) and (4.30).

4.6 Polynomial model:
a) Based on the modeling process discussed in Section 4.7, find the parameters to represent the polynomial model.
b) Plot the polynomial model output with the ISDM output for comparison. Assess the performance using the indices defined in Section 4.2.

References

Huang PH, Xiao W, Peng JC, Kirtley and J. 2016 Comprehensive parameterization of solar cell: Improved accuracy with simulation efficiency. *Industrial Electronics, IEEE Transactions on* **63**(3), 1549–1560.

Mahmoud Y, Xiao W and Zeineldin H 2012 A simple approach to modeling and simulation of photovoltaic modules. *Sustainable Energy, IEEE Transactions on* **3**(1), 185–186.

Mahmoud Y, Xiao W and Zeineldin H 2013 A parameterization approach for enhancing PV model accuracy. *Industrial Electronics, IEEE Transactions on* **60**(12), 5708–5716.

Romero-Cadaval E, Spagnuolo G, Franquelo L, Ramos-Paja C, Suntio T and Xiao W 2013 Grid-connected photovoltaic generation plants: components and operation. *IEEE Industrial Electronics Magazine* **7**(3), 6–20.

Shockley W 1949 The theory of p-n junctions in semiconductors and p-n junction transistors. *Bell System Technical Journal* **28**(3), 435–489.

Xiao W 2007 *Improved control of photovoltaic interfaces* PhD thesis University of British Columbia.

Xiao W, Edwin FF, Spagnuolo G and Jatskevich J 2013 Efficient approaches for modeling and simulating photovoltaic power systems. *Photovoltaics, IEEE Journal of* **3**(1), 500–508.

Xiao W, Lind MG, Dunford WG and Capel A 2006 Real-time identification of optimal operating points in photovoltaic power systems.*Industrial Electronics, IEEE Transactions on* **53**(4), 1017–1026.

5

Power Conditioning

Power conditioning is required to interface the PV source circuit for power processing. The key functional blocks of a typical two-stage conversion system for AC grid interconnection is shown in Figure 5.1.

The PV link is the interface between the PV source circuit and the PV-side converter (PVSC). It provides a filtering function to maintain a steady voltage at the link. The PVSC is a DC/DC power interface, the input of which is coupled to the PV link and is usually controlled by the maximum power point tracking (MPPT) algorithm so that maximum energy harvesting is achieved. Throughout the book, it is occasionally referred to the "PV-side power interface." The grid link is the interface between the grid and the grid-side converter (GSC). It provides a filtering function to guarantee the power quality required by the grid. A transformer can be implemented at this stage for the purpose of galvanic isolation and voltage conversion. The GSC is the power interface between the DC link and the AC grid link. It converts DC to AC for grid interconnection. The DC link is commonly formed by capacitors, which maintain a steady DC-link voltage between the two conversion stages.

In systems with single-stage conversion, one centralized power converter is used, as shown in Figure 5.2. Thanks to the simplicity of its configuration, the system gives higher efficiency and lower costs than a two-stage-conversion system. To avoid confusion, the DC/AC converter for single-stage conversion is considered the GSC. The PV link is merged with the DC link, as shown in Figure 5.2. The design and analysis of the GSC in single-stage conversion systems is generally the same as in two-stage conversion systems.

Shown in Figure 1.16 and described in Section 1.11, a DC microgrid is composed of the power interfaces for energy storage units. Bidirectional DC/DC converters are usually required to interface batteries with a DC bus. Therefore, one section in this chapter is about the battery-side converters (BSCs) that are required for the charge and discharge operations of battery storage units.

The inductance and capacitance parameters in the following design examples are theoretically sized and based on the ideal converter system. They are considered as minimal values for proof of concept and simulation. They are also the reference and the starting point for practical designs in which all constraints are considered. When all the parameters are re-tuned to meet all requirements, they can be again used for simulation to prove the upgraded design concept. The modeling and simulations in this chapter are based on the basic functions of Simulink rather than circuit-based modeling methods, such as Simscape Power Systems.

Photovoltaic Power System: Modeling, Design, and Control, First Edition. Weidong Xiao.
© 2017 John Wiley & Sons Ltd. Published 2017 by John Wiley & Sons Ltd.
Companion Website: www.wiley.com/go/xiao/pvpower

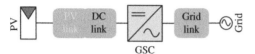

Figure 5.1 Block diagram of a grid-connected PV system with two-stage power conversion. PVSC, PV-side converter; GSC, grid-side converter.

Figure 5.2 Block diagram of grid-connected PV system with single-stage power conversion. GSC, grid-side converter.

5.1 PV-side Converters

The PVSC is the DC/DC power-conditioning circuit, the input of which is connected to the PV link. In standalone systems, the PVSC can be used as the power interface to charge batteries or supply power to local loads. The common DC/DC topologies used for PVSCs can include buck, boost, buck–boost, flyback, tapped-inductor, and full-bridge isolated DC/DC converters. It is important to design a topology that gives maximum power yield, without increasing the circuit complexity. Figure 5.3 illustrates the procedure that is recommended to design, simulate, and evaluate the PVSC in order

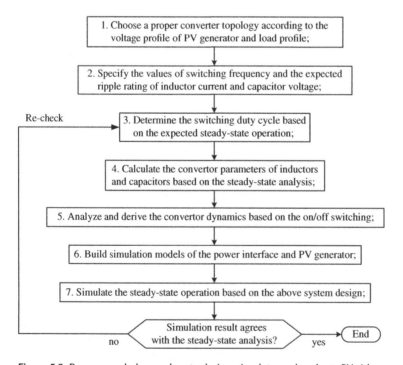

Figure 5.3 Recommended procedure to design, simulate, and evaluate PV-side power interface.

to meet the system specification and give the correct performance rating. The procedure follows the sequence of

- topology selection
- specification
- steady-state analysis
- design
- simulation modeling
- simulation evaluation.

Simulation is an effective tool and widely used to prove the design concepts of PVSCs and the effectiveness of models.

5.1.1 PV Module for Case Study

In the following study, one specific PV module is used to demonstrate PVSC design and simulation. Its electrical characteristics at STC are shown in Figure 5.4. The PV module is constructed from 72 multi-crystalline cells and is used for demonstration purposes only. The simulation model is based on the ISDM that is introduced in Section 4.1. The short-circuit current and open-circuit voltage are 8.34 A and 44.17 V respectively. The maximum power point (MPP) is at (37.0 V, 7.79 A). The peak power is 288.3 W, as indicated in the plot.

5.1.2 Buck Converter

A buck converter circuit, as shown in Figure 5.5, can be used for the PVSC. The converter is controlled by pulse width modulation (PWM), of which the switching duty cycle is the control-input variable. A step-down topology should be considered if the converter-output voltage is never higher than the PV terminal voltage (v_{pv}) at the MPP, V_{MPP}, when the normal voltage variation of both sides has been considered. The condition can be expressed as: $V_{O(max)} \leq V_{MPP(min)}$. The lowest value of the PV terminal voltage at the MPP, $V_{MPP(min)}$, can be estimated from the highest ambient temperature

Figure 5.4 I–V and P–V curves of PV module output.

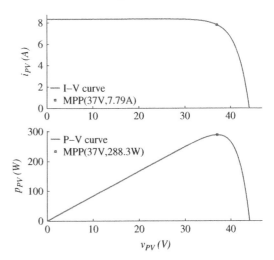

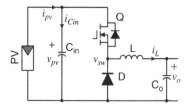

Figure 5.5 Buck converter used for PV power interface.

and the minimum irradiance for the converter to operate. The highest value of the output voltage, $V_{O(max)}$, can be determined from the load profile. For battery-charging applications, the battery voltage becomes the converter-output voltage, v_o, which varies from the cut-off voltage at 0% state of charge (SOC) to the highest level: the open-circuit voltage at 100% SOC. In this case, the value of $V_{O(max)}$ is equivalent to the open-circuit voltage of the battery at 100% SOC.

One additional concern of the buck topology is that the input current is always pulsing or "chopped," so significant input capacitance is required to smooth the PV-link voltage. This generally affects the dynamics of the PV-link voltage and causes significant ripple in the capacitor current, i_{Cin}. The output current is smooth in the buck topology because the inductor appears on the output side.

At STC and the predefined switching frequency, f_{sw}, the inductor ripple current and ripple voltage at the PV link should be specified by the peak-to-peak values, ΔI_L and ΔV_{PV}, respectively. Steady-state analysis can determine the duty cycle at the nominal load operating condition. At STC, the PV source circuit should be operated at the MPP, which is represented by V_{MPP} and I_{MPP}. The duty cycle can be calculated from:

$$D_o = \frac{V_{O-NOM}}{V_{MPP}} \tag{5.1}$$

with the assumption that the converter is operated in continuous conduction mode (CCM). The symbol V_{O-NOM} represents the nominal output voltage, which can be specified from the load profile. The value of the inductance, L, and the capacitance, C_{in}, can be calculated from (5.2) and (5.3), respectively.

$$L = \frac{V_{O-NOM}(1 - D_o)}{\Delta I_L f_{sw}} \tag{5.2}$$

$$C_{in} = \frac{I_{MPP}(1 - D_o)}{\Delta V_{PV} f_{sw}} \tag{5.3}$$

In this system, the capacitor C_{in} represents the PV link between the PV source circuit and the PVSC.

It should be noted that the thermal characteristics should be evaluated and considered for capacitor selection. Significant current ripple can be expected for filtering the pulsing current at the input side of the buck converter. Each capacitor is constrained by the allowable ripple current. The intrinsic equivalent series resistance (ESR) results in power loss when significant ripple current is conducted. High temperature is related to the early failure of capacitors. To evaluate the power loss and estimate the core temperature, the RMS value of the ripple current in i_{Cin} should be determined (see discussion in Section 5.6). In general, thermal analysis is considered as one of the most important aspects of the selection of capacitors and the design of power electronics.

When switch Q is "on," the system dynamics can be represented as in (5.4) and (5.5).

$$i_L = \frac{1}{L} \int (v_{pv} - v_o)dt \tag{5.4}$$

$$v_{pv} = \frac{1}{C_{in}} \int (i_{pv} - i_L)dt \tag{5.5}$$

When Q is "off," the system dynamics are as in (5.6) and (5.7).

$$i_L = \frac{1}{L} \int (-v_o)dt \tag{5.6}$$

$$v_{pv} = \frac{1}{C_{in}} \int i_{pv}dt \tag{5.7}$$

From the on/off states and the integral operation, a simulation model can be built by Simulink, as shown in Figure 5.6a. The model is based on the ideal buck converter, in which loss is not considered. Two single-pole–double-throw (SPDT) switches are utilized in the Simulink model for the switching between the on-state, (5.4) and (5.5), and the off-state, (5.6) and (5.7). The model includes three inputs:

- the pulse width modulation (PWM) command signal for the switches
- the output voltage determined by the load condition (v_o)
- the PV output current (i_{pv}).

It outputs two signals: the inductor current (i_L) and the PV-link voltage (v_{pv}). The dynamic interaction is simulated by the integral operation, which is defined by (5.4)–(5.7). The saturation signs are shown in the Simulink blocks for integration, which constrains the inductor current (i_L) and the PV-link voltage (v_{pv}) to be always positive. All components can be packed together to form a single block to represent the buck converter, as shown in Figure 5.6b. The correspondence of the PV-link voltage (v_{pv})

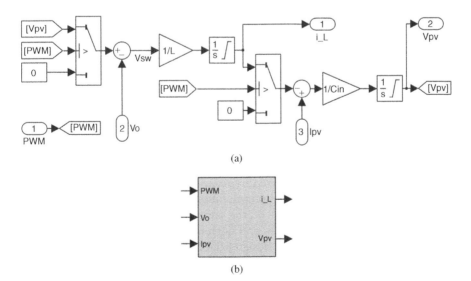

(a)

(b)

Figure 5.6 Simulink model of buck converter used for PV power interface: (a) model composition; (b) integrated block.

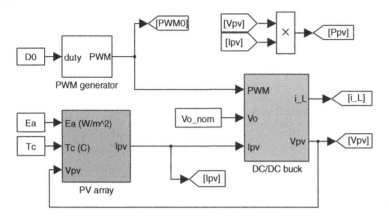

Figure 5.7 Simulink system using the buck converter for PV-side converter.

and the PV output current (i_{pv}) should follow the PV source circuit model discussed in Chapter 4 and shown in Figure 4.25. The steady-state analysis and design can be simulated and verified using the Simulink model.

An example is given of charging a battery module with a nominal voltage of 24 V. The open-circuit voltage is 28 V when the battery module is fully charged. The power source is the PV module that was introduced in Section 5.1.1. Since the V_{MPP} is significantly higher than 28 V, a buck converter is selected for the power conditioning. The switching frequency is designed to be 50 kHz; the peak-to-peak ripple voltage of the PV module is specified as 0.2 V; and the peak-to-peak ripple of the inductor current is specified as 1 A. Based on the nominal battery voltage (24 V) and the STC of the PV module, the nominal duty cycle of the PWM can be calculated as 64.9% according to (5.1). Then the values of L and C_{in} can be calculated as 167 µH and 272 µF, by following (5.2) and (5.3), respectively. The system simulation model, as shown in Figure 5.7, is constructed together with the blocks of the PV module and the PWM generator.

The simulation result is shown in Figure 5.8, including the waveforms of i_L, v_{pv}, i_{pv}, and p_{pv}. When PWM signals are applied to the converter with its nominal duty cycle of 64.9%, the value of v_{pv} drops from the open-circuit voltage, 44.17 V, to the value of V_{MPP}, which is 37.0 V at the steady state. The signal of i_{pv} increases from zero to the value of I_{MPP}, which is 7.79 A at the steady state. The PV power (p_{pv}) reaches the value of the MPP with the corresponding voltage and current, the same as the rating shown in Figure 5.8.

Figure 5.9 provides a zoom-in look at the waveform that was presented in Figure 5.8. The peak-to-peak ripple of the inductor current is measured as 1 A, corresponding to the specification. The peak-to-peak ripple of the PV-link voltage is also measured to match the specified value of 0.2 V. The fixed step-size of the numerical solver is chosen as 20 ns, which indicates that the simulation resolution is 1000 sampling points in each PWM switching cycle. The MPP is 288.3 W, the same as the PV module specification. The simulation verifies the system design and demonstrates the model's performance in the steady state. The same sizing and design principles can be applied to cases with different electrical ratings from the example.

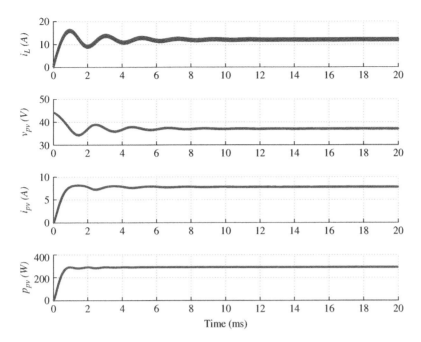

Figure 5.8 Simulated waveforms of the PV power charger in the steady state.

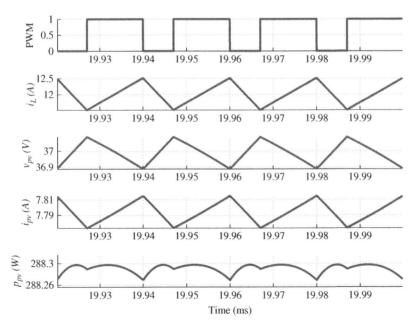

Figure 5.9 Simulated waveforms of the PV power charger in the steady state.

5.1.3 Full-bridge Isolated Transformer DC/DC Converter

One typical topology is the full-bridge transformer isolated DC/DC converter, the circuit of which is shown in Figure 5.10. The transformer can be constructed with a winding-turn ratio of 1:N. The circuit has a higher component count than another isolated topology, the flyback converter (see Section 5.1.7). Therefore, the converter is mainly used for higher power capacities than flyback topologies.

The H bridge at the left-hand side of the transformer produces a high-frequency AC (HFAC) signal, v_T, which can be transmitted through the transformer. The winding-turn ratio of the transformer provides options for either stepping up or down the input voltage into the output terminal. The HFAC signal is rectified into a pulsed DC form at the right-hand side of the transformer, labeled v_S in the circuit diagram. The pulsed DC signal is filtered by the inductor–capacitor (LC) circuit to produce a steady DC voltage at the output terminal. Under the nominal CCM operation, the output current, i_L, and voltage, v_o, are expected to be smooth DC signals with low switching ripples.

Depending on the transformer design, the topology can be configured in two forms. The simple transformer design, as shown Figure 5.10a, requires a full bridge to form the rectifier. When the center tap is applied in the transformer design, the rectifier can be composed of two diodes, as shown in Figure 5.10b. Therefore, the total component count is reduced, with a slightly increased complexity in the transformer design.

Even though a full-bridge transformer isolated converter can step up the voltage from the input to the output port, the operating principle follows the same analysis as for the non-isolated buck converter. When the winding-turn ratio of the transformer is

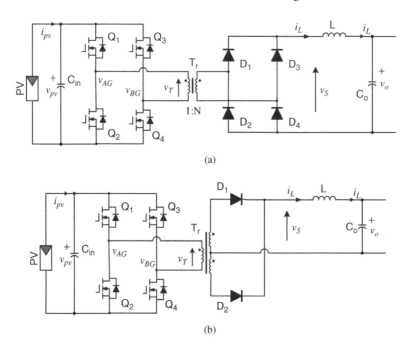

(a)

(b)

Figure 5.10 Circuit of full-bridge DC/DC converter for PV-side converter: (a) full-bridge rectifier; (b) two-diode based rectifier.

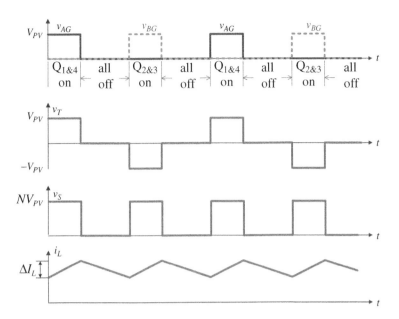

Figure 5.11 Typical steady-state waveforms in full-bridge transformer isolated DC/DC converter.

reset to 1:1, the converter always produces a DC output voltage lower in magnitude than the DC input voltage.

The inductor current can be expressed as

$$L\frac{di_L}{dt} = v_s - v_o \tag{5.8}$$

The waveforms in the steady state are shown in Figure 5.11, including the HFAC signal, v_T, the switching signal, v_S, and the inductor current, i_L. The voltage potentials in the two legs are shown as v_{AG} and v_{BG}, as indicated in Figure 5.10. The pulse widths of v_{AG} and v_{BG} determine the voltage-conversion ratio between the input and output terminals. The duty cycle of the PWM can be the control variable in regulating the voltage conversion.

When either Q_1 and Q_4 or Q_2 and Q_3 are in an on-state, the system dynamics can be represented as

$$i_L = \frac{1}{L}\int(Nv_{pv} - v_o)dt \tag{5.9}$$

$$v_{pv} = \frac{1}{C_{in}}\int(i_{pv} - Ni_L)dt \tag{5.10}$$

When all switches are in an off-state, the system dynamics is given by

$$i_L = \frac{1}{L}\int(-v_o)dt \tag{5.11}$$

$$v_{pv} = \frac{1}{C_{in}}\int i_{pv}dt \tag{5.12}$$

Equations 5.9–5.12 show similarities in the switching operation to the analysis for the buck converter. The difference is that the winding-turn ratio appears in the equations.

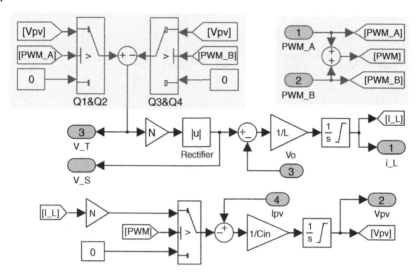

Figure 5.12 Simulink model of full-bridge DC/DC converter for PV power interface.

Using Simulink, the simulation model for the full-bridge isolated DC/DC converter can be derived (see Figure 5.12) when it is used for the PVSC. The symbol N represents the winding-turn ratio of the isolating transformer. Four inputs are included in the simulation model: the PWM signals for Q_1, Q_4, Q_2, and Q_3, the output voltage, and the current of the PV source circuit. There are four output signals: the inductor current, i_L, the voltage of v_{PV}, v_T, and v_S.

The sizing of the LC filter can follow the principle of the non-isolated buck converter. At STC and the predefined switching frequency, f_{sw}, the inductor ripple current and the ripple voltage at the PV link should be specified by the peak-to-peak values, ΔI_L and ΔV_{PV}. The symbol f_{vs} represents the frequency of the signal v_S, which is twice that of v_T. Steady-state analysis can determine the duty cycle at the nominal load operating condition. At STC, the PV source circuit is expected to output V_{MPP} and I_{MPP}, which represent the MPP. Referring to the switching waveforms shown in Figure 5.11, the percentage of the positive pulse in v_S can be calculated from

$$D_{vs} = \frac{V_{O-NOM}}{N V_{MPP}} \tag{5.13}$$

when the converter is operated in continuous conduction mode (CCM). The symbol V_{O-NOM} represents the nominal output voltage, which can be specified by the load profile or the DC bus rating. The value of the inductance, L, and the capacitance C_{in} can be calculated from (5.14) and (5.15), respectively.

$$L = \frac{V_{O-NOM}(1 - D_{vs})}{\Delta I_L f_{vs}} \tag{5.14}$$

$$C_{in} = \frac{I_{MPP}(1 - D_{vs})}{\Delta V_{PV} f_{vs}} \tag{5.15}$$

It should be noted that the switching frequency (f_{sw}) of Q_1, Q_4, Q_2, and Q_3 is half of the predefined frequency, f_{vs}. For practical implementations, the switch-on duty cycle is also the half of the value of D_{vs}.

An example is given of supplying a DC bus with a nominal voltage of 48 V. The power source is the PV module, of which the electrical characteristics at STC are shown in Figure 5.4. Since V_{MPP} at STC is 37 V, the transformer turns ratio is designed as $N = 2$. The frequency of v_S is specified as 50 kHz. The peak-to-peak ripple voltage of the PV module is specified as 0.2 V, and the peak-to-peak ripple of the inductor current is specified as 1 A. Based on the nominal battery voltage at STC, the percentage of the positive pulse in v_S can be calculated as 64.9% according to (5.13). Then the values of L and C_{in} can be calculated as 337 μH and 274 μF, following (5.14) and (5.15), respectively. The Simulink model, as shown in Figure 5.12, includes the simulation blocks of the PV module and the PWM generator.

The simulation result is shown in Figure 5.8, including the waveforms of i_L, v_{pv}, i_{pv}, and p_{pv}. When the duty cycle of 64.9% is applied to the converter, the value of v_{pv} drops from the open-circuit voltage to the value of V_{MPP} and then reaches a steady state. The signal of i_{pv} increases from zero to the value of I_{MPP} and a steady state. The PV power (p_{pv}) reaches the MPP value at the corresponding voltage and current, the same as the rating shown in Figure 5.13. Except for the waveform of i_L, the rest of waveforms are the same as those in Figure 5.8.

Figure 5.14 zooms in on the waveforms, showing that the peak-to-peak ripple of the inductor current is about 1 A, corresponding to the design specification. The peak-to-peak ripple of the PV-link voltage is also measured, and matches the specified value of 0.2 V. The simulation verifies the system design and demonstrates the model performance in a steady state. For practical implementation, the PWM signals for Q_1, Q_4, Q_2, and Q_3 are 32.5% in duty cycle and 25 kHz in switching frequency. Comparing to the non-isolated buck converter, the topology provides galvanic isolation and flexibility in conversion ratios thanks to the HFAC transformer.

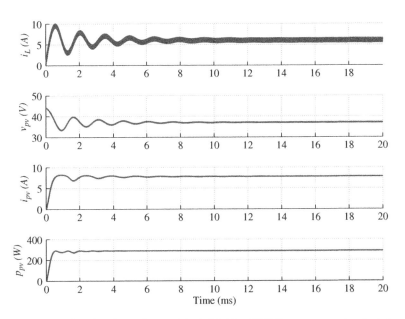

Figure 5.13 Simulated waveforms of the full-bridge DC/DC converter at steady state.

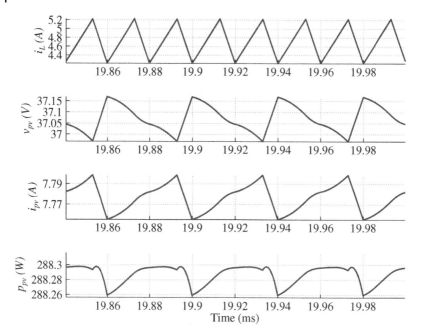

Figure 5.14 Simulated waveforms of the full-bridge DC/DC converter at steady state: zoomed-in view to illustrate the ripple magnitude.

The full-bridge operation can also be modulated by the technique of phase shifting instead of PWM. The converter can be fully controlled by the phase shift and produce the required waveforms. The operational principle is demonstrated in Figure 5.15. All symbols refer to the definitions in Figure 5.10. The switches of Q_1 and Q_4 are modulated by a constant 50% duty cycle, which generates a signal v_{AG}. Meanwhile, the switches of Q_2 and Q_3 are modulated by the constant 50% duty cycle, generating a signal v_{BG}. Both v_{AG} and v_{BG} share the same switching frequency, but are different in phase. The phase shifting allows generation of the same waveform as v_T, as shown in Figure 5.11.

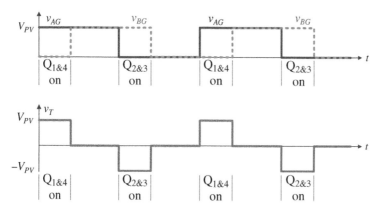

Figure 5.15 Typical steady-state waveforms in full-bridge transformer isolated DC/DC converter.

Figure 5.16 Boost converter for PV power interface.

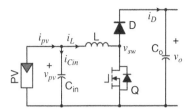

In contrast to PWM, which is controlled by the switching duty cycle, there is one state in which Q1 and Q3 are turned on, which results in zero voltage across the transformer input terminal. The phase-shift technique for the full-bridge DC/DC converter has the advantage of a soft-switching capability, which can reduce switching losses and increase the conversion efficiency.

5.1.4 Boost Converter

A boost converter circuit for the PVSC is illustrated in Figure 5.16. A step-up topology should be considered if the converter-output voltage is never lower than the PV-link voltage at MPP, V_{MPP}, with consideration of the normal voltage variation of both terminals. The condition can be expressed as: $V_{O(min)} \geq V_{MPP(max)}$. The highest value of the PV-link voltage of the MPP, $V_{MPP(max)}$, can be estimated for a combination of the lowest ambient temperature and the highest irradiance. The load profile can be used to determine the value of $V_{O(min)}$. For battery-charging applications, the value of $V_{O(min)}$ can be selected to be the same as the cut-off voltage at 0% SOC. One advantage of the boost topology is that the input current is smooth, since the inductor appears on the input side. This feature requires less capacitance at the PV link in comparison with the buck topology. Under the same conditions, the dynamics at the PV link is faster than with the application of buck or buck–boost converters (Xiao et al. 2007). However, the output current is always pulsating or "chopped," which must be considered in relation to the load profile.

At STC and the predefined switching frequency, f_{sw}, the inductor ripple current and ripple voltage at the PV link should be specified by the peak-to-peak values, ΔI_L and ΔV_{PV}, respectively. Steady-state analysis can determine the duty cycle at the nominal load operating condition. At STC, the PV source circuit is expected to output V_{MPP} and I_{MPP}, which represent the MPP. The duty cycle can be calculated as

$$D_o = 1 - \frac{V_{MPP}}{V_{O-NOM}} \tag{5.16}$$

when the converter is operated at CCM. The symbol V_{O-NOM} represents the output voltage of the nominal operating condition, which should be determined by the load profile. The value of the inductance, L, can be calculated as

$$L = \frac{V_{MPP}D_o}{\Delta I_L f_{sw}} \tag{5.17}$$

At steady state, the inductor current waveform is composed of the DC component, I_{MPP}, and the linear ripple current with a peak-to-peak value of ΔI_L, as shown in Figure 5.17. All variables are defined according to the circuit diagram shown in Figure 5.16. The capacitor, C_{in}, provides filtering of the switching ripple to minimize

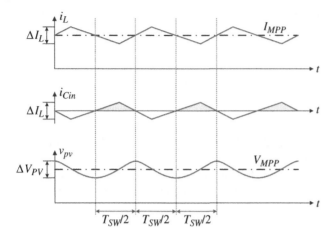

Figure 5.17 Typical steady-state waveforms in boost DC/DC converter.

the coupling with the PV output current, i_{pv}. The variation of the capacitor voltage, v_{pv}, depends on the charge and discharge current through it. At steady state, the charge and discharge period to be distributed is half of the switching cycle. During the charging period, as marked in Figure 5.17, the voltage, v_{pv} increases by ΔV_{PV}, which is from the minimum to the maximum value. The charge during the half cycle can be calculated as $q = C_{in}\Delta V_{PV}$, which is equivalent to the integration of the current waveform between its zero crossing. The equilibrium can be expressed as

$$C_{in}\Delta V_{PV} = \frac{1}{2}\frac{\Delta I_L}{2}\frac{T_{SW}}{2} \tag{5.18}$$

which includes the integration of the triangular area of the current waveform. Therefore, the capacitance of C_{in} can be determined according the ripple specification, ΔV_{PV}, as

$$C_{in} = \frac{\Delta I_L}{8\Delta V_{PV} f_{sw}} \tag{5.19}$$

where T_{SW} and f_{SW} represent the switching period and switching frequency, respectively.

When Q is at the on-state, the system dynamics can be represented as

$$i_L = \frac{1}{L}\int v_{pv}dt \tag{5.20}$$

$$v_{pv} = \frac{1}{C_{in}}\int (i_{pv} - i_L)dt \tag{5.21}$$

When Q is at the off-state, the system dynamics is given by

$$i_L = \frac{1}{L}\int (v_{pv} - v_o)dt \tag{5.22}$$

$$v_{pv} = \frac{1}{C_{in}}\int (i_{pv} - i_L)dt \tag{5.23}$$

Based on the on/off switching, a simulation model can be built using Simulink, as shown in Figure 5.18a. The model is based on the ideal boost converter without

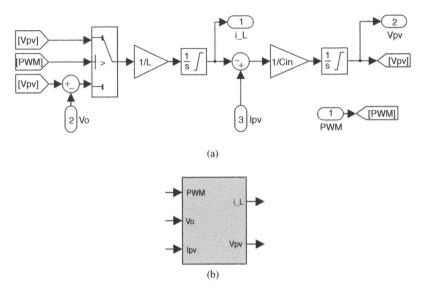

(a)

(b)

Figure 5.18 Simulink model of boost converter for PV power interface: (a) model composition; (b) integrated block.

the consideration of power loss. It includes three inputs: the PWM signal for the switches, the output voltage determined by the load condition (v_o), and the PV output current (i_{pv}). It also outputs two signals: the inductor current (i_L) and the PV-link voltage (v_{pv}). The dynamic interaction is simulated by the integral operation, which is defined in (5.20)–(5.23). The saturation signs are shown in the Simulink integration blocks. These ensure that the inductor current (i_L) and the PV-link voltage (v_{pv}) are always positive. All components can be packed together to form a single block to represent the converter model, as shown in Figure 5.18b. The relation of the PV-link voltage (v_{pv}) and the PV output current (i_{pv}) should follow the PV source circuit model discussed in Chapter 4. The steady-state analysis and design can be simulated and evaluated by the developed model.

An example is the charging of a battery module with a nominal voltage of 48 V. The open-circuit voltage is 40 V when the battery module is fully discharged. The power source is a PV module, the STC electrical characteristics of which are shown in Figure 5.4. Since V_{MPP} is expected to be lower than 40 V, the boost topology is selected. The switching frequency is designed to be 50 kHz, the peak-to-peak ripple voltage of the PV module is specified as 0.2 V, and the peak-to-peak ripple of the inductor current is specified as 1 A. Based on the nominal battery voltage and STC, the nominal duty cycle of the PWM can be calculated as 22.9% according to (5.16). Then the values of L and C_{in} can be calculated as 170 μH and 12.5 μF, by following (5.17) and (5.19), respectively.

The simulation result is shown in Figure 5.19, including the waveforms of i_L, v_{pv}, i_{pv}, and p_{pv}. When PWM signals with a nominal duty cycle of 22.9% are applied to the converter, the value of v_{pv} drops from the open-circuit voltage to the value of V_{MPP} and then reaches the steady state. The signal of i_{pv} increases from zero to the value of I_{MPP} and maintains the steady state. The PV power (p_{pv}) reaches the value of MPP at the

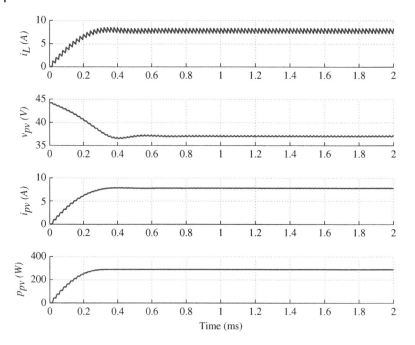

Figure 5.19 Simulated waveforms of the PV power charger at steady state.

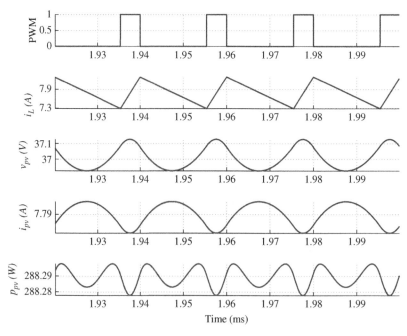

Figure 5.20 Simulated waveforms of the PV power charger at steady state. Zoomed-in view to illustrate ripple.

Figure 5.21 Tapped-inductor converter used for PV power interface.

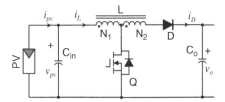

corresponding voltage and current, just as the rating shown in Figure 5.19. Figure 5.20 provides a zoom-in look at the waveforms, showing that the peak-to-peak ripple of the inductor current is about 1 A, which corresponds to the specification. The peak-to-peak ripple of the PV-link voltage also matches the specified value of 0.2 V. The fixed step size of the numerical solver is chosen as 20 ns, which indicates that the simulation resolution is 1000 sampling points in each switching cycle. The simulation verifies the system design and demonstrates the model's operation.

5.1.5 Tapped-inductor Boost Topology

Figure 5.21 illustrates the circuit of a tapped-inductor boost converter used as a PVSC. The circuit shows that the tapping of the inductor is connected to a power switch, Q. The operation follows the same principles as for the analysis of the boost topology. However, due to the the the tapping connection, different inductances should be considered for the on- and off-states of Q. This topology can be selected if the converter-output voltage is significantly higher than the PV-link voltage at MPP (V_{MPP}) and if galvanic isolation is not required. The topology suits module-integrated parallel converter (MIPCs) applications due to the need for high conversion ratios, as discussed in Section 2.4.2. The topology has shown the potential for high conversion efficiency (Cheng 2006). Efficiency of 98.3% has been reported in an MIPC application (Krzywinski 2015).

When Q is switched on, current is drawn from the PV source circuit and the input capacitor to charge the first section of the inductor. When Q is off, the energy stored in the magnetic core forces the current to flow and supply the output side through the whole inductor. The split ratio of the tapped inductor is given by $N_1 : N_2$. The topology has the same advantage as the boost topology: the current is smooth at the input port due to the presence of the inductor, as shown in Figure 5.21.

The dynamics of the current, i_L, can be expressed as (5.24) and (5.25) in response to the switch on and off, respectively.

$$\left(\frac{N_1}{N_1 + N_2}\right) L \frac{di_L}{dt} = v_{pv} \tag{5.24}$$

$$L \frac{di_L}{dt} = v_{pv} - v_o \tag{5.25}$$

At steady state, the magnitude of the increasing inductor current at the on-state should be equal to that of the decreasing magnitude in every switching cycle, which is expressed as

$$\Delta I_{L(on)} + \Delta I_{L(off)} = 0 \tag{5.26}$$

which can be rearranged to

$$\left(\frac{N_1 + N_2}{N_1}\right) \frac{V_{PV}}{Lf_{sw}} d = \frac{V_{PV} - V_O}{Lf_{sw}}(1 - d) \tag{5.27}$$

from which the steady-state equivalence can be found and expressed as

$$\frac{V_O}{V_{PV}} = \frac{1 + (N_2/N_1)d}{1 - d} \tag{5.28}$$

where V_{PV} and V_O are the input and output voltages at the steady state, and f_{sw} and d are the switching frequency and switching duty ratio. When $N_2 = 0$, the tapped-inductor topology is equivalent to a conventional boost converter. When $N_2 > 0$, the voltage-conversion ratio can be increased by the split ratio of the tapped inductor. Figure 5.22 shows the voltage-conversion ratio, which is significantly raised by the split ratios.

At STC and the predefined switching frequency, f_{sw}, the inductor ripple current and ripple voltage at the PV link can be specified by the peak-to-peak values as ΔI_L and ΔV_{PV}, respectively. Steady-state analysis can determine the duty cycle at the nominal-load operating condition. At STC, the output of the PV source circuit is expected to be located at the V_{MPP} and I_{MPP}, which represent the MPP. The duty cycle can be calculated from

$$D_0 = \frac{V_{O-NOM}/V_{MPP} - 1}{N_2/N_1 + V_{O-NOM}/V_{MPP}} \tag{5.29}$$

which is derived from (5.28) when the converter is operating at CCM. The symbol V_{O-NOM} represents the output voltage of the nominal operating condition, which can be determined from the load profile. The value of the inductance, L, and the capacitance C_{in} can be calculated as

$$L = \frac{(V_{O-NOM} - V_{MPP})(1 - D_0)}{\Delta I_L f_{sw}} \tag{5.30}$$

$$C_{in} = \frac{\Delta I_L}{8 \Delta V_{PV} f_{sw}} \tag{5.31}$$

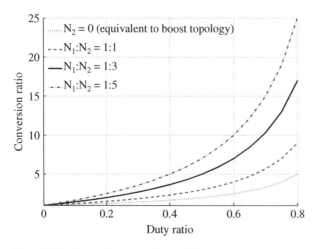

Figure 5.22 Conversion ratio of tapped-inductor topology.

When Q is at the on-state, the system dynamics according to the PWM operation of the converter circuit can be represented by

$$i_L = \frac{N_1 + N_2}{N_1 L} \int v_{pv} dt \tag{5.32a}$$

$$v_{pv} = \frac{1}{C_{in}} \int (i_{pv} - i_L) dt \tag{5.32b}$$

When Q is in the off-state, the system dynamics is given by

$$i_L = \frac{1}{L} \int (v_{pv} - v_o) dt \tag{5.33a}$$

$$v_{pv} = \frac{1}{C_{in}} \int (i_{pv} - i_L) dt \tag{5.33b}$$

Based on the on/off states shown in (5.32) and (5.33), a simulation model can be built using Simulink, as shown in Figure 5.23a. The model includes three inputs: the PWM signal for the switches, the output voltage determined by the load condition (v_o), and the PV output current (i_{pv}). It also outputs two signals: the inductor current (i_L) and the PV-link voltage (v_{pv}). The dynamic interaction is simulated by the integral operation, which is defined in (5.32)–(5.33). The saturation signs shown in the Simulink blocks of integration constrain the inductor current (i_L) and the PV-link voltage (v_{pv}) to always be positive. All components can be packed together to form a single block to represent the converter model, as shown in Figure 5.23b.

A design example is a DC microgrid that has a DC bus voltage of 380 V. The power source is a PV module, of which the electrical characteristics at STC are shown in Figure 5.4. Comparing the source and load voltage profiles, the step-up conversion

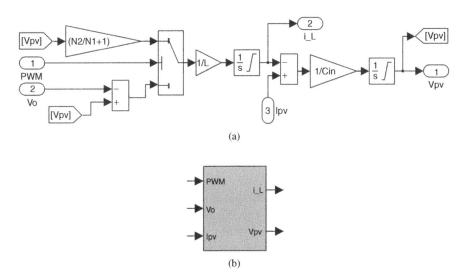

(a)

(b)

Figure 5.23 Simulink model of tapped-inductor boost converter for PV power interface: (a) model composition; (b) integrated block.

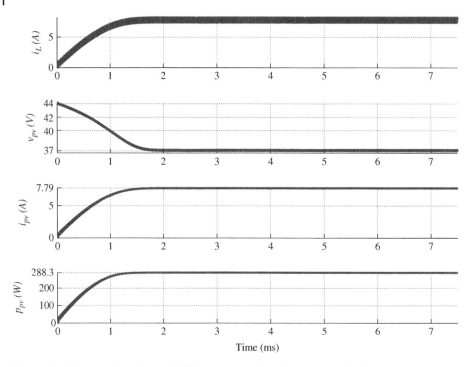

Figure 5.24 Simulated waveforms of PV power system based on a tapped-inductor converter.

ratio is higher than 1:10, which can be supported by a tapped-inductor boost topology. The split ratio of the tapped inductor is designed to be (N_1 : N_2 = 1 : 6). The switching frequency is 50 kHz, the peak-to-peak ripple voltage of the PV module is 0.2 V, and the peak-to-peak ripple of the magnetic inductor current is 1 A. At the nominal DC bus voltage of 380 V and STC, the nominal duty cycle of PWM can be calculated as 57.0% from (5.29). Then the values of L and C_{in} can be calculated as 3.6 mH and 12.5 µF, from (5.30) and (5.31), respectively.

The simulation results are shown in Figure 5.24, including the waveforms of i_L, v_{pv}, i_{pv}, and p_{pv}. When PWM signals at a nominal duty cycle of 57.0% are applied to the converter, the value of v_{pv} drops from the open-circuit voltage to the value of V_{MPP} (37 V) and reaches the steady state. The signal of i_{pv} increases from zero to the value of I_{MPP} and then stays at the steady state. The PV power (p_{pv}) reaches the MPP value, at corresponding voltage and current the same as the ratings shown in Figure 5.4. Figure 5.25 provides a zoom-in on the waveforms, showing that the peak-to-peak ripple of inductor current is about 1 A, matching the design specification. The peak-to-peak ripple of the PV-link voltage is also matches the specified values, at 0.2 V. The tapped-inductor boost converter has the potential for high voltage-conversion ratios when used as a PVSC.

5.1.6 Buck–Boost Converter

When PV power is used for battery charging, the converter-output voltage and the PV-link voltage of the MPP (V_{MPP}) cannot be distinguished as either step-up or step-down, because of the significant variation of the PV-link and converter-output

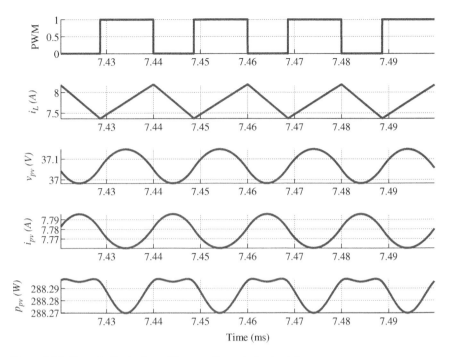

Figure 5.25 Zoom-in illustration of the waveform at steady state.

Figure 5.26 Buck–boost converter used as PV power
interface.

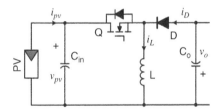

voltages. Under these conditions, a buck–boost topology can be considered as the
PVSC when opposite polarity of the output voltage is not a concern.

Figure (5.26) is the circuit diagram of a buck–boost converter used as a PV-side
power interface. One disadvantage of the buck–boost topology is that both the input
and output current are chopped, which usually requires significant filtering at both
ports to smooth the input and output voltages. As the PV power interface, the dynamics
of the PV link is mainly affected by the applied filtering. Meanwhile, the output current
is always pulsing, which should be considered in relation to the load profile.

At STC and the predefined switching frequency, f_{sw}, the inductor ripple current
and ripple voltage at the PV link should be specified by the peak-to-peak values,
ΔI_L and ΔV_{PV}, respectively. Steady-state analysis can determine the duty cycle at the
nominal-load operating condition. At STC, the PV source circuit is expected to output
V_{MPP} and I_{MPP}, which represent the MPP. The duty cycle can be calculated as

$$D_o = \frac{V_{O-NOM}}{V_{O-NOM} + V_{MPP}} \tag{5.34}$$

when the converter is operated at CCM. The symbol V_{O-NOM} represents the output voltage of the nominal operating condition, which can be determined by the load profile. The value of the inductance, L, and the capacitance C_{in} can be calculated from (5.35) and (5.36), respectively:

$$L = \frac{V_{MPP}D_o}{\Delta I_L f_{sw}} \tag{5.35}$$

$$C_{in} = \frac{I_{MPP}(1 - D_o)}{\Delta V_{PV} f_{sw}} \tag{5.36}$$

When Q is in the on-state, the system dynamics can be given by

$$i_L = \frac{1}{L} \int v_{pv} dt \tag{5.37}$$

$$v_{pv} = \frac{1}{C_{in}} \int (i_{pv} - i_L) dt \tag{5.38}$$

When Q is in the of- state, the system dynamics can be given by

$$i_L = \frac{1}{L} \int (-v_o) dt \tag{5.39}$$

$$v_{pv} = \frac{1}{C_{in}} \int i_{pv} dt \tag{5.40}$$

Based on the repeatable operation of on/off states, a simulation model can be built using Simulink, as shown in Figure 5.27a. The model is based on the ideal buck–boost converter, without consideration of power loss. It includes three inputs: the PWM signal for the switches (PWM), the output voltage determined by the load condition (v_o), and

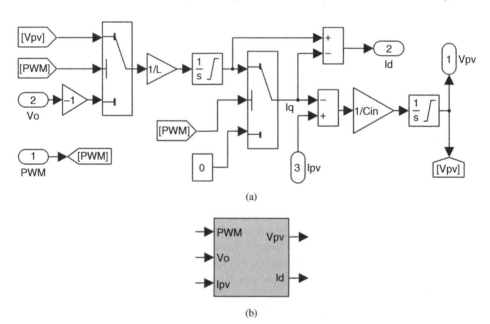

(a)

(b)

Figure 5.27 Simulink model of buck–boost converter for PV power interface: (a) model composition; (b) integrated block.

the PV output current (i_{pv}). It also outputs two signals: the diode current (i_D) and the PV-link voltage (v_{pv}). The dynamic interaction is simulated by the integral operation, which is defined in (5.37)–(5.40). The saturation signs are shown in the Simulink integration blocks. This constrains the inductor current (i_L) and the PV-link voltage (v_{pv}) to be DC signals. All components can be packed together to form a single block to represent the converter model for simulation, as shown in Figure 5.27b. The relation of the PV-link voltage (v_{pv}) and the PV output current (i_{pv}) can be determined from the PV source circuit model, as discussed in Chapter 4.

A design example is the charging of a battery module with a nominal voltage of 36 V. The battery voltage ranges from 30 V to 42 V depending on the level of SOC. The power source is a PV module, and its electrical characteristics at STC are shown Figure 5.4. By comparing the source and load voltage profiles, the voltage conversion can be either step-up or step-down depending on the status of both the PV source circuit and the battery. Therefore, a buck–boost converter is chosen. The switching frequency is designed to be 50 kHz, the peak-to-peak ripple voltage of the PV module is specified as 0.2 V, and the peak-to-peak ripple of the inductor current is specified as 1 A. At the nominal battery voltage and STC, the nominal duty cycle of the PWM can be calculated as 49.3% from (5.34). Then the values of L and C_{in} can be calculated as 365 µH and 395 µF, following (5.35) and (5.36), respectively. The simulink model, as shown in Figure 5.27 is utilized with the simulation models of the PV module and PWM generator.

The simulation results are shown in Figure 5.28, including the waveforms of i_L, v_{pv}, i_{pv}, and p_{pv}. When PWM signals with the nominal duty cycle of 49.3% are applied to the

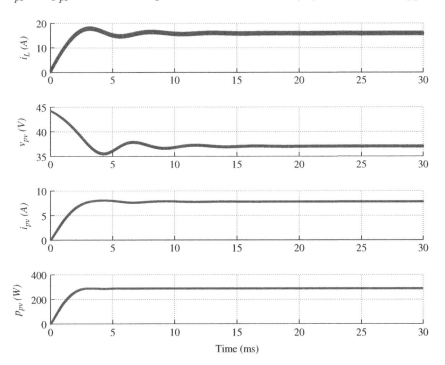

Figure 5.28 Simulated waveforms of a PV power system based on buck–boost topology.

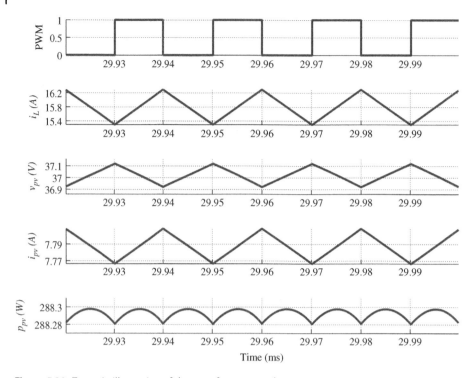

Figure 5.29 Zoom-in illustration of the waveform at steady state.

converter, the value of v_{pv} drops from the open-circuit voltage to the value of V_{MPP} and reaches the steady state. The signal of i_{pv} increases from zero to the value of I_{MPP} and a steady state. The PV power (p_{pv}) reaches the value of MPP at a voltage and current that is the same as the rating shown in Figure 5.4.

Figure 5.29 provides a zoom-in on the waveforms, showing that the peak-to-peak ripple of inductor current is about 1 A, matching the specification. The peak-to-peak ripple of the PV-link voltage is also measured, and matches the specified value of 0.2 V. The fixed step size of the numerical solver is chosen as 20 ns, which indicates that the simulation resolution is 1000 sampling points in each switching cycle of 20 μs. The simulation verifies the system design and demonstrates the system operation.

5.1.7 Flyback Converter

A flyback topology should be considered if the converter-output voltage is either significantly higher or significantly lower than the PV-link voltage at the MPP (V_{MPP}), and galvanic isolation is required. A high conversion ratio can be achieved through the winding-turn ratio of the flyback transformer. The analysis of the flyback converter can be derived from the principle of the buck–boost topology, as illustrated in Figure 5.30 (Erickson and Maksimovic 2001). An ideal transformer with a $1:n$ winding ratio can be added to the buck–boost converter to provide galvanic isolation or extend the voltage-conversion ratio, as shown in Figure 5.30b. A flyback transformer can be designed and constructed to accommodate both the inductance and the voltage transformer, as illustrated in Figure 5.30c. L_m is the magnetic inductance of the flyback

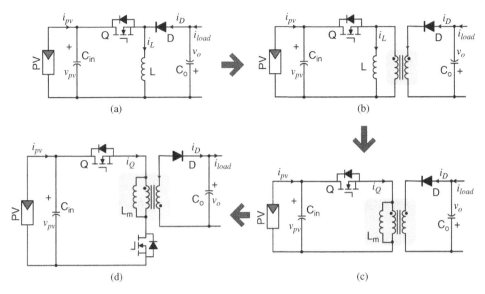

Figure 5.30 Evolution of flyback topology from the buck–boost topology: (a) buck-boost; (b) Isolated buck-boost; (c) integrated magnetic inductor into transformer; (d) flyback converter.

transformer, which is equivalent to the symbol L in the steady-state analysis for the buck–boost converter. The location of the power switch can be moved to the low side to simplify the driving circuit. The polarity of the output voltage can be corrected by reassigning the polarity of the transformer output terminal. Therefore, a standard flyback topology can be formed, as shown in Figure 5.30d, in which the conversion ratio for CCM operation is expressed as:

$$\frac{V_O}{V_{PV}} = n\frac{D}{1-D} \tag{5.41}$$

where D is the switching duty cycle for the on-state and n refers to the $1:n$ winding-turn ratio.

For PVSC applications, the discontinuous conduction mode (DCM) of the flyback topology is used. It should be noted that the inductor current cannot be directly measured in the flyback topology because the magnetic inductor is embedded inside the flyback transformer. The topology has the same disadvantage as the buck–boost topology, namely that both the input current and output current are always pulsing. The voltage-conversion ratio in steady state and DCM can be determined from

$$\frac{V_O}{V_{PV}} = n\frac{T_{up}}{T_{down}} \tag{5.42}$$

where T_{up} is to the time period that the inductor current takes to rise from zero to the peak and T_{down} is the period taken to drop back to zero.

In DCM, there is a zero state of the inductor current in each switching cycle, in contrast to CCM operation. Designing a flyback converter can start with defining the winding-turn ratio of the flyback transformer to achieve the required voltage-conversion ratio. Other steps follow the design of the buck–boost converter.

The design procedure can specify the current ripple value in order to calculate the magnetic inductance and therefore design the flyback transformer. At STC and the predefined switching frequency, f_{sw}, the inductor ripple current and ripple voltage at the PV link should be specified by the peak-to-peak values, ΔI_L and ΔV_{PV}, respectively. Steady-state analysis can determine the duty cycle at the nominal operating condition. The PV source circuit is expected to output V_{MPP} and I_{MPP}, which represent the MPP. When the converter is operated at CCM, the duty cycle can be calculated as

$$D_o = \frac{V_{O-NOM}}{V_{O-NOM} + nV_{MPP}} \tag{5.43}$$

The symbol V_{O-NOM} represents the output voltage of the nominal operating condition, which can be determined from the load profile. Following the same derivation as the buck–boost converter, the value of the magnetic inductance, L_m, and the capacitance, C_{in}, can be calculated from (5.44) and (5.45), respectively.

$$L_m = \frac{V_{MPP}D_o}{\Delta I_L f_{sw}} \tag{5.44}$$

$$C_{in} = \frac{I_{MPP}(1 - D_o)}{\Delta V_{PV} f_{sw}} \tag{5.45}$$

A minor modification of the Simulink model of the buck–boost converter is required to simulate the flyback topology. This is shown in Figure 5.31. A new parameter, n, is included in the model to represent the winding-turn ratio of the flyback transformer. When $n > 1$, high step-up voltage ratios can be achieved, which is desirable

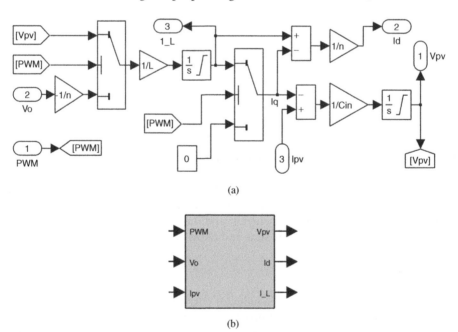

(a)

(b)

Figure 5.31 Simulink model of flyback converter for PV power interface: (a) model composition; (b) integrated block.

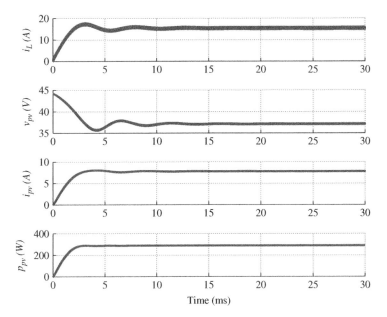

Figure 5.32 Simulated waveforms of a PV power system based on the flyback topology.

for distributed MPPT systems at the module and submodule levels, as discussed in Sections 2.4 and 2.5, respectively.

A design example is the use of a flyback topology for a DC microgrid with a DC bus voltage of 380 V. The power source is a PV module, with STC electrical characteristics as shown in Figure 5.4. Comparing the source and load voltage profiles, the step-up conversion ratio is higher than 1:10, which supports use of the flyback topology. The winding-turn ratio of the flyback transformer is designed as 1:10 ($n = 10$). The switching frequency is specified as 50 kHz, the peak-to-peak ripple voltage of the PV module is specified as 0.2 V, and the peak-to-peak ripple of the magnetic inductor current is specified as 1 A, which indicates CCM. At the nominal DC bus voltage and STC, the nominal duty cycle of the PWM can be calculated as 50.7% from (5.43). The values of L_m and C_{in} can be calculated as 375 µH and 384 µF, according to (5.44) and (5.45), respectively.

The simulation results are shown in Figure 5.32, including the waveforms of i_L, v_{pv}, i_{pv}, and p_{pv}. When the PWM signals with the nominal duty cycle of 50.7% are applied to the converter, the value of v_{pv} drops from the open-circuit voltage to the value of V_{MPP} and reach the steady state. The signal of i_{pv} increases from zero to the value of I_{MPP} and enters the steady state. The PV power (p_{pv}) reaches the MPP value at voltage and current values the same as the ratings shown in Figure 5.4. Figure 5.33 provides a zoom-in on the waveforms, showing that the peak-to-peak ripple of the inductor current is about 1 A, matching the specification. The peak-to-peak ripple of the PV-link voltage also matches the specified values of 0.2 V.

The fixed step size of the numerical solver is chosen as 20 ns, which indicates that the simulation resolution is 1000 sampling points in each switching cycle of 20 µs. Oversampling is unnecessary since it slows down the simulation. The simulation result proves the system design and verifies the operation. It can be seen that the

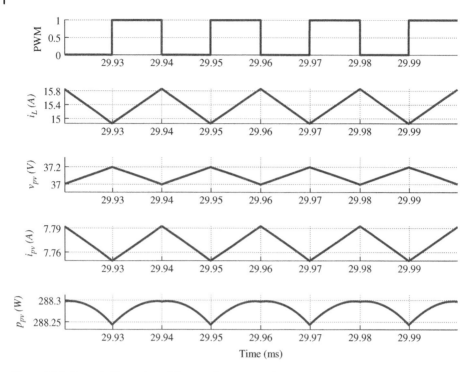

Figure 5.33 Zoom-in illustration of the waveform at steady state.

waveforms in Figures 5.32 and 5.33 are similar to those in Figures 5.29 and 5.28. This is because both flyback and buck–boost topology share the same operating principle, as illustrated in Figure 5.30, and they are based on very similar specifications except for the winding-turn ratio of 10.

5.2 Battery-side Converter for DC/DC Stage

Batteries are critical in standalone PV systems, as discussed in Section 1.9. In a DC microgrid, as shown in Figure 1.16, bidirectional power flow is required to exchange power between the DC bus and the energy storage system. The battery bank contributes power to the DC bus when the load demand is higher than the PV power generation. Current flows in the opposite direction to charge the battery when additional PV power is available. One solution for bidirectional conversion is to use two DC/DC converters, one for the charging operation and one for discharge. Since the charge and discharge for one battery storage unit do not happen at the same time, a bidirectional DC/DC converter that switches operating mode as required can also be used.

5.2.1 Introduction to Dual Active Bridges

One special DC/DC topology has drawn a great deal of recent research attention. This is the dual active bridge (DAB) DC/DC converter. This topology suits applications of battery power interfaces and solid-state transformers thanks to its capability for

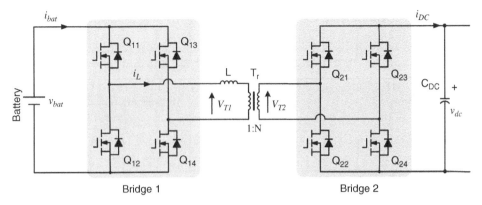

Figure 5.34 Bidirectional DC/DC for battery power interface based on dual active bridge topology.

bidirectional power flow, its high power density, the controllability of the power flow, and the inherent soft switching (Wen et al. 2013, 2014). Figure 5.34 shows a schematic of a conventional DAB, with two H bridges, connected through an inductor, L, and a transformer, T_r.

The transformer provides galvanic isolation and the winding-turn ratio (1:N) gives the flexibility to achieve high voltage-conversion ratios between the battery and the DC link. The diagonal devices in both bridges are paired and controlled by the same gate signal. Shown in Figure 5.34, the on/off operations of Q_{11}, Q_{12}, Q_{21}, and Q_{22} are the same as those of Q_{14}, Q_{13}, Q_{24}, and Q_{23}, respectively. The operation creates two square waves in AC format, V_{T1} and V_{T2}. The voltage difference between V_{T1} and V_{T2} appears across the inductor, L. Limited by the inductance in L, the energy can be transferred in both directions depending on the leading and lagging phases between the square waveforms of V_{T1} and V_{T2}.

5.2.2 Discharge Operation

The discharge operation is when power is delivered to the DC bus from the battery bank. A typical waveform is shown in Figure 5.35. The power-exchange level is determined by the degree of phase shift between the two HFAC signals. This is the control variable for regulating the flow of active power. The degree of phase shift between V_{T1} and V_{T2} is represented by the symbol T_1 in the time domain, as shown in Figure 5.35. Both bridges are switched at the same switching frequency and a constant 50% duty cycle.

The dynamics of the inductor current is given by

$$i_L = \frac{1}{L} \int (V_{T1} - V_{T2})dt \tag{5.46}$$

Thus, the Simulink model of the DAB can be built by following the circuits of Figures 5.34 and (5.46). Besides the control signals for the eight power switches, the input- and output-voltage signals are the other model inputs. Shown in Figure 5.36, the model will output the signals v_{T1}, v_{T2}, and i_L, as defined in Figure 5.34. The signals of the delivered power and the input current are also computed for the output variables.

From T_0 to T_1, the inductor current increase, as shown in Figure 5.35, is

$$i_L(t) = i_L(0) + \frac{V_{bat} + V_{dc}/N}{L}t \tag{5.47}$$

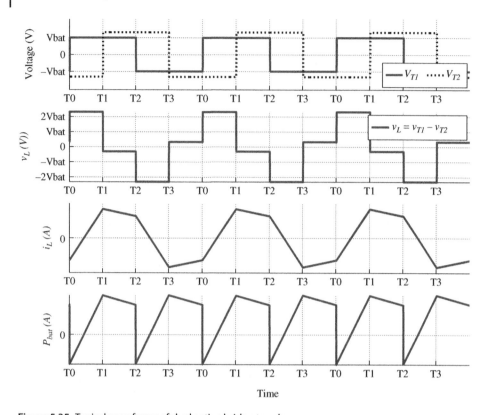

Figure 5.35 Typical waveforms of dual active bridge topology.

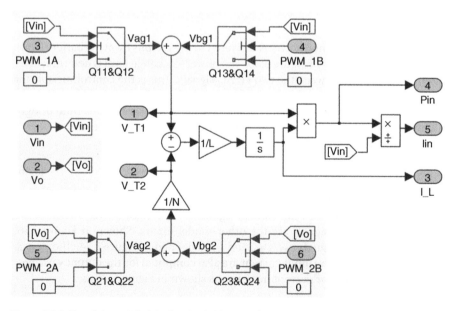

Figure 5.36 Simulink model of dual active bridge topology.

where V_{bat} and V_o symbolize the steady-state values of the input and output voltages respectively. The initial value of the inductor current is $i_L(T_0)$, which indicates the beginning of the switching cycle. At T_1, the current value is

$$i_L(T_1) = i_L(T_0) + \frac{V_{bat} + V_{dc}/N}{L} T_1 \tag{5.48}$$

From T_1 to T_2, the inductor current varies according to the difference between V_{bat} and V_{dc} and is expressed as

$$i_L(t) = i_L(T_1) + \frac{V_{bat} - V_{dc}/N}{L}(t - T_1) \tag{5.49}$$

At T_2, the current value is

$$i_L(T_2) = i_L(T_1) + \frac{V_{bat} - V_{dc}/N}{L}(T_2 - T_1) \tag{5.50}$$

Time T_2 is also the indicator of the half cycle of the waveform, I_L, as shown in Figure 5.35. At steady state, the magnitude of $i_L(0)$ and $i_L(T_2)$ should be identical, to form the symmetrical waveform of i_L. Therefore, by following (5.48) and (5.50), the equilibrium can be expressed as

$$-i_L(T_0) = i_L(T_0) + \frac{V_{bat} + V_{dc}/N}{L} T1 + \frac{V_{bat} - V_{dc}/N}{L}(T_2 - T_1) \tag{5.51}$$

Then, the initial value of the inductor current, $i_L(0)$, can be determined as

$$i_L(T_0) = -\frac{T_2 V_{bat} + 2T_1 V_{dc}/N - T_2 V_{dc}/N}{2L} \tag{5.52}$$

The waveforms of v_{T1}, v_{T2}, and i_L are fixed at the constant switching frequency, f_{sw}, as shown in Figure 5.35. The angular frequency (ω) that is expressed as $\omega = 2\pi f_{sw}$ can be applied to (5.52). The current initial condition can be expressed as

$$i_L(T_0) = -\frac{T_2 \omega V_{bat} + 2T_1 \omega V_{dc}/N - T_2 \omega V_{dc}/N}{2\omega L}$$
$$= -\frac{\pi V_{bat} + 2\varphi V_{dc}/N - \pi V_{dc}/N}{2\omega L} \tag{5.53}$$

where the symbol φ is the phase shift with units of radians. It should be noted that the value of $i_L(T_0)$ should be considered as the initial condition to start system simulation. The value can be pre-calculated by (5.53) as the initial state and the implemented into the integrator block of the Simulink model, as shown in Figure 5.36. This can avoid initial-value errors in simulations, which cause a DC bias in the waveform of i_L.

The power waveform shows the repeated cycle from T_0 to T_2, as illustrated in Figure 5.35. Without considering power loss, the power flow level that is transferred between the battery and the DC link is

$$P_{bat} = \frac{1}{T_2} \int_0^{T_2} v_{bat} i_L(t) dt \tag{5.54}$$

The integration in the half cycle of both v_{T1} and i_L can be separated into two parts, and divided by the phase shift moment of T_1:

$$P_{bat} = \frac{V_{bat}}{T_2} \left[\int_0^{T_1} i_L(t) dt + \int_{T_1}^{T_2} i_L(t) dt \right]. \tag{5.55}$$

The power flow equation at steady state can be derived as

$$P_{bat} = \frac{V_{bat}V_o\varphi(\pi - \varphi)}{\pi\omega LN} \tag{5.56}$$

which then gives the battery current equation

$$I_{bat} = \frac{V_o\varphi(\pi - \varphi)}{\pi\omega LN} \tag{5.57}$$

The maximum power flow can be determined by the extremum theory:

$$\frac{dp_{bat}}{dt} = 0 \tag{5.58}$$

which then gives

$$1 - \frac{2\varphi}{\pi} = 0 \tag{5.59}$$

A phase shift of 90° (or $\pi/2$ radians) gives the highest power delivery:

$$P_{bat}(max) = \frac{V_{bat}V_o}{8f_{sw}LN} \tag{5.60}$$

This is the highest power capacity at which the DAB can be operated for the specific ratings of the switching frequency, the inductance, and the input and output voltages.

For the predefined maximum power level, the inductance should be satisfied by the following condition (5.61), which is commonly used for circuit design.

$$L \le \frac{V_{bat}V_o}{8f_{sw}NP_{bat}(max)} \tag{5.61}$$

A design example is a power interface to link a battery bank with a DC bus. The nominal voltage of the battery bank is 48 V, which is a common level in the telecommunications industry. The DC bus is rated at 380 V, which is standard for low-voltage DC systems. Due to the high voltage-conversion ratio and bidirectional power flow required, the DAB topology is selected for the power interface. The winding-turn ratio of the transformer is constructed as $N = 6$. Due to the nature of the zero voltage switching of the DAB, the switching frequency is selected as 200 kHz. The maximum power capacity is 1 kW, based on a phase shift of 90°. According to (5.61), for the maximum power delivery, the inductance can be calculated and rated as 1.9 μH. All parameters are summarized in Table 5.1.

Table 5.1 Specification of dual active bridge.

Parameters	Value
Nominal battery voltage	48 V
Nominal DC bus voltage	380 V
Transformer winding-turns ratio	$N = 6$
Switching frequency	200 kHz
Inductance	$L = 1.9\,\mu H$
Maximum power capacity	1000 W
Nominal power rating	750 W

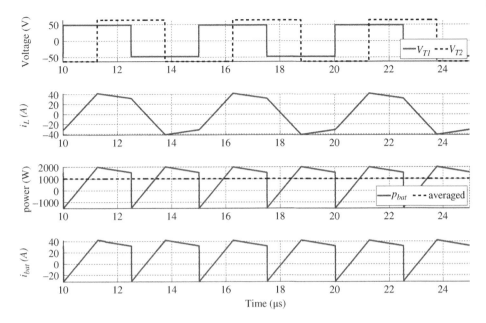

Figure 5.37 Simulated waveforms of the DAB topology at full power.

The simulation results are shown in Figure 5.37, including the waveforms of v_{T1}, v_{T2}, i_L, p_{bat}, and i_{bat}, as defined in Figure 5.34. At 90° phase shift, the averaged power output of the battery bank is 1000 W, which matches the design specification and proves the concept design.

Following (5.56), the power output is 750 W when the phase shift is 45°. This is defined as the nominal power rating. For verification, the simulation is shown in Figure 5.38, which indicates that the power level is controlled to be 750 W. Since the phase of v_{T1} leads that of v_{T2}, the active power flows from the battery into the DC bus, which is the discharge operation.

The reactive power and circulating current can be seen in Figures 5.37 and 5.38, since a portion of power is delivered back to the source. The portion of power is also referred to as "nonactive" power in the literature (Wen et al. 2014). It is clear that the circulating current should be minimized since it always adds an energy loss in conducting elements. Furthermore, the circulating current also limits the range of zero-voltage switching (ZVS), and therefore introduces more switching loss. This results from the difference between the magnitudes of v_{T1} and v_{T2} in the steady state.

5.2.3 Charging Operation

When the phase of v_{T2} leads the phase of v_{T1}, the power flow is from the DC link to the battery. The power exchange level is determined by the degree of phase shift between the two AC signals. The waveforms at steady state are illustrated in Figure 5.39 indicating the charge operation.

T_1 is no longer the indicator of the phase shift in the time domain since the phase of v_{T2} is leading. This can be seen in

$$\omega T_1 = \pi + \varphi \tag{5.62}$$

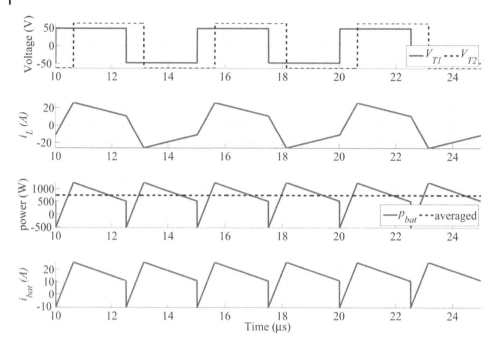

Figure 5.38 Simulated waveforms of the DAB topology when the phase shift is 45°.

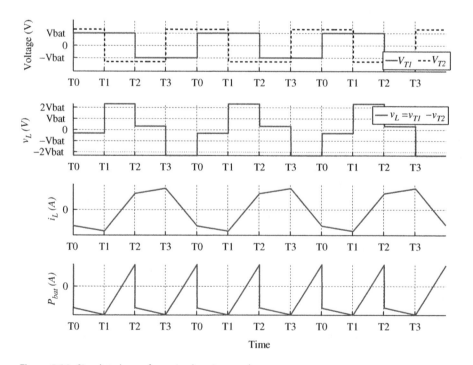

Figure 5.39 Simulated waveforms in charging mode.

where the phase angle (φ) is negative. From 0 to T_1, the change of the inductor current follows the difference between the magnitudes of v_{T2} and v_{T1} and can be expressed as

$$i_L(t) = i_L(T_0) + \frac{V_{bat} - V_{dc}/N}{L}t \tag{5.63}$$

The initial value is shown as $i_L(T_0)$, which indicates the beginning of the switching cycle. At T_1, the current value is shown as

$$i_L(T_1) = i_L(T_0) + \frac{V_{bat} - V_{dc}/N}{L}T_1 \tag{5.64}$$

From T_1 to T_2, the inductor current increases due to the difference between V_{bat} and V_{dc} and is expressed as

$$i_L(t) = i_L(T_1) + \frac{V_{bat} + V_{dc}/N}{L}(t - T_1) \tag{5.65}$$

At T_2, the current value is

$$i_L(T_2) = i_L(0) + \frac{V_{bat} - V_{dc}/N}{L}T_1 + \frac{V_{bat} + V_{dc}/N}{L}(T_2 - T_1) \tag{5.66}$$

T_2 reflects the moment of the half cycle of the waveform, v_{T1}. In the steady state, the magnitudes of $i_L(T_0)$ and $i_L(T_2)$ are identical so as to form the symmetrical waveform of i_L. Therefore, the symmetry can be used to compute the initial value of the inductor current, $i_L(0)$:

$$i_L(T_0) = -\frac{T_2 V_{bat} - 2T_1 V_{dc}/N + T_2 V_{dc}/N}{2L} \tag{5.67}$$

The waveforms of v_{T1}, v_{T2}, and i_L are fixed to the constant switching frequency, f_{sw}, as shown in Figure 5.39. The angular frequency (ω) that is expressed as $\omega = 2\pi f_{sw}$ can be applied to (5.67). The current initial condition can be expressed as (5.68), (5.69), and then (5.70)

$$i_L(T_0) = -\frac{T_2 \omega V_{bat} - 2T_1 \omega V_{dc}/N + T_2 \omega V_{dc}/N}{2\omega L} \tag{5.68}$$

$$i_L(T_0) = -\frac{\pi V_{bat} - 2(\pi - \varphi)V_{dc}/N + \pi V_{dc}/N}{2\omega L} \tag{5.69}$$

$$i_L(T_0) = -\frac{\pi V_{bat} - 2\varphi V_{dc}/N - \pi V_{dc}/N}{2\omega L} \tag{5.70}$$

where φ represents the lagging phase angle, which has a negative value for the equations. The initial value of the inductor current, $i_L(T_0)$ should be implemented in the integrator block of the Simulink model, as shown in Figure 5.36. It should be noted that the expression in (5.70) is different from the initial inductor current, as expressed in (5.53). The initial condition of the inductor current can be expressed in a general format by (5.53) and (5.70), which is expressed (5.71) for both the charge and discharge condition:

$$i_L(T_0) = -\frac{\pi V_{bat} + 2|\varphi|V_{dc}/N - \pi V_{dc}/N}{2\omega L} \tag{5.71}$$

The operation status – either charge or discharge – should be determined before the initial value is computed. This can avoid any initial value error, which would cause a DC bias in the waveform of i_L. The power flow equation in the steady state can be derived from

$$P_{bat} = \frac{V_{bat}}{T_2}\left[\int_0^{T_1} i_L(t)dt + \int_{T_1}^{T_2} i_L(t)dt\right] \qquad (5.72)$$

since the power waveform is repeated in every cycle between T_0 to T_2. This can be expanded to

$$P_{bat} = \frac{V_{bat}}{T_2}\int_0^{T_1}\left[i_L(0) + \frac{V_{bat} - V_{dc}/N}{L}t\right]dt$$
$$+ \frac{V_{bat}}{T_2}\int_{T_1}^{T_2}\left[i_L(T_1) + \frac{V_{bat} + V_{dc}/N}{L}(t - T_1)\right]dt \qquad (5.73)$$

The power flow equation in the steady state can be derived as

$$P_{bat} = \frac{V_{bat}V_o\varphi(\pi + \varphi)}{\pi\omega LN} \qquad (5.74)$$

where φ is a negative value. The expression is different from the computation for the discharge state, as expressed in (5.56). The value of the power flow can be expressed in a general format by (5.56) and (5.74) for both charging and discharging:

$$P_{bat} = \frac{V_{bat}V_o\varphi(\pi - |\varphi|)}{\pi\omega LN} \qquad (5.75)$$

Following (5.75), the power output is $-750\,\text{W}$ when the phase shift becomes $-45°$ or $-\pi/4$, which indicates the phase of v_{T1} is lagging that of v_{T2}. For verification, the simulation is shown in Figure 5.40, which indicates the power level is averaged as $-750\,\text{W}$.

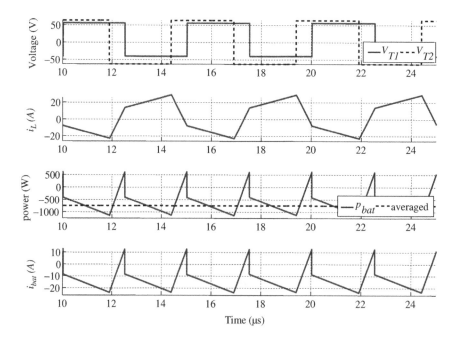

Figure 5.40 Simulated waveforms in charging mode.

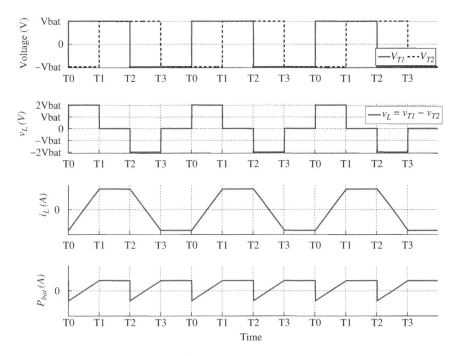

Figure 5.41 Simulated waveforms of the DAB topology when the magnitudes of v_{T1} and v_{T2} are equal.

The active power flow is from the DC bus to the battery. The simulation proves the theoretical design and analysis of the DAB converter as a battery-side power interface.

5.2.4 Zero Voltage Switching

Study has shown that zero voltage switching (ZVS) can be always maintained if the amplitudes of v_{T1} and v_{T2} are equal in the steady state (Wen et al. 2014). This is considered the ideal operating condition since ZVS can be maintained for the full phase-shift range from 0 to 90°. The condition is illustrated in Figure 5.41, where the flat-top shape of the inductor current is noticeable in the discharge operation. The on/off switching happens at T_0, T_1, T_2, and T_3, and repeatably so. The switching scheme is summarized in Table 5.2. It shows that all turn-on operations are ZVS since the current

Table 5.2 Switching operation of dual active bridge.

Time	Current	Switching	Soft switching
T_1	$i_L > 0$; $i_{DC} > 0$	Q_{11} and Q_{14} are conducting; Q_{22} and Q_{23} are switched off; Q_{21} and Q_{24} are switched on.	ZVS
T_2	$i_L > 0$; $i_{DC} > 0$	Q_{21} and Q_{24} are conducting; Q_{11} and Q_{14} are switched off; Q_{12} and Q_{13} are switched on	ZVS
T_3	$i_L < 0$; $i_{DC} < 0$	Q_{12} and Q_{13} are conducting; Q_{21} and Q_{24} are switched off; Q_{22} and Q_{23} are switched on.	ZVS
T_0	$i_L < 0$; $i_{DC} < 0$	Q_{22} and Q_{23} are conducting; Q_{12} and Q_{13} are switched off; Q_{11} and Q_{14} are switched on.	ZVS

direction allows the diode to conduct before the the MOSFET switches, as shown in Figure 5.34. The turn-off operation is always hard switching. The circulating power is noticeable in the power waveform, p_{bat}, since it appears to be negative at certain moments.

Table 5.2 shows that ZVS can be achieved for turning on either Q_{12} and Q_{13} or Q_{21} and Q_{24} when $i_L > 0$. The turning-on switch of either Q_{11} and Q_{14} or Q_{22} and Q_{23} requires $i_L < 0$ for ZVS. The condition can be preserved if the magnitudes of v_{T1} and v_{T2} are equal in the steady state since the inductor current is always in the correct direction, as shown in Figure 5.41.

However, the flat-top condition is difficult to maintain for a battery power interface since the battery voltage varies over a large range, depending on the SOC and the temperature. The analysis can be based on the same example discussed in the previous sections. When the system is in discharge mode, the battery voltage can drop to 44 V, which shows that the magnitude of v_{T1} is 19.33 V lower than that of v_{T2}. When the phase-shift angle is in a low range, say 10°, the discharge power can be calculated as 210 W from (5.75). The simulation results are shown in Figure 5.42, which represents half of the switching cycle. Compared with the switching operation shown in Table 5.2, the inductor current direction is different at the switching moments T_0 and T_2, which indicates the loss of ZVS. The details are summarized in Table 5.3 based on the waveforms. The ZVS is lost for the turn-on switching of Q_{12} and Q_{13}, and Q_{11} and Q_{14}.

To maintain ZVS of all switches, the inductor current, i_L, should not be positive at the T_0:

$$i_L(T_0) \leq 0 \tag{5.76}$$

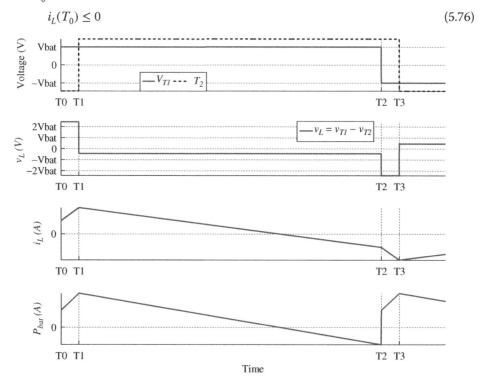

Figure 5.42 Simulated waveforms of the DAB topology when losing ZVS.

Table 5.3 Switching operation of dual active bridge.

Time	Current	Switching	Soft switching
T_1	$i_L > 0$; $i_{DC} > 0$	Q_{11} and Q_{14} are conducting; Q_{22} and Q_{23} are switched off; Q_{21} and Q_{24} are switched on.	ZVS
T_2	$i_L < 0$; $i_{DC} < 0$	Q_{21} and Q_{24} are conducting; Q_{11} and Q_{14} are switched off; Q_{12} and Q_{13} are switched on	ZVS lost
T_3	$i_L < 0$; $i_{DC} < 0$	Q_{12} and Q_{13} are conducting; Q_{21} and Q_{24} are switched off; Q_{22} and Q_{23} are switched on.	ZVS
T_0	$i_L > 0$; $i_{DC} > 0$	Q_{22} and Q_{23} are conducting; Q_{12} and Q_{13} are switched off; Q_{11} and Q_{14} are switched on.	ZVS lost

Using (5.53), the constraint can be derived as

$$-\frac{\pi V_{bat} + 2|\varphi|V_{dc}/N - \pi V_{dc}/N}{2\omega L} \leq 0 \tag{5.77}$$

The minimal phase angle for the ZVS of all switches can be determined from (5.78), where the unit of φ is radians:

$$|\varphi| \geq \frac{\pi(V_{dc} - NV_{bat})}{2V_{dc}} \tag{5.78}$$

The value becomes lower with decreasing difference between V_{bat} and V_{dc}/N and reaches zero when they are equal.

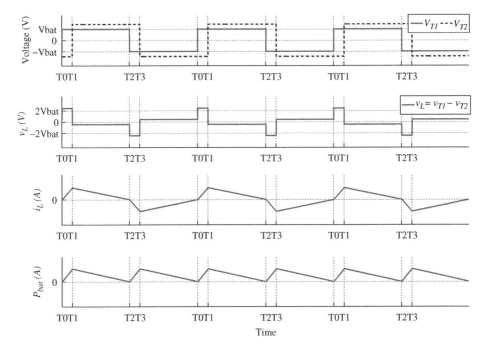

Figure 5.43 Simulated waveforms of the DAB topology to maintain both ZVS and zero circulating of active power.

The case study has a battery voltage of 44 V, a DC bus voltage of 380 V, and a transformer turn ratio, $N = 6$. The minimal phase angle can be calculated as 27.47° for ZVS at the switch-on moments. Figure 5.43 shows the simulation results, with the zero crossing located at the switching moments of T_0 and T_2 to maintain ZVS for all switches when the phase shift is 27.47° or 0.48 radians. It should be noted that the circulating power is also eliminated for the operation when the zero crossing is utilized at the switching moment, which is noticeable in the power waveform. The operating condition is desirable since both the ZVS and zero circulating current are achieved.

5.3 DC Link

The DC link is important to interface the PVSC and the GSC in a two-stage conversion system, as shown in Figure 5.44a. In a single-stage conversion system, the DC link is merged with the PV link acting as the interface between the PV source circuit and the GSC, as shown Figure 5.44b. The DC-link analyses are the same, but the design and specifications of a single-stage conversion system and a double-stage one are different. In the single-stage conversion system, the voltage of the DC link is the same as the voltage of the PV terminal. Special attention should be given to the trade-off of the voltage ripple and the dynamic response, which influences the MPPT performance.

The voltage variation of the DC link is caused by the interaction between the injected current and the extracted current corresponding to the capacitance of the DC link. The dynamics can be represented by

$$C_{dc}\frac{dv_{dc}}{dt} = i_{dc} - i_{inv} \tag{5.79}$$

where v_{dc} is expressed by the differential equation with the coefficient, C_{dc}, and two variables, i_{dc} and i_{inv}. It can be derived in an integral form

$$v_{DC} = \frac{1}{C_{dc}}\int(i_{dc} - i_{inv})dt \tag{5.80}$$

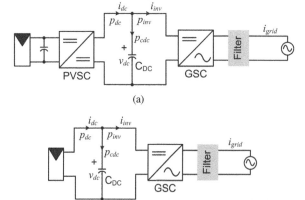

(a)

(b)

Figure 5.44 DC link in grid-connected PV power systems: (a) two-stage conversion; (b) single-stage conversion.

Figure 5.45 Simulink model of DC link
with the current input.

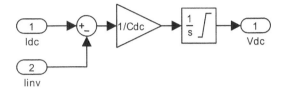

which is useful for simulation implementation. Based on the capacitor-based DC link, the Simulink model can be constructed and is shown in Figure 5.45. All variables and symbols refer to the same definition, as shown in Figure 5.44.

The injected (i_{dc}) and extracted current (i_{inv}) might not be directly sensed in practical systems. Alternatively, the dynamics of the DC-link voltage can also be represented by the interaction with the input power (p_{dc}) and output power (p_{inv}) of the DC link:

$$C_{dc}\frac{dv_{dc}}{dt} = \frac{p_{dc} - p_{inv}}{v_{dc}} \tag{5.81}$$

This can be derived in the integral form

$$v_{dc} = \frac{1}{C_{dc}}\int \left(\frac{p_{dc} - p_{inv}}{v_{dc}}\right) dt \tag{5.82}$$

for the simulation implementation. The Simulink model can be built, as shown in Figure 5.46.

In a practical PV power system, the power measurements at the PV side and grid side are available for the purpose of MPPT and current regulation. These can also be used to estimate the values of the input power (p_{dc}) and output power (p_{inv}).

5.3.1 DC Link for Single-phase Grid Interconnection

To illustrate the steady-state operation of a single-phase grid-connected system, Figure 5.47 plots the waveforms of v_{grid}, p_{dc}, p_{inv}, p_{cdc}, and v_{dc}, which correspond to the signals indicated in Figure 5.44. In the steady state, the DC power injection can be considered as a constant value, P_{dc}, as shown in Figure 5.47. The averaged value of the extracted power (p_{inv}) should be equivalent to P_{dc}, so that the DC-link voltage can be maintained at a steady value, shown as V_{dc}. For reference, the grid voltage (v_{grid}) is plotted in Figure 5.47, showing one cycle from 0 to 2π, as expressed in as

$$v_{grid} = V_{mag}\sin(\omega_b t) \tag{5.83}$$

where ω_b corresponds to the grid line frequency in units of rad/s. Since a unity power factor is considered, the grid injection current is expressed as

$$i_{grid} = I_{mag}\sin(\omega_b t) \tag{5.84}$$

$$p_{inv} = P_{dc} - P_{dc}\cos(2\omega_b t) \tag{5.85}$$

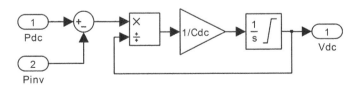

Figure 5.46 Simulink model of DC link with the power input.

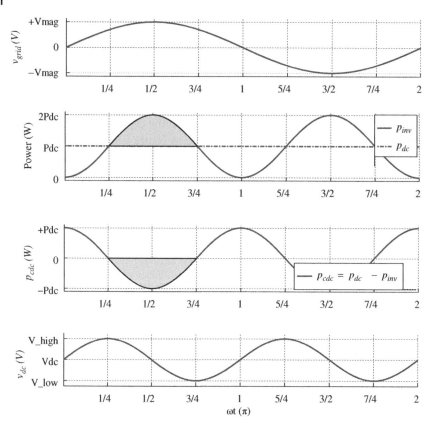

Figure 5.47 Balanced power flow at the DC link in steady state for single-phase AC grid connection.

The DC-link capacitor (C_{dc}) provides an energy buffer to accommodate instant power differences between p_{dc} and p_{inv}, as expressed in (5.86) and shown in Figure 5.47.

$$p_{cdc} = p_{dc} - p_{inv} = +P_{dc}\cos(2\omega_b t) \tag{5.86}$$

When the instantaneous power of p_{inv} is higher than P_{dc}, the deficit is compensated for by the capacitor energy in the period from $\pi/4$ to $3\pi/4$, which is highlighted in Figure 5.47. The voltage of v_{dc} drops from the peak value (V_{high}) to the minimum (V_{low}), to compensate the deficit by supplying the stored energy to the DC link. The compensation can be derived as

$$\frac{1}{2}C_{in}(V_{high}^2 - V_{low}^2) = -\int_{\pi/4\omega_b}^{3\pi/4\omega_b} p_{cdc}dt \tag{5.87}$$

$$\frac{1}{2}C_{in}(V_{high}^2 - V_{low}^2) = -\int_{\pi/4\omega_b}^{3\pi/4\omega_b} P_{dc}\cos(2\omega_b t)dt \tag{5.88}$$

$$C_{dc}\underbrace{\frac{V_{high} + V_{low}}{2}}_{V_{dc}}\underbrace{(V_{high} - V_{low})}_{\Delta V_{dc}} = \frac{P_{dc}}{\omega_b} \tag{5.89}$$

When the peak-to-peak voltage ripple (ΔV_{dc}) is specified for the steady state, the DC-link capacitance (C_{dc}) can be rated for the single-phase grid interconnection by:

$$C_{dc} = \frac{P_{dc}}{\omega_b V_{dc} \Delta V_{dc}} \qquad (5.90)$$

The theoretical value is commonly considered the minimum capacitance for practical implementations since all capacitors are non-ideal.

In single-phase grid interconnections, the calculation can be applied to both the single- and two-stage conversion systems since the ripple voltage is dominated by the double-line-frequency ripple. However, the specification of the peak-to-peak voltage ripple (ΔV_{dc}) is different for the two systems. For single-stage conversion, the DC link becomes the PV link, in which the voltage ripple directly influences the level of deviation from the MPP. The value of ΔV_{dc} should be specified by the tradeoff of MPP accuracy versus the volume of C_{dc}. In the two-stage conversion system, the DC-link voltage ripple has an indirect influence on the deviation of the MPP. However, the double-line frequency ripple in the voltage also induces a series of harmonics at the point of common coupling of the grid side (Du et al. 2015).

In Figure 5.47, the peak of P_{inv} is equivalent to $2P_{dc}$ and can be approximated by $V_{mag}I_{mag}$ when the DC to AC conversion loss is neglected. The values of I_{mag} and V_{mag} are constant in the steady state, representing the amplitudes of the grid current, i_{grid}, and the grid voltage, v_{grid}, respectively. Therefore, the value of I_{mag} can be estimated by (5.91), which can be utilized as the reference for the grid current regulation.

$$I_{mag} \approx \frac{2P_{dc}}{V_{mag}} \qquad (5.91)$$

5.3.2 DC Link for Three-phase Grid Interconnections

When all three phases are balanced and operated at a unity power factor, the power distribution at the steady state is as shown in Figure 5.48, where the three-phase

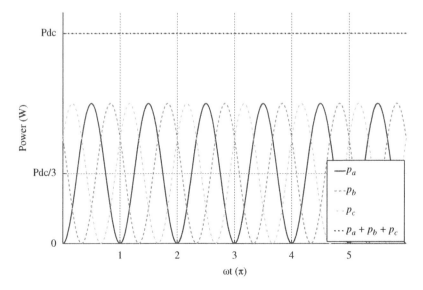

Figure 5.48 Balanced power flow in steady state at the DC link for three-phase AC grid connection.

power forms a constant DC value. Since it shows no variation between the DC power and the accumulation of the AC power, the DC-link capacitor is not rated the same as the single-phase case, which focuses mainly on the double-line-frequency ripples.

The volume of the DC-link capacitor is therefore sized by the filtering effect for high-frequency switching and unexpected power variations. In the two-stage conversion system, as shown in Figure 5.44a, both the PVSC and the GSC should be considered for sizing the capacitor. The DC-link energy buffer is designed according to the dominant ripple component. Most power electronic converters are switched at frequencies higher than 1 kHz, so the need for filtering capacitance is less than for a single-phase grid connection.

Since the PV power source is intermittent, another important factor is the power variation capacity and frequency, which are difficult to accurately predict. The variation can have many causes, such as the annual and daily weather changes. For PV power systems, the power mismatch between the DC and AC stages can be estimated as ΔP_{max} and ΔT_r, which represents the power difference and the time period of the variation (Wittig et al. 2009). Therefore, the DC-link capacitance can be sized by either (5.92) or (5.93):

$$C_{dc} = \frac{\Delta P_{max} \Delta T_r}{2 V_{dc} \Delta V_{dc}} \tag{5.92}$$

$$C_{dc} = \frac{\Delta P_{max} \Delta T_r}{V_{dc} \Delta V_{dc} + \Delta V_{dc}^2 / 2} \tag{5.93}$$

where V_{dc} and ΔV_{dc} are the nominal voltage of the DC link and the specified ripple voltage, respectively.

An example of the determination of a DC-link capacitance is now outlined. The system specification is as summarized in Table 5.4. The peak-to-peak voltage ripple is allowed to be 1 V at the DC link. Ten percent of the rated power and half cycle of the AC signal is considered as the variation energy volume, the capacitance of C_{dc} is 2.6 mF and 5.1 mF according to (5.92) and (5.93), respectively.

Another method can be used that is based on the rating method for the DC to single-phase AC conversion. First, the capacitance value is computed by (5.90), and then reduced by six for the three-phase implementation. Based on the case study shown above, the capacitance is computed as 16.4 mF using (5.90), which is based on the single-phase AC grid. The final value is appropriately 2.7 mF for the three-phase grid interconnection.

Table 5.4 Specification of three-phase DC/AC conversion.

Parameters	Value
Nominal AC grid phase-to-neutral voltage	$V_{LN} = 220$ V RMS
Nominal DC-link voltage	$V_{dc} = 560$ V
Nominal AC frequency	$f_b = 50$ Hz
Rated power from PV generation	$P_{dc} = 2883$ W

5.4   Grid-side Converter for DC/AC Stage

GSCs perform DC/AC conversion in order to transform DC power for AC grid interconnections. Multiple functions should be implemented to address the safety and power quality of the grid connection. The common topology for GSCs includes the H bridge for single-phase grid interconnections and a current source inverter (CSI) for three-phase interconnection.

5.4.1   DC to Single-phase AC Grid

For single-phase grid interconnections, a H bridge or so-called "full-bridge," is commonly used as the GSC for DC-to-AC conversion. A schematic is shown in Figure 5.49, which is built by two half bridge legs, A and B. For the two-stage conversion topology, the PV-generated power is processed by the PVSC and transmitted to the GSC through the DC link, as shown in Figure 5.49a. The input port of the GSC is the DC link, as discussed in Section 5.3. The output is coupled to the grid through the grid link, also known as the AC filter.

For a single-stage topology, a GSC converts the PV-generated power directly to AC and supplies the grid, as illustrated in Figure 5.49b. The analysis and design of the DC link is generally the same as for two-stage conversion systems, as described in Section 5.3.1.

It should be noted that there are always the paired operations Q_1/Q_4 and Q_2/Q_3 to produce an AC waveform. When Q_1/Q_4 pair is turned on, the DC voltage is applied to increase i_{grid} in the positive direction. On the other hand, the DC voltage is reversibly applied, which leads i_{grid} in the opposite direction when the Q_2/Q_3 pair is turned on.

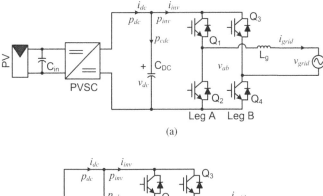

(a)

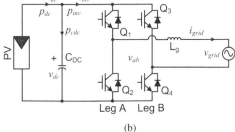

(b)

Figure 5.49 H bridge used for grid-side converter in two-stage conversion topology: (a) two-stage conversion; (b) single-stage conversion.

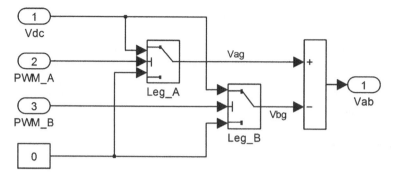

Figure 5.50 Simulink model of H bridge for grid-side converter.

The Simulink model of the H bridge can be built accordingly and is shown in Figure 5.50. The legs A and B as defined as in the circuit definition, Figure 5.49. PWM should be applied to operate the H-bridge converter and transform from DC power to active AC power.

In control theory, the bang-bang controller has proven a simple and robust solution. It is a feedback controller that switches abruptly between two states: ON and OFF. It is also called the on-off controller, or the hysteresis controller, and is widely used in power electronics. For GSCs, the output current can be regulated by the hysteresis controller in order to produce a sinusoidal waveform in phase with the grid voltage. The switching signals can be generated by the hysteresis controller through comparison of the current error in a predefined hysteresis band ($\pm\epsilon$), as demonstrated in Figure 5.51. If the injected current exceeds the upper hysteresis band (HB) in the positive cycle, the upper switch of leg A (Q_1) is turned off and the lower switch of leg A (Q_2) is turned on. As the current decays and crosses the lower HB in the positive cycle, the lower switch of leg A (Q_2) is turned off and the upper switch (Q_1) is turned on.

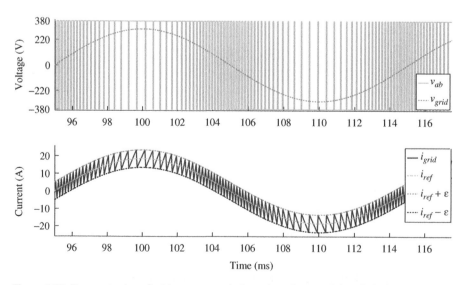

Figure 5.51 Demonstration of grid current regulation using a hysteresis band of ±5 A.

Figure 5.52 Simulink model of hysteresis controller for single-phase grid current modulation.

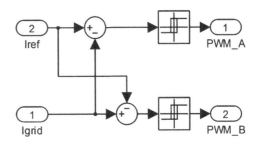

The hysteresis controller for the current regulation can be implemented by Simulink blocks, as shown in Figure 5.52. The value of ϵ should be assigned in the two hysteresis blocks, which can force the injected current (i_{grid}) to follow the command signal (i_{ref}) within the hysteresis band ($\pm\epsilon$).

An example system is constructed to continue the study for the operation, modulation, and control of the H-bridge converter for grid interconnections. The system specification is summarized in Table 5.5. The PV source circuit is constructed from ten solar modules in series connection. The output characteristics of the individual PV modules are as shown in Figure 5.4. Because of the significant capacitance across the DC link, the design and analysis can be based on the DC link and the GSC, which are separate in the analysis of the PVSC in the two-stage conversion system.

The simulated waveform in Figure 5.51 gives a detailed look at the hysteresis operation. The hysteresis envelope is assigned as ± 5 A in the Simulink model for demonstration purposes. The command signal to regulate the grid current is shown as i_{ref}. The grid current, i_{grid}, should follow the command signal within the predefined hysteresis band ($\pm\epsilon$). The grid link is represented by the AC filter, which is the L-type, as shown in Figure 5.49. The value of L_g is 3 mH.

It can be seen that the waveform of i_{grid} has a significant current ripple due to the relatively high hysteresis band, which results in low power quality. The current control technique using a fixed hysteresis band has one major disadvantage in that the PWM frequency varies dramatically because the peak-to-peak current ripples are the same at every point of the fundamental frequency wave, as shown in Figure 5.51. It is noticeable that the switching frequency becomes higher when the regulated current approaches the zero crossing. The switching frequency is 4.2 kHz on average, which results in 22% total harmonic distortion (THD) in the current. The example is only for illustration purposes, since the THD value is too high for practical grid connections.

The current harmonics can be reduced by lowering the value of ϵ. It is recommended to assign a value of ϵ lower than 5% of the magnitude of the rated grid current for low

Table 5.5 Specification of the DC to single-phase AC conversion.

Parameters	Value
Nominal AC grid RMS voltage	220 V
Nominal DC-link voltage	380 V
Nominal AC frequency	$f_b = 50$ Hz
Rated power from PV generation	$P_{dc} = 2883$ W

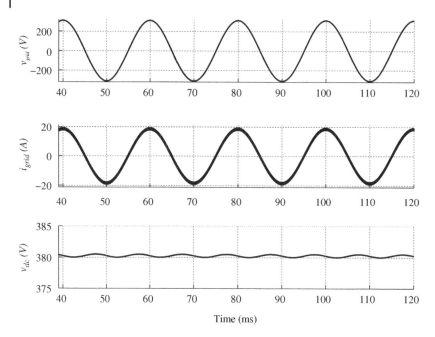

Figure 5.53 Demonstration of grid current regulation using a hysteresis band of ±0.74 A.

harmonic injection. The switching frequency can be determined by the values of ϵ and the parameters of the AC filter, as discussed in Section 5.5.

For the case in Table 5.5, according to the grid interconnection code, as discussed in Section 3.2, the upper limit of the current THD is assigned to be 4%. The peak-to-peak ripple voltage of the DC link is specified as 1 V. The capacitance C_{dc} can be specified as 24 mF according to the rated ripple value and (5.90). The amplitude of the command signal, i_{ref}, is assigned as 18.53 A according to (5.91). The hysteresis envelope is assigned as ±0.74 A, which is equivalent to 4% of the rated current magnitude. The system operation is simulated using the developed models, as shown in Figures 5.50 and 5.52. The simulated signals of v_{grid}, i_{grid}, and v_{dc} are shown in Figure 5.53.

A unity power factor is applied so that the signal of v_{grid} is in phase with i_{grid}. The tradeoff is clear by comparing the waveforms in Figures 5.51 and 5.53. The current THD is measured as 3.3%, thanks to the relatively low rate of ϵ. The switching frequency is significantly increased to 28 kHz, which corresponds to the same value of L_g (3 mH) as the previous case. If the switching frequency is too high for practical implementation, it can be reduced by re-sizing the AC filters, as discussed in Section 5.5. It should be noted that a current of 7.59 A should be injected to the DC link in order to maintain a steady DC-link voltage and balance the injection and extraction powers.

For AC grid connections, the H bridge can also be operated for current unfolding in two-stage conversion systems, as shown in Figure 5.54. In contrast to conventional two-stage power conversion systems (Figure 5.49a), the significant energy buffer at the DC link is removed. The PVSC outputs DC current with significant double-line-frequency ripples, i_{inv}, as shown in Figure 5.55. In the second stage, the H bridge is operated at a line frequency to unfold the current from DC to AC form. As shown in Figure 5.55, the waveform of i_{inv} is DC, but that of i_{grid} becomes

Figure 5.54 Current unfolding circuit using H bridge for the DC to single-phase AC conversion.

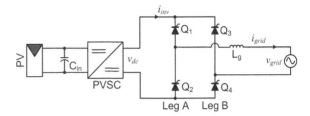

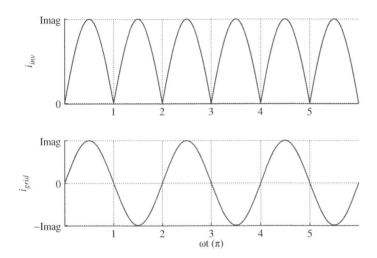

Figure 5.55 Simulated waveform to demonstrate the current unfolding for the DC to single-phase AC conversion.

AC, folded by the H bridge. High-frequency switching is no longer required, so that low-cost silicon-controlled rectifiers (SCRs) can be used as the power switches in the H bridge. This arrangement is frequently found in MIPIs, as discussed in Section 2.4.1. Edwin et al. (2012) noted that a PVSC with the flyback topology could step up from a low PV-module voltage and produce DC current waveforms, i_{inv}. The signal is unfolded by the H bridge to an AC current for grid interconnection.

5.4.2 DC to Three-phase AC Grid

Current source inverters and voltage source inverters (CSIs and VSIs) are relevant mainly for DC to three-phase AC applications. Sometimes, the terms are confusing to readers. The names were originally defined when inverters were used to drive loads, such as motors. For such applications, a CSI is for DC/AC conversion, with an inductor as the DC-link buffer. The VSI uses capacitor banks as the DC link, when the AC voltage needs to be regulated.

In grid interconnections for PV power systems, the DC link is usually formed by a capacitance, as shown in Figure 5.56. To supply power to grid, the power interface acts as a current source rather than a voltage source. The same approach is used for VSIs to supply a three-phase load where regulation of the output voltage is required. The CSI and VSI share the same six-switch bridge, but are different in their control operation.

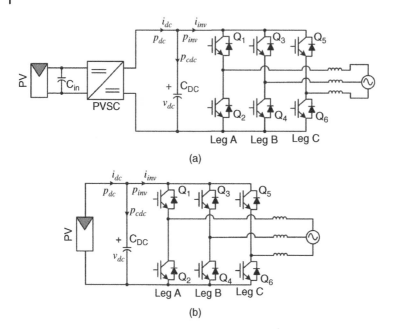

Figure 5.56 Current source inverter used as grid-side converter: (a) two-stage conversion; (b) single-stage conversion.

The AC filters for grid connection are also different. A VSI is commonly applied with an LC filter. The CSI is coupled with either an L or LCL filter, for grid interconnection.

In the two-stage conversion system shown in Figure 5.56a, the DC link is used to couple the PVSC and the CSI. In the single-stage conversion system, the PVSC is neglected, and therefore the DC link becomes the PV link, as shown Figure 5.56b.

The Simulink model of the CSI can be built accordingly and is shown in Figure 5.57. The legs A, B, and C are defined according to the switching circuit with six power switches shown in Figure 5.56. The model requires the PWM control signals to convert from DC power to three-phase AC. Three signals are required to modulate the three legs, indicated by the multiplex signal "PWM 3P." The DC-link voltage should be applied and is labeled V_{dc}. The outputs are the three-phase signals v_{as}, v_{bs}, and v_{cs}, which are multiplexed as one signal, V_{abcs}. The use of multiplexing makes the Simulink circuit simple and readable.

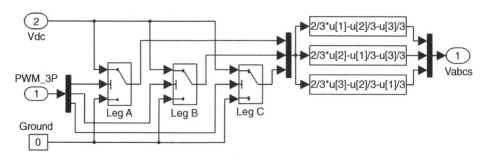

Figure 5.57 Simulink model for three-phase DC-to-AC conversion.

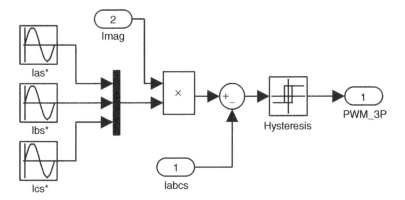

Figure 5.58 Simulink model for the hysteresis controller in regulating three-phase current source inverter.

For a CSI, the output current can be regulated by a hysteresis controller in order to produce a sinusoidal waveform in phase with the grid voltage. The switching signals can be generated by the hysteresis controller through the comparison of the current error in a predefined hysteresis band ($\pm\epsilon$). This is similar to the approach for the single-phase application, as illustrated in Figure 5.51. The current regulation forces the current vector in three phases according to a reference trajectory. The hysteresis controller for current modulation can be implemented by the Simulink blocks, as shown in Figure 5.58. The value of ϵ is assigned in the hysteresis block, which conducts the comparison and outputs the PWM for the DC to three-phase AC conversion. The command is formed by the three-phase current reference, shown as i_{as}^*, i_{bs}^*, and i_{cs}^*, with the assigned amplitude of I_{mag}. The signal, i_{abcs}, represents the three-phase current, which is multiplexed into one symbol. The three-phase signals of the PWM are shown as "PWM 3P" and are output to control the CSI.

The accumulated current appearing at the DC link is shown as i_{inv} and can be simulated by the model, as shown in Figure 5.59.

5.4.3 Reactive Power

The previous section focuses on injecting grid current in phase with the grid voltage. Recent grid-connected PV systems have been flexibly designed to support grid networks by adjusting the power factor at the grid link to be either leading or lagging. This can be achieved by modifying the reference signals by including the desired phase shift. Using the same hysteresis control technique to regulate the grid current, a reactive power contribution, either leading or lagging, can be achieved. The simulation implementation process is the same as that discussed previously.

Figure 5.59 Simulink model for the computation of i_{inv}.

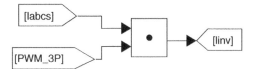

5.5 Grid Link

The THD of the voltage and current is an important measurement to characterize the quality of the grid power injection. THD is the ratio of the root mean square (RMS) of the harmonic content to the RMS value of the fundamental quantity. An AC filter is required at the grid link in order to ensure the power quality. One goal of research is to design effective AC filters that will minimize injection of harmonics. At the rated power level, the value of THD should be lower than 5% according to the IEEE 1547 standard, which was discussed in Section 3.2. Another term that is commonly used for the regulation of power quality is the TDD, which was also defined and explained in Section 3.2.

5.5.1 L-type for Single-phase Grid Connections

The AC filter illustrated in Figure 5.49, is the inductive type (L-type). The mathematical model of L-type filters is expressed as

$$L_g \frac{di_{Lg}}{dt} = v_{ab} - v_{grid} \tag{5.94}$$

where v_{grid} is the grid voltage and v_{ab} is the output voltage of the H bridge, consisting of a fundamental component and higher-order harmonics. Using (5.94), the Simulink model for a single-phase grid connection based on an L filter is as shown in Figure 5.60.

When the hysteresis current controller is applied, the switching frequency is high at the zero crossing, but low at the peak, as shown in Figure 5.51. The ON time of Q_1 and Q_4 is equal to their OFF time, and can be expressed as

$$\Delta T_{on-zero} = \Delta T_{off-zero} = \frac{2\epsilon L_g}{V_{dc}} \tag{5.95}$$

when the grid voltage is zero crossing. Since the values of ϵ and V_{dc} are defined by the system specification, the value of $(\Delta T_{on-zero} + \Delta T_{off-zero})$ can determine the highest switching frequency of the switches. At the current peak, the ON time of Q_1 and Q_4 is the maximum of

$$\Delta T_{on-peak} = \frac{2\epsilon L_g}{V_{dc} - V_{mag}} \tag{5.96}$$

where V_{mag} is the peak value of the grid voltage, at steady state. Meanwhile, the OFF time can be derived as

$$\Delta T_{off-peak} = \frac{2\epsilon L_g}{V_{dc} + V_{mag}} \tag{5.97}$$

The lowest switching frequency can be calculated by the time period $(\Delta T_{on-peak} + \Delta T_{off-peak})$.

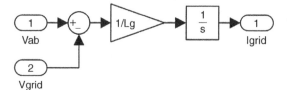

Figure 5.60 Simulink model for L filter for single-phase applications.

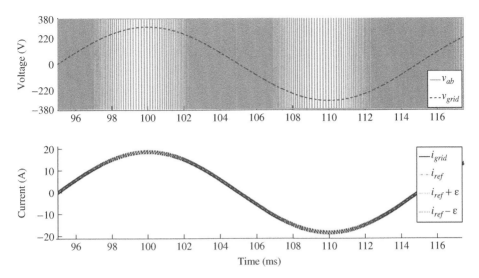

Figure 5.61 Simulated waveform of grid voltage and current.

Based on the upper limit of the switching frequency, the parameters for the L-type filter can be determined from the above information. To follow the example discussed in Section 5.4.1, the amplitude of the reference signal (i_{ref}) is kept the same, at 18.53 A. Meanwhile, the hysteresis envelope is also the same, at ± 0.74 A. A new constraint is applied, namely that the switching frequency should be not more than 20.0 kHz. The rating of L_g should be re-sized by making $\Delta T_{on-zero} + \Delta T_{off-zero} = 50\,\mu s$, which results in 6.4 mH in value according to (5.95).

The simulated waveform is as shown in Figure 5.61, with 3.3% THD in the grid current. The lowest switching frequency can be calculated from (5.96) and (5.97) as 6.6 kHz. The average value of the switched signal of v_{ab} is 13.1 kHz, which matches the estimate from averaging the highest and lowest frequencies. Comparing with the design in Section 5.4.1, the switching frequency can be constrained by a proper design of the filter inductance. The example demonstrates that the hysteresis controller and AC filter can be designed to meet the system specification

5.5.2 L-type for Three-phase Grid Interconnections

Sizing an L-type filter for a three-phase grid tied system can follow the same process as for single phase. The RMS value of the phase current is

$$I_{RMS} = \frac{P_{dc}}{3V_{LN}} \tag{5.98}$$

where V_{LN} is the RMS value of the line-to-neutral voltage. It assumes that all phases are balanced. The amplitude of each phase current is then expressed as

$$I_{mag} = \frac{P_{dc}\sqrt{2}}{3V_{LN}} \tag{5.99}$$

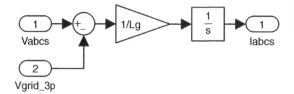

Figure 5.62 Simulink model for L filter for three-phase applications.

Following the design case discussed in Section 5.3.2, the amplitude of each phase current can be computed as 6.18 A from (5.99). The value is assigned to the hysteresis controller for the current regulation of the three phases. The hysteresis envelope is set to be 4% of I_{mag}, which is 0.25 A. The upper band of the switching frequency is specified as 20 kHz. According to (5.95), the AC filter inductance is 28 mH.

The Simulink model to represent the L-type filter for the three-phase system is shown in Figure 5.62. Vabcs, Vgrid 3P, and Iabcs, represent the multiplex signals for the three-phase representation.

The simulation results in Figure 5.63 include the grid voltage, the injected three-phase current, the extracted current from the DC link, and the DC-link voltage. The THD of the current in phases A, B, and C is measured as 3.3%. When the amplitude of the AC current is regulated to 6.18 A, the DC-link voltage is regulated to 560 V according to the specification. The result verifies the design and modeling process, as described in Sections 5.3.2 and 5.4.2.

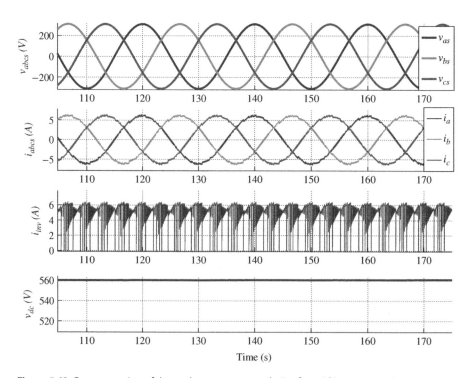

Figure 5.63 Demonstration of three-phase current regulation for grid interconnections.

Figure 5.64 LCL filter circuits: (a) undamped LCL filter; (b) LCL filter with damping resistor.

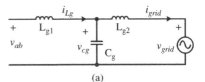

(a)

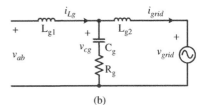

(b)

5.5.3 LCL-type Filters

LCL-type filters are also commonly used for grid connections, as shown in Figure 5.64. The key components are L_{g1}, L_{g2}, and C_g. In Figure 5.64b, a resistor is applied in series with the filter capacitor in order to add damping into the circuit. Therefore, the circuit in Figure 5.64a is called the "undamped" LCL filter. The difference between the two circuits lies in the damping resistor. Compared to the L-type filter, it is claimed that an LCL filter can attenuate high-frequency harmonics more effectively and requires less total inductance to achieve the same performance, which makes them more cost effective (Hanif et al. 2014). However, the circuit has high-order dynamics that result in self-oscillation. This then requires suitable damping.

The input port of the AC filter is represented by v_{ab}, which is the modulated voltage signal from the GSC. The output port is the grid connection, showing as the grid voltage and current, represented by v_{grid} and i_{grid}, respectively. The mathematical expressions for the LCL filter without a damping resistor are

$$L_{g1} \frac{di_{Lg}}{dt} = v_{ab} - v_{cg} \tag{5.100a}$$

$$C_g \frac{dv_{cg}}{dt} = i_{Lg} - i_{grid} \tag{5.100b}$$

$$L_{g2} \frac{di_{grid}}{dt} = v_{cg} - v_{grid} \tag{5.100c}$$

which can also be written as

$$i_{Lg} = \frac{1}{L_{g1}} \int (v_{ab} - v_{cg}) dt \tag{5.101a}$$

$$v_{cg} = \frac{1}{C_g} \int (i_{Lg} - i_{grid}) dt \tag{5.101b}$$

$$i_{grid} = \frac{1}{L_{g2}} \int (v_{cg} - v_{grid}) dt \tag{5.101c}$$

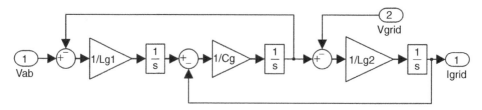

Figure 5.65 Simulink model for LCL filter without damping resistor.

for simulation purposes. The Simulink model for the LCL filter without damping resistor, based on (5.101), is shown in Figure 5.65.

From (5.100), the undamped LCL filter can be expressed in the state-space format:

$$
\begin{bmatrix} \dfrac{di_{Lg}}{dt} \\[2mm] \dfrac{dv_{cg}}{dt} \\[2mm] \dfrac{di_{grid}}{dt} \end{bmatrix} =
\begin{bmatrix} -\dfrac{R_g}{L_{g1}} & -\dfrac{1}{L_{g1}} & 0 \\[2mm] \dfrac{1}{C_g} & 0 & -\dfrac{1}{C_g} \\[2mm] 0 & \dfrac{1}{L_{g2}} & 0 \end{bmatrix}
\begin{bmatrix} i_{Lg} \\[2mm] v_{cg} \\[2mm] i_{grid} \end{bmatrix} +
\begin{bmatrix} \dfrac{1}{L_{g1}} & 0 \\[2mm] 0 & 0 \\[2mm] 0 & -\dfrac{1}{L_{g2}} \end{bmatrix}
\begin{bmatrix} v_{ab} \\[2mm] v_{grid} \end{bmatrix}
\tag{5.102}
$$

where i_{grid} is the variable that should be regulated. The model inputs include the signal, v_{ab} and v_{grid}. To represent the third-order dynamics, three eigenvalues of the dynamic matrix can be derived and expressed as

$$\lambda_1 = +j\omega_n \tag{5.103a}$$

$$\lambda_2 = -j\omega_n \tag{5.103b}$$

$$\lambda_3 = 0 \tag{5.103c}$$

where $\lambda_{1,2,3}$ are the eigenvalues. Two eigenvalues are located on the imaginary axis in the complex plane, which indicate the resonant characteristics of the LCL circuit. One eigenvalue is equal to zero, indicating the integral characteristics of the filter circuit.

The undamped natural frequency, ω_n, is determined by

$$\omega_n = \sqrt{\dfrac{L_{g1} + L_{g2}}{L_{g1}L_{g2}C_g}} \tag{5.104}$$

Damping can be added to the undamped LCL circuit by inserting a damping resistor in series with the filter capacitor, as shown in Figure 5.64b. This is called a "damped LCL filter" hereafter. The system dynamics can be expressed as either (5.105) or (5.106) for simulation purposes.

$$L_{g1}\dfrac{di_{Lg}}{dt} = -R_g i_{Lg} - v_{cg} + R_g i_{grid} + v_{ab} \tag{5.105a}$$

$$C_g\dfrac{dv_{cg}}{dt} = i_{Lg} - i_{grid} \tag{5.105b}$$

$$L_{g2}\dfrac{di_{grid}}{dt} = R_g i_{Lg} + v_{cg} - R_g i_{grid} - v_g rid \tag{5.105c}$$

$$i_{Lg} = \frac{1}{L_{g1}} \int [v_{ab} - v_{cg} - R_g(i_{Lg} - i_{grid})]dt \tag{5.106a}$$

$$v_{cg} = \frac{1}{C_g} \int (i_{Lg} - i_{grid})dt \tag{5.106b}$$

$$i_{grid} = \frac{1}{L_{g2}} \int [v_{cg} + R_g(i_{Lg} - i_{grid}) - v_{grid}]dt \tag{5.106c}$$

The Simulink model for the damped LCL filter is illustrated in Figure 5.66.

From (5.105), the damped LCL filter can be represented in a state-space format:

$$\begin{bmatrix} \dfrac{di_{Lg}}{dt} \\ \dfrac{dv_{cg}}{dt} \\ \dfrac{di_{grid}}{dt} \end{bmatrix} = \begin{bmatrix} -\dfrac{R_g}{L_{g1}} & -\dfrac{1}{L_{g1}} & \dfrac{R_g}{L_{g1}} \\ \dfrac{1}{C_g} & 0 & -\dfrac{1}{C_g} \\ \dfrac{R_g}{L_{g2}} & \dfrac{1}{L_{g2}} & -\dfrac{R_g}{L_{g2}} \end{bmatrix} \begin{bmatrix} i_{Lg} \\ v_{cg} \\ i_{grid} \end{bmatrix} + \begin{bmatrix} \dfrac{1}{L_{g1}} & 0 \\ 0 & 0 \\ 0 & -\dfrac{1}{L_{g2}} \end{bmatrix} \begin{bmatrix} v_{ab} \\ v_{grid} \end{bmatrix} \tag{5.107}$$

where i_{grid} is the variable that should be regulated for grid connection. The model inputs include the signal, v_{ab} and v_{grid}. The eigenvalues of the dynamic matrix can be derived as

$$\lambda_1 = -\omega_r + j\omega_i \tag{5.108a}$$

$$\lambda_2 = -\omega_r - j\omega_i \tag{5.108b}$$

$$\lambda_3 = 0 \tag{5.108c}$$

where ω_r and ω_i are computed as

$$\omega_r = \frac{R_g(L_{g1} + L_{g2})}{2L_{g1}L_{g2}} \tag{5.109a}$$

$$\omega_i = \frac{\sqrt{(L_{g1} + L_{g2})(R_g^2 C_g^2 L_{g1} + R_g^2 C_g^2 L_{g2} - 4L_{g1}L_{g2}C_g)}}{2L_{g1}L_{g2}C_g} \tag{5.109b}$$

This shows that the resistance, R_g, adds damping to the filter. One eigenvalue is equal to zero, indicating the integral characteristics of the filter circuit.

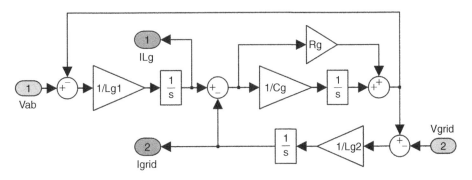

Figure 5.66 Simulink model for damped LCL filter.

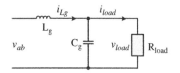

Figure 5.67 LC filter for AC loads.

When the hysteresis current controller is applied, the switching frequency is the highest when the grid voltage is zero. At each zero crossing of the grid voltage, v_{grid}, the transfer function can be derived as

$$\frac{I_{grid}(s)}{V_{ab}(s)} = \frac{R_g C_g s + 1}{L_{g1} L_{g2} C_g s^3 + R_g C_g (L_{g1} + L_{g2}) s^2 + (L_{g1} + L_{g2}) s} \tag{5.110}$$

The passive damping technique causes a power loss in the added resistance element and reduces the overall efficiency. Active damping control techniques are required to reduce the oscillation when the undamped LCL is utilized. The design of LCL filters and active damping control techniques are described in a recent publication by Hanif et al. (2014), and are not repeated here.

5.5.4 LC-type Filters

An LC filter is mainly used in VSI applications to mitigate harmonics in the output voltage signal. Figure 5.67 shows the type of LC filter commonly used for standalone PV power systems supplying local AC loads. The DC/AC converter acts as a VSI, the output voltage of which is regulated to meet the load requirement. When the ripple voltage and current are specified, the filter parameters L_g and C_g can be determined. The derivation follows the mathematical expression shown in (5.111) and (5.112) for the DC to single-phase AC conversion.

$$L_g \frac{di_{Lg}}{dt} = v_{ab} - v_{load} \tag{5.111}$$

$$C_g \frac{dv_{load}}{dt} = i_{Lg} - \frac{v_{load}}{R_{load}} \tag{5.112}$$

The design and simulation involved apply not only to PV power systems but to all applications when DC/AC conversion is required. Therefore, the details are not covered here. Interested readers can find them in other books on power electronics.

5.6 Loss Analysis

Designers of power converters should consider the tradeoffs of cost, efficiency, and size. Loss models are important to improve the conversion efficiency of power interfaces. It is important to identify the losses in order to identify efficiency improvements. The total power loss of converters is the sum of the individual power losses of all components – conduction losses and switching losses.

5.6.1 Conduction Loss

The conduction losses of capacitors and inductors are mainly from the ESR, which results in joule heating. The values can be calculated as

$$P_{loss} = ESR \times I_{RMS}^2 \tag{5.113}$$

where I_{RMS} is the RMS current value. The loss model can be applied to determine the conduction loss of MOSFETs, which show the effects of ESR, which dissipates power as current is conducted through the device. The equivalent resistance accumulated by several resistive components during conduction is commonly referred as the "on-state resistance" $R_{ds(on)}$.

The ESR value depends on the applied frequency, which is occasionally available from the manufacturer's datasheet. It can also be measured by an LCR meter or impedance analyzer when certain frequency signals are applied. It should be noted that the accurate value of ESR can be difficult to determine since it changes with temperature. It is very important to apply temperature coefficients based on either experimental tests or theoretical estimations.

The general definition over a time period (T) is

$$I_{RMS} = \sqrt{\frac{1}{T} \int_0^T i^2(t)dt} \tag{5.114}$$

In the steady state, the RMS value over all time of a periodic waveform is equal to the RMS value over one cycle.

In switching-mode converter systems, the trapezoidal current waveforms are as shown in Figure 5.68. These commonly refer to the current waveform through power semiconductors, such as MOSFETs, IGBTs, and diodes. The calculation is

$$I_{RMS} = \sqrt{\frac{T_{ON}}{3T}(I_1^2 + I_1 I_2 + I_2^2)} \tag{5.115}$$

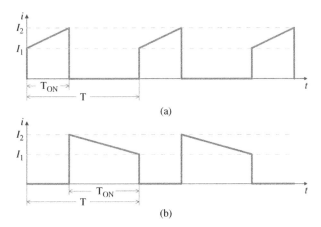

Figure 5.68 Common current waveforms in switching-mode power converters: (a) trapezoidal waveform with rising top; (b) trapezoidal waveform with dropping top.

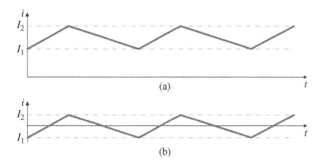

Figure 5.69 Common current waveforms in switching-mode power converters: (a) triangular waveform with DC offside; (b) triangular waveform without DC offside.

In CCM, the triangular waveform with DC component shown in Figure 5.69a commonly represents the inductor current in DC/DC converters, and can be calculated as

$$I_{RMS} = \sqrt{\frac{1}{3}(I_1^2 + I_1 I_2 + I_2^2)} \tag{5.116}$$

The capacitor current can be the triangular waveform without a DC component, as shown in Figure 5.69b. Since $|I_1| = |I_2|$, the RMS value can be determined from

$$I_{RMS} = \frac{I_1}{\sqrt{3}} \tag{5.117}$$

The sine wave with a DC offset shown in Figure 5.70 can be expressed as

$$i(t) = I_{AVG} + I_p \sin(\omega t) \tag{5.118}$$

The RMS value can be determined from

$$I_{RMS} = \sqrt{I_{AVG}^2 + \frac{I_p^2}{2}} \tag{5.119}$$

The appendix in the book by Erickson and Maksimovic (2001), covers more formulas for waveforms found in electronic power converters.

In CCM, the RMS value of the inductor current can also be approximated by the average value if the ripple is insignificant. The calculation of the average current is

$$I_{AVR} = \frac{1}{T} \int_0^T i(t)dt \tag{5.120}$$

Another conduction loss is from the non-ideal factor of power semiconductors in terms of voltage drop in forward bias conditions. Power devices that show a voltage

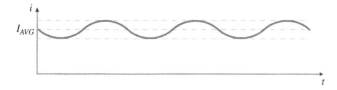

Figure 5.70 Common current waveform in switching-mode power converters.

drop during conduction include diodes, SCRs, and IGBTs. The average value (I_{AVR}) can be used to calculate such losses:

$$P_{loss} = I_{AVR} \times V_{drop} \tag{5.121}$$

5.6.2 High-frequency Loss

Modern power converters have relatively high-frequency switching of the power semiconductors. Switching loss, or high-frequency loss, is introduced due to the state transition. It becomes more and more significant as power electronics pushes to higher switching frequencies. High-frequency conversion is one way to reduce the size of energy-storage components and make converters compact and low cost.

The turn-on loss happens in the short moment when the power switch starts to conduct current without dropping the voltage to zero or the minimum level. Meanwhile, the turn-off loss occurs when the power switch starts to build voltage without shutting down the current to zero. The overlap of these losses results in the switching loss, which is caused by the intrinsic parasitic capacitors of power switches. The capacitors store and dissipate energy and delay any transition during each switching cycle.

Figure 5.71 demonstrates the model of an IGBT and a MOSFET with consideration of the parasitic capacitance. For the IGBT, they are represented as a gate-emitter capacitance (C_{GE}), a collector-emitter capacitance (C_{CE}), and a gate-collector capacitance (C_{GC}). For the MOSFET, there is a gate-source capacitance (C_{GS}), a drain-source capacitance (C_{DS}), and a gate-drain capacitance (C_{GD}).

Calculation of the switching loss is based on the overlap time of the voltage across and the current through the IGBT or MOSFET. One example is shown in Figure 5.72, where the MOSFET turn-on and turn-off are demonstrated. The energy dissipation during the overlap can be determined by the overlap time and the voltage and current ratings. The overlap times during switch-on and switch-off are given by T_{ON} and T_{OFF} respectively. The variables V_{DS} and I_{DS} represent the voltage across the MOSFET drain and source and the current through the drain and source at the point of switching. The switching loss can determined by

$$P_{loss-SW} = \frac{V_{DS} \times I_{DS}}{2}(T_{ON} + T_{OFF})F_{SW} \tag{5.122}$$

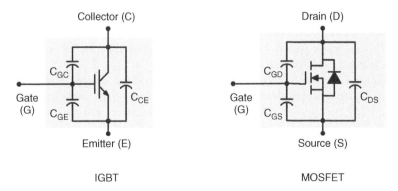

Figure 5.71 IGBT and MOSFET models with consideration of parasitic capacitance.

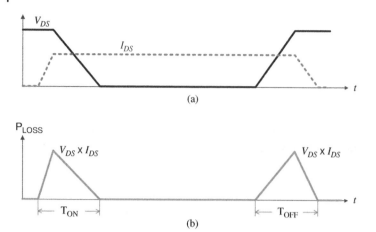

Figure 5.72 Switching loss illustration for MOSFET turn-on state and turn-off state.

where F_{SW} is the switching frequency. The loss is proportional to the switching frequency and the overlap time. The switching loss can be reduced by minimizing the duration of T_{ON} and T_{OFF}. A dedicated gate driver that can supply a significant driving current is commonly used to charge and discharge the capacitance quickly in order to shorten the turn-on and turn-off times. Soft-switching techniques aim to eliminate or reduce the switching loss, by reducing the value of either V_{DS} or I_{DS} to either zero or a low value during each switching. From (5.122), the switching loss can be either eliminated or significantly reduced when soft switching is applied. The technique of ZVS was briefly introduced in Section 5.2.4.

The same calculation can be applied for switching loss in IGBTs. Determining the overlap times of T_{ON} and T_{OFF} is based on the charge and discharge operations of the parasitic capacitance. Low values of parasitic capacitance are preferred for low switching losses. The loss also depends on the gate driving capability, the applied voltage (V_{DS}), and the through current (I_{DS}). The voltage and current can be either measured or estimated from a steady-state analysis. However, an accurate value of the time periods is very difficult to determine due to the non-linear nature of the parasitic capacitance. The parameters of the parasitic capacitors depend on many factors, such as temperature, voltage, current, and operating frequency. Even though the values can be found in the datasheets of the power semiconductor devices, they generally refer to a very specific testing condition. Methods for estimation of the overlap time can be found in many industrial white papers and application notes, but accurate estimation of switching losses in power semiconductors is never easy.

The inductor component is commonly considered to cause a conduction loss. It should be noted that high-frequency losses should be considered for inductors. These include the hysteresis loss, eddy currents, the skin effect, and the proximity effect. The loss from hysteresis and eddy currents is commonly referred to as "core loss". It is known that the B–H curve[1] of magnetic cores has a loop shape, which represents the hysteresis loss. Eddy currents causing joule heating are induced within the core by the

---

1 A B–H curve is used to illustrate the nonlinear relationship between magnetic flux density (B) and magnetic field strength (H).

changing magnetic fields. Loss of winding can be caused by the skin and proximity effects because of the applied frequency. In general, high-frequency losses should be considered when selecting the correct magnetic core material and designing the winding configuration. It is challenging to derive an accurate model for high-frequency losses due to the complexity of core materials and the difficulties of parameter identification. However, high-frequency loss can be evaluated through experimental tests in combination with the conduction loss. The temperature of the magnetic core and winding is an indicator for loss analysis.

5.7 Conversion Efficiency

The conversion efficiency is considered the most important measure of PV power interfaces. It is the ratio in percentage terms of the input to output power. A converter efficiency is not uniformly distributed at all power levels. Conventionally, peak efficiency is used as the performance index, because it represents the optimal operating point for the best conversion efficiency. However, the operation of PV power systems depends on the environmental conditions. The efficiency at a single operating condition will not be the best index to represent performance. Therefore, weighted values are introduced to evaluate the conversion efficiency of PV power interfaces.

The European efficiency and CEC efficiency measures are commonly applied and are expressed in (5.123) and (5.124), respectively.

$$\eta_{eu} = 0.03 \times \eta_{5\%} + 0.06 \times \eta_{10\%} + 0.13 \times \eta_{20\%} + 0.10 \times \eta_{30\%}$$
$$+ 0.48 \times \eta_{50\%} + 0.20 \times \eta_{100\%} \tag{5.123}$$

$$\eta_{cec} = 0.04 \times \eta_{10\%} + 0.05 \times \eta_{20\%} + 0.12 \times \eta_{30\%} + 0.21 \times \eta_{50\%}$$
$$+ 0.53 \times \eta_{75\%} + 0.05 \times \eta_{100\%} \tag{5.124}$$

The term "CEC" refers to the California Energy Commission. The coefficients indicate the importance of the efficiency of each level, based on assumptions about how often the PV converter will function at that level. The weighted values correspond to the climate in central Europe and California, USA. The symbol $\eta_{x\%}$ represents the efficiency tested at x percent of the rated power level. The symbol $\eta_{100\%}$ refers to the efficiency value at the rated power level.

Comparing the two performance indices, the European measure gives more weight at 50% of the inverter power rating, meanwhile, the CEC measure gives more at 70% of the conversion capacity. This reflects that the solar resource in California is greater than in central Europe.

5.8 Wide Band-gap Devices for Future Power Conversion

In general, the conversion efficiency in PV power systems has been pushed to a higher level than in other power electronic applications. The DC/AC converters that are applied at the string and array levels for grid connections have reached 97% or even higher efficiency. The DC/DC converters used for PV-side power interfaces are usually more than than 98% efficient. MIPIs have relatively lower conversion efficiencies of between

94% and 97% because of the high conversion ratio of the voltage from DC to single-phase AC. Conventional power interfaces are mainly constructed by power switches; either MOSFETs or IGBTs. For relatively low voltage levels of 150 V or less, a MOSFET is preferred since the conduction loss is determined by the ESR or $R_{ds}(on)$, which is relatively low for the latest technologies. IGBTs are mainly used for higher voltages – 400 V or more – because the conduction loss is mainly a function of the voltage drop. Both MOSFETs and IGBTs are silicon-based and are reaching the limits of this technology. Therefore, even though various new converters have been proposed in recent years, higher efficiencies are becoming difficult to achieve through improved circuit topologies. A breakthrough in power semiconductor technology is required if higher efficiencies are to be achieved for the PV power industry.

Researchers have started to use wide-bandgap semiconductors to fabricate power switches, leading to high efficiencies in power conversion. Silicon carbide (SiC) and gallium nitride (GaN) technologies have drawn significant attention in both industry and academia and have exhibited outstanding performance in terms of small losses and small sizes. SiC-based diodes have very low reverse recovery charge in order to improve the PV inverter efficiency (Islam and Mekhilef 2016). Experiments have given 98.5% peak efficiency and 98.3% European efficiency, indicating the effectiveness of SiC diodes in the system design.

SiC MOSFETs also have the advantage of low parasitic capacitances, resulting in lower switching losses than conventional IGBTs. They are used in DC/AC converters for grid connections, with improved performance (Barater et al. 2016). Such MOSFETs are available in 650 V and 1200 V planar types. Meanwhile, industry is pushing to higher-voltage models for even broader applications.

Figure 5.73 Submodule-integrated DC/DC converters constructed from GaN FETs.

Another wide-bandgap technology is gallium nitride (GaN), which has been used for a PV power system (Khan et al. 2016). The application is based on a traditional synchronous buck converter, but exhibits high efficiency gains thanks to the low parasitic capacitance and low $R_{ds}(on)$ of GaN power switches. This allows for high switching frequencies (>200 kHz) without significant losses due to hard switching. The application focuses on low-voltage DC/DC conversion for subMISCs. A prototype circuit is shown in Figure 5.73, including three synchronous DC/DC converters for the subMISCs. Six GaN FETs and three drivers are shown in the circuit. Each converter is rated at 100 W since the small size of GaN devices allows for a very compact design and high switching frequency. The circuit is designed to be integrated inside the junction box of a PV panel, as shown in Figure 5.73, in order to maximize the power generation of each submodule.

A recent development shows the potential of a 600-V GaN FET based on so-called GaN-on-Si technology (Stubbe et al. 2014). The voltage level allows it to be used for grid-connected inverters and many other applications. Studies show that the efficiency can be pushed to 99% and higher when GaN devices are used. The device size can also be significantly reduced due to its high switching frequency and the small size of GaN components.

5.9 Summary

This chapter focuses on electronic technology for PV power conversion. It looks at PV-side converters (PVSCs), the DC link, grid-side converters (GSCs), and grid links in order to design a complete grid-connected PV power system. The study covers power interface design, steady-state analysis, simulation modeling, evaluation, and proofs of concept for AC grid-connected systems and DC microgrid systems.

In the steady-state analysis, the parameters of the passive components, control inputs, voltage-conversion ratios, ripple currents, and ripple voltages are considered and designed. The values of the inductance and capacitance of the passive components are determined through steady-state analysis. Since these are based on a theoretical analysis, the values can be considered as the basis of practical system implementations.

Table 5.6 summarizes the DC/DC converter topologies used for PV-side converters. It provides the information on the modulation methods, galvanic isolation, and selection criteria. The selection criteria should be the highest conversion efficiency and simplest circuit design. The buck, boost, buck–boost, and tapped-inductor topologies are non-isolated DC/DC converters. When galvanic isolation is mandatory, the common topologies are the flyback and full-bridge isolated buck converter, which can also be distinguished by the power capacity for practical implementation.

Table 5.7 summarizes the common converter topologies used for grid-side converters. For single-phase power conversion, the H bridge can be modulated by hysteresis current control, sine-triangle, and unfolding for AC grid connections. Hysteresis current modulation has been demonstrated for DC to three-phase AC conversion. Other technologies can also be explored for three-phase grid connections, including space vector modulation.

The chapter describes topology analysis, circuit design, and simulation, focusing on PV power systems. A brief introduction and analysis of LCL filters for grid links is

Table 5.6 Common converter topologies for PV-side converters.

Topology	Modulation mean	Galvanic isolation	Selection criteria
Buck	PWM	no	$V_{O(max)} \leq V_{MPP(min)}$ and $\frac{V_{MPP}}{V_O} \leq 5$
Boost	PWM	no	$V_{O(min)} \geq V_{MPP(max)}$ and $\frac{V_O}{V_{MPP}} \leq 5$
Buck–boost	PWM	no	Either buck or boost is not an option; no common ground for input and output.
Full-bridge isolated buck converter	PWM or phase shift	yes	Commonly utilized for high power capacity and galvanic isolation. The voltage-conversion ratio can be flexible due to the freedom of the transformer design.
Flyback	PWM	yes	$\frac{V_{MPP}}{V_O} > 3$ or $\frac{V_O}{V_{MPP}} > 3$; commonly for low power capacity requiring high voltage conversion and galvanic isolation.
Tapped inductor	PWM	no	For high step-up conversion: $\frac{V_O}{V_{MPP}} > 3$

V_O and V_{MPP} refer to the nominal output voltage and voltage of maximum power point.

Table 5.7 Common converter topologies for AC grid-side power interfaces.

Topology	Modulation	Application
H bridge	Hysteresis, sine-triangle, and unfolding	Single-phase grid connection
CSI	Hysteresis, sine-triangle, and SVM	Three-phase grid connection

CSI, current source inverter.

given. Since the majority involve a common topology that is widely covered in power electronics books, the details are not presented.

The simulation models presented in this chapter are based on the ideal operating condition, and are mainly for proving design concepts. In some cases, detailed models are required to identify loss elements, including non-ideal factors, in order to improve the conversion efficiency. It is always important to evaluate the design by simulation. The design and simulation model can verify each other if the model output matches the design specification,

Maximum power point tracking can be achieved with full knowledge of the PV model, load profile, and environmental conditions. However, it is difficult to achieve this in reality, because of the time-variant nature of the PV models and the load conditions. Moreover, the identification of the environment condition in terms of irradiance and cell temperature can be costly and it can be difficult to achieve the required accuracy. To achieve maximum power point tracking, a dedicated algorithm is needed, and this is discussed in the following chapters. Although all the design examples involve relatively low power capacities, the design and simulation principles can be applied to high-capacity systems too.

One specific section focuses on the battery-side converter. One topology, the DAB, has been analyzed and demonstrated to be suitable for the battery-side power interface. The features of bidirectional power operation and zero voltage switching are attractive, not only for battery-side converters but also for solid-state transformers and electric vehicles. The concerns of DAB lie in the loss of zero voltage switching and the significance of circulating power, which affects conversion efficiency.

Problems

5.1 Duplicate the simulation results in design cases in this chapter. The process is valuable in becoming familiar with the principle of design, analysis, modeling, evaluation, simulation evaluation, and proof of concept for the various PV-side and grid-side converters. It is also helpful in understanding the operation and design of the DC and grid links.

5.2 Design a PV-side converter to supply a constant-voltage bus rated at 12 V DC. One of the non-isolated topologies should be used.

a) Choose a crystalline-based PV module to be the power source. The PV module should be constructed from either 60 or 72 cells in series connection. Formulate a simulation model to represent its output characteristics.

b) At STC, the PV module is operated at the MPP, which is represented by V_{MPP} and I_{MPP}. Based on the rated bus voltage, calculate the duty cycle at steady state with the assumption of CCM.

c) The switching frequency is designed to be 40 kHz, the peak-to-peak ripple voltage of the PV module is specified as 1% of V_{MPP}, and the peak-to-peak ripple of the inductor current is specified as 20% of I_{MPP}. Find the rating of the inductor (L) and capacitor (C_{in}) to meet the above requirements.

d) Build and show the simulation model to represent the system operation.

e) Apply the specified switching duty cycle to the PVSC and simulate the system operation. Plot the waveforms of i_L, v_{pv}, i_{pv}, and p_{pv} to prove the converter design, operation, and simulation.

f) Plot the zoom-in waveforms of i_L, v_{pv}, i_{pv}, and p_{pv} to check the ripple voltage and current and verify the design specification in the steady state.

5.3 Design a PVSC to supply a constant-voltage bus rated at 60 V DC. One of the non-isolated topologies should be used.

a) Choose a crystalline-based PV module to be the power source. The PV module should be constructed from either 60 or 72 cells in series connection. Formulate a simulation model to represent its output characteristics.

b) At STC, the PV module is operated at the MPP, which is represented by V_{MPP} and I_{MPP}. Based on the rated bus voltage, calculate the duty cycle at steady state with the assumption of CCM.

c) The switching frequency is designed to be 40 kHz; the peak-to-peak ripple voltage of the PV module is specified as 1% of the value, V_{MPP}; and the peak-to-peak ripple of the inductor current is specified as 10% of the value,

I_{MPP}. Find the rating of the inductor (L) and capacitor (C_{in}) to meet the above requirement.

d) Build and show the simulation model to represent the system operation.

e) Apply the specified switching duty cycle to the PVSC and simulate the system operation. Plot the important waveforms of i_L, v_{pv}, i_{pv}, and p_{pv} to prove the converter design, operation, and simulation.

f) Plot the zoom-in waveforms of i_L, v_{pv}, i_{pv}, and p_{pv} to check the ripple voltage and current and verify the design specification in the steady state.

5.4 Design a PVSC to supply a constant-voltage bus rated at 300 V DC. One of the non-isolated topologies should be used.

a) Choose a crystalline-based PV module to be used as the power source. The PV module should be constructed from either 60 or 72 cells in series connection. Formulate a simulation model to represent its output characteristics.

b) At STC, the PV module is operated at the MPP, which is represented by V_{MPP} and I_{MPP}. Based on the rated bus voltage, calculate the duty cycle at steady state with the assumption of CCM.

c) The switching frequency is set at 40 kHz, the peak-to-peak ripple voltage of the PV module is specified as 1% of V_{MPP}, and the peak-to-peak ripple of the inductor current is specified as 10% of I_{MPP}. Find the rating of the inductor (L) and capacitor (C_{in}) to meet the above requirements.

d) Build and show the simulation model to represent the system operation.

e) Apply the specified switching duty cycle to the PVSC and simulate the system operation. Plot the important waveforms i_L, v_{pv}, i_{pv}, and p_{pv} to prove the converter design, operation, and simulation.

f) Plot the zoom-in waveforms of i_L, v_{pv}, i_{pv}, and p_{pv} to check the ripple voltage and current and verify the design specification in the steady state.

5.5 Design a PVSC to supply a constant-voltage bus rated at 380 V DC. One of the isolated topologies should be used.

a) Choose a crystalline-based PV module to be used as the power source. The PV module should be constructed from either 60 or 72 cells in series connection. Formulate a simulation model to represent its output characteristics.

b) At STC, the PV module is operated at the MPP, which is represented by V_{MPP} and I_{MPP}. Based on the rated bus voltage, calculate the duty cycle at steady state with the assumption of CCM.

c) The switching frequency is designed to be 40 kHz; the peak-to-peak ripple voltage of the PV module is specified as 1% of the value, V_{MPP}; and the peak-to-peak ripple of the inductor current is specified as 10% of the value, I_{MPP}. Find the rating of the inductor (L) and capacitor (C_{in}) to meet the above requirements.

d) Build and show the simulation model to represent the system operation.

e) Apply the specified switching duty cycle to the PVSC and simulate the system operation. Plot the important waveforms i_L, v_{pv}, i_{pv}, and p_{pv} to prove the converter design, operation, and simulation.

f) Plot the zoom-in waveforms of i_L, v_{pv}, i_{pv}, and p_{pv} to check the ripple voltage and current and verify the design specification in the steady state.

References

Barater D, Concari C, Buticchi G, Gurpinar E, De D and Castellazzi A 2016 Performance evaluation of a three-level ANPC photovoltaic grid-connected inverter with 650-V SiC devices and optimized PWM. *Industry Applications, IEEE Transactions on* **52**(3), 2475–2485.

Cheng K 2006 Tapped inductor for switched-mode power converters *Power Electronics Systems and Applications, 2006. ICPESA '06. 2nd International Conference on*, pp. 14–20.

Du Y, Lu D, Chu G and Xiao W 2015 Closed-form solution of time-varying model and its applications for output current harmonics in two-stage PV inverter. *Sustainable Energy, IEEE Transactions on* **6**(1), 142–150.

Edwin F, Xiao W and Khadkikar V 2012 Topology review of single phase grid-connected module integrated converters for PV applications *IECON 2012 – 38th Annual Conference on IEEE Industrial Electronics Society*, pp. 821–827.

Erickson R and Maksimovic D 2001 *Fundamentals of Power Electronics*. Springer.

Hanif M, Khadkikar V, Xiao W and Kirtley J 2014 Two degrees of freedom active damping technique for an LCL filter based grid connected PV systems. *Industrial Electronics, IEEE Transactions on* **61**(6), 2795–2803.

Islam M and Mekhilef S 2016 Efficient transformerless MOSFET inverter for a grid-tied photovoltaic system. *Power Electronics, IEEE Transactions on* **31**(9), 6305–6316.

Khan O, Xiao W and Zeineldin H 2016 Gallium nitride based submodule integrated converters for high-efficiency distributed maximum power point tracking PV applications. *Industrial Electronics, IEEE Transactions on* **63**(2), 966–975.

Krzywinski G 2015 Integrating storage and renewable energy sources into a DC microgrid using high gain DC DC boost converters *DC Microgrids (ICDCM), 2015 IEEE First International Conference on*, pp. 251–256.

Stubbe T, Mallwitz R, Rupp R, Pozzovivo G, Bergner W, Haeberlen O and Kunze M 2014 GaN power semiconductors for PV inverter applications – opportunities and risks *Integrated Power Systems (CIPS), 2014 8th International Conference on*, pp. 1–6.

Wen H, Su BR and Xiao W 2013 Design and performance evaluation of a bidirectional isolated DC-DC converter with extended dual-phaseshift scheme. *Power Electronics, IET* **6**(5), 914–924.

Wen H, Xiao W and Su B 2014 Nonactive power losses minimization in a bidirectional isolated DC-DC converter for distributed power system. *Industrial Electronics, IEEE Transactions on* **61**(12), 6822–6831.

Wittig B, Franke WT and Fuchs FW 2009 Design and analysis of a DC/DC/AC three phase solar converter with minimized DC link capacitance *Power Electronics and Applications, 2009. EPE '09. 13th European Conference on*, pp. 1–9.

Xiao W, Ozog N and Dunford WG 2007 Topology study of photovoltaic interface for maximum power point tracking. *Industrial Electronics, IEEE Transactions on* **54**(3), 1696–1704.

6

Dynamic Modeling

This chapter discusses dynamic modeling and voltage regulation in PV power systems. For two-stage conversion systems, the dynamics of the PV and DC links should be analyzed to achieve the required control performance. For single-stage conversion, the PV link is merged with the DC link. The objective is to develop a mathematical model combining both PV generators and power interfaces that can be analyzed by well-established control theory.

The output of a PV generator is nonlinear, as discussed in Chapter 4. Power-conditioning circuits are mainly constructed from switching-mode power converters that are nonlinear due to the on-off switching and other nonlinear features. From the dynamic expression of the switching on/off operation, an averaged model can be derived using the state-space averaging method. Linearization is required to derive small-signal linear models for dynamic analysis and controller synthesis.

6.1 State-space Averaging

When switching-mode power converters are used for PV power interfaces, the state-space averaging technique is required to derive mathematical models. The state-space concept was first introduced at the Power Electronics Specialists Conference in 1976 (Middlebrook and Cuk 1976). Since then, the method has been widely used for various converter topologies in order to derive mathematical models for control analysis.

The averaging approach is based on the condition that the switching frequency is much higher than the system's critical dynamics, which are formed by energy-storage components, such as the inductor and capacitor. Under this condition, the nonlinear switching dynamics can be neglected for dynamic analysis and controller synthesis. Over one switching cycle, the system state-space model can be derived from the on-state and off-state of the power switch. The averaged value of continuous signals can be used to form a dynamic model without the need to represent the switching ripples. If the averaged model exhibits linearity, a large-signal model is developed, which can be used for control-loop analysis and design. If there is nonlinearity, a linearization process is required to derive a mathematical model, which can then be used through the linear control theorem.

Photovoltaic Power System: Modeling, Design, and Control, First Edition. Weidong Xiao.
© 2017 John Wiley & Sons Ltd. Published 2017 by John Wiley & Sons Ltd.
Companion Website: www.wiley.com/go/xiao/pvpower

6.2 Linearization

If $f(x)$ is an infinitely differentiable function, the Taylor series expansion with regard to an equilibrium point (x_0) can be expressed as

$$f(x) = \sum_{n=0}^{\infty} \frac{f^{(n)}(x_0)}{n!}(x-x_0)^n \tag{6.1}$$

where $n!$ denotes the factorial of n and $f^{(n)}(x_0)$ denotes the nth derivative evaluated at point x_0. Based on the Taylor series, a small-signal model can be derived from the first-order approximation by neglecting higher-order terms:

$$f(x) \approx f(x_0) + \underbrace{\left.\frac{df(x)}{dx}\right|_{x=x_0}}_{C_1}(x-x_0) \tag{6.2}$$

where C_1 is considered as a constant parameter. Applying the approximation, a nonlinear differential equation $\dot{x} = f(x)$ can be expressed as

$$\frac{dx}{dt} \approx f(x_0) + C_1(x-x_0) \tag{6.3}$$

If the perturbation is defined as $\tilde{x} = x - x_0$, the equation can be expressed as

$$\frac{d(x_0 + \tilde{x})}{dt} \approx f(x_0) + C_1(\tilde{x}) \tag{6.4}$$

The linearization can be finalized and shown as

$$\frac{d\tilde{x}}{dt} = C_1(\tilde{x}) \tag{6.5}$$

due to the equilibrium at the point of x_0. It should be noted that the linear model is only valid near x_0.

The same approximation can be used for a nonlinear differential equation with multiple states and inputs. A nonlinear differential equation including two state variables (x_1, x_2) and one input (u) is given as

$$\frac{dx_1}{dt} = f(x_1, x_2, u) \tag{6.6a}$$

$$\frac{dx_2}{dt} = g(x_1, x_2, u) \tag{6.6b}$$

Applying the linearization process, the linearized dynamics of the two state variables can be expressed as

$$\frac{d\tilde{x}_1}{dt} = \underbrace{\left[\left.\frac{\partial f(x_1, x_2, u)}{\partial x_1}\right|_{X_2, U}\right]}_{a_{11}}\tilde{x}_1 + \underbrace{\left[\left.\frac{\partial f(x_1, x_2, u)}{\partial x_2}\right|_{X_1, U}\right]}_{a_{12}}\tilde{x}_2 + \underbrace{\left[\left.\frac{\partial f(x_1, x_2, u)}{\partial u}\right|_{X_1, X_2}\right]}_{b_1}\tilde{u}$$

$$\tag{6.7a}$$

$$\frac{d\tilde{x}_2}{dt} = \underbrace{\left[\left.\frac{\partial g(x_1, x_2, u)}{\partial x_1}\right|_{X_2, U}\right]}_{a_{21}}\tilde{x}_1 + \underbrace{\left[\left.\frac{\partial g(x_1, x_2, u)}{\partial x_2}\right|_{X_1, U}\right]}_{a_{22}}\tilde{x}_2 + \underbrace{\left[\left.\frac{\partial g(x_1, x_2, u)}{\partial u}\right|_{X_1, X_2}\right]}_{b_2}\tilde{u}$$

$$\tag{6.7b}$$

where $\tilde{x}_1 = x_1 - X_1$, $\tilde{x}_2 = x_2 - X_2$, and $\tilde{u} = u - U$. The system can also be expressed in the state-space format as

$$\begin{bmatrix} \dfrac{d\tilde{x}_1}{dt} \\[2mm] \dfrac{d\tilde{x}_2}{dt} \end{bmatrix} = \begin{bmatrix} a_{11} & a_{12} \\ a_{21} & a_{22} \end{bmatrix} \begin{bmatrix} \tilde{x}_1 \\ \tilde{x}_2 \end{bmatrix} + \begin{bmatrix} b_1 \\ b_2 \end{bmatrix} \tilde{u} \tag{6.8}$$

where a_{11}, a_{12}, a_{21}, and a_{22} are the parameters of the dynamic matrix, while the constants of b_1 and b_2 form the control matrix. A linear small-signal model is derived, which is only valid near the steady states in terms of X_1, X_2, and U.

6.3 Dynamics of PV Link

There is always a tradeoff in defining the ripple voltage at the PV link. The low value of the voltage ripple at the PV link can enhance MPPT accuracy at the steady state. However, it results in a high capacitance appearing at the PV link, which slows down the dynamics. The value of the ripple voltage should be specified by the loss that is caused by the deviation of the MPP.

Due to the nonlinear behavior of on/off switching, averaging is required to derive the model for power interfaces. Capacitors are required across the PV link. Inductors are also needed to construct the PVSC, as discussed in Section 5.1. The system dynamics can be expressed in a general form showing the dynamics of the inductor current, i_L, and the PV-link voltage, v_{pv}, which is controlled by the switching duty cycle d:

$$\frac{di_L}{dt} = f(i_L, v_{pv}, d) \tag{6.9a}$$

$$\frac{dv_{pv}}{dt} = g(i_L, v_{pv}, d) \tag{6.9b}$$

The piecewise-linear or small-signal model can be further derived by a linearization process:

$$\frac{d\tilde{i}_L}{dt} = \frac{\partial f}{\partial v_{pv}}\bigg|_{SS} \tilde{v}_{pv} + \frac{\partial f}{\partial i_L}\bigg|_{SS} \tilde{i}_L + \frac{\partial f}{\partial d}\bigg|_{SS} \tilde{d} \tag{6.10a}$$

$$\frac{d\tilde{v}_{pv}}{dt} = \frac{\partial g}{\partial v_{pv}}\bigg|_{SS} \tilde{v}_{pv} + \frac{\partial g}{\partial i_L}\bigg|_{SS} \tilde{i}_L + \frac{\partial g}{\partial d}\bigg|_{SS} \tilde{d} \tag{6.10b}$$

where $\tilde{i}_L$, $\tilde{v}_{pv}$, and $\tilde{d}$ represent the small signals of the PV module voltage v_{pv}, the inductor current i_L, and the switching duty cycle d, respectively and SS denotes the steady state. The small-signal model characterizes the system dynamics and is important for model-based controller design. In the following section, the modeling process will be demonstrated using buck, full-bridge transformer isolated, boost, buck–boost, flyback, and tapped-inductor topologies.

6.3.1 Linearization of PV Output Characteristics

The PV output characteristics were introduced in Chapter 4. They are expressed by nonlinear equations, and represented as equivalent circuits. The partial derivative $\partial i_{pv}/\partial v_{pv}$ should give the small-signal model through the linearization process in (6.10). Based on the ideal single-diode model (ISDM), a PV array can be expressed as

$$i_{pv} = N_P[i_{ph} - i_s(e^{\frac{qv_{pv}}{N_S k T_c A}} - 1)]$$ (6.11)

with N_S and N_P being the numbers of PV cells in series and parallel connection. The dynamic conductance can be derived by partial differentiation and expressed by

$$G_{pv}(v_{pv}) = \frac{\partial i_{pv}}{\partial v_{pv}} = -\frac{N_P q i_s}{N_S k T_c A} e^{\frac{qv_{pv}}{N_S k T_c A}}$$ (6.12)

This is a variable that is a function of v_{pv} and is influenced by the environmental conditions. Consequently, the dynamic resistance can be derived from (6.12) and expressed as

$$R_{PV}(v_{pv}) = \frac{1}{G_{PV}(v_{pv})}$$ (6.13)

In the modeling process, the steady state is the maximum power point (MPP) at STC, which is available from the product datasheet. The dynamic resistance is R_{PV}, which is a constant to represent the linearized correlation of slight deviation of i_{pv} and v_{pv} from the steady state. It is given by

$$\tilde{i}_{pv} = \frac{\tilde{v}_{pv}}{R_{PV}}$$ (6.14)

The electrical characteristics of the PV module shown in Figure 5.4 can be expanded to include information about the dynamic conductance and resistance for dynamic modeling purposes, as shown in Figure 6.1, where the absolute value of the dynamic resistance changes with the operating point in the I–V curve. The absolute value becomes higher when the operating point deviates from the MPP and approaches the left-hand side, which can be called the current source region. It becomes lower when the operating point lies to the right of the MPP. The area surrounding the MPP is defined as the power source region (Xiao and Zhang 2013; Xiao et al. 2007). This three-zone definition is shown in Figure 6.2, which is based on a normalized I–V curve.

6.3.2 Buck Converter as the PV-link Power Interface

The design, analysis, and simulation of the buck topology were discussed in Section 5.1.2. Based on the schematic in Figure 5.5, the system dynamics can be discovered for CCM operation. When the PV-link voltage is the control variable, the output voltage, V_o, is assumed to be constant for dynamic modeling.

Q on-state dynamics:

$$L\frac{di_L}{dt} = v_{pv} - V_0$$ (6.15a)

$$C_{in}\frac{dv_{pv}}{dt} = i_{pv} - i_L$$ (6.15b)

Q off-state dynamics:

$$L\frac{di_L}{dt} = -V_o$$ (6.16a)

$$C_{in}\frac{dv_{pv}}{dt} = i_{pv}$$ (6.16b)

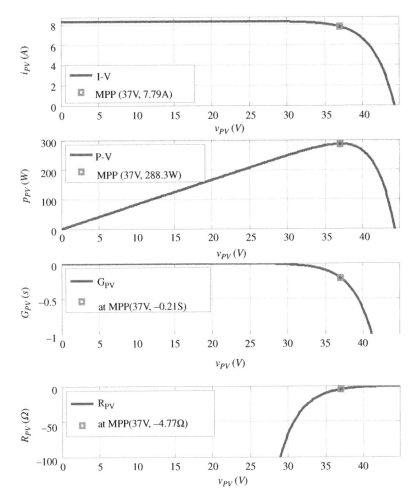

Figure 6.1 Electrical characteristics of the PV module showing dynamic conductance and resistance.

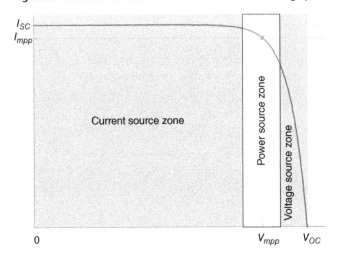

Figure 6.2 Three-zone definition based on I–V curve.

Averaging:

$$\frac{di_L}{dt} = \underbrace{\frac{1}{L}[dv_{pv} - V_o]}_{f(v_{pv},d,i_L)} \tag{6.17a}$$

$$\frac{dv_{pv}}{dt} = \underbrace{\frac{1}{C_{in}}[i_{pv} - di_L]}_{g(v_{pv},d,i_L)} \tag{6.17b}$$

where d is the switching duty cycle and the control variable.

Due to the nonlinear characteristics in (6.17), linearization is required to derive the small-signal model at the nominal operating condition. Based on the predefined steady state, the small-signal model can be derived using (6.10) and expressed in the state-space form as

$$\begin{bmatrix} \dfrac{d\tilde{i}_L}{dt} \\[2mm] \dfrac{d\tilde{v}_{pv}}{dt} \end{bmatrix} = \begin{bmatrix} 0 & \dfrac{D}{L} \\[2mm] -\dfrac{D}{C_{in}} & \dfrac{1}{R_{PV}C_{in}} \end{bmatrix} \begin{bmatrix} \tilde{i}_L \\[2mm] \tilde{v}_{pv} \end{bmatrix} + \begin{bmatrix} \dfrac{V_{pv}}{L} \\[2mm] -\dfrac{I_L}{C_{in}} \end{bmatrix} \tilde{d} \tag{6.18}$$

where D, V_{PV}, and I_L represent the switching duty cycle, the PV terminal voltage, and the inductor current, which are considered to be constant in the steady state. The signals $\tilde{i}_L$ and $\tilde{v}_{pv}$ are the state variables, and $\tilde{d}$ represents the control variable in the small-signal model. It can be transformed and expressed in the transfer function:

$$\frac{\tilde{v}_{pv}(s)}{\tilde{d}(s)} = \frac{-\dfrac{I_L}{C_{in}}s - \dfrac{DV_{pv}}{LC_{in}}}{s^2 - \left(\dfrac{1}{R_{PV}C_{in}}\right)s + \dfrac{D^2}{LC_{in}}} \tag{6.19}$$

This is a second-order system with two poles and one minimal-phase zero, which can be standardized by

$$G_0(s) = \frac{K_0(\beta s + 1)}{s^2 + 2\xi\omega_n s + \omega_n^2} \tag{6.20}$$

where the undamped natural frequency and damping factor are expressed as ω_n and ξ, respectively. To represent (6.19), the following coefficients can be derived:

$$\omega_n = \frac{D}{\sqrt{LC_{in}}} \tag{6.21a}$$

$$\xi = -\frac{\sqrt{L}}{2DR_{pv}\sqrt{C_{in}}} \tag{6.21b}$$

$$K_0 = -\frac{DV_{pv}}{LC_{in}} \tag{6.21c}$$

$$\beta = \frac{I_L L}{DV_{pv}} \tag{6.21d}$$

The gain K_0 is negative because the change of the PV terminal voltage is in the opposite direction to the duty cycle. It should be noted that the absolute value of R_{pv} – the operating point of PV generator – affects only the damping ratio. When the value is higher, the damping becomes lighter, which causes more oscillation. It has been shown that the damping becomes critical when the operating point enters the current source region, as defined in Figure 6.2.

To continue the case study in Section 5.1.2, the small-signal model is developed using the same system parameters and is expressed in the state-space form:

$$
\begin{bmatrix} \dfrac{d\tilde{i}_L}{dt} \\ \dfrac{d\tilde{v}_{pv}}{dt} \end{bmatrix} = \begin{bmatrix} 0 & 3846 \\ -2370 & -765 \end{bmatrix} \begin{bmatrix} \tilde{i}_L \\ \tilde{v}_{pv} \end{bmatrix} + \begin{bmatrix} 200038 \\ -38919 \end{bmatrix} \tilde{d} \tag{6.22}
$$

This can be transferred to a single-input, single-output (SISO) transfer function, which represents the small-signal dynamics between $\tilde{v}_{pv}$ and the duty cycle, $\tilde{d}$:

$$
\frac{\tilde{v}_{pv}(s)}{\tilde{d}(s)} = \frac{-(43878s + 519994944)}{s^2 + 765s + 9113887} \tag{6.23}
$$

The undamped natural frequency is 3019 rad/s and the damping ratio is 0.13. The location of the poles and zero can be plotted in the complex plane, as shown in Figure 6.3. Both the low value of the damping factor and the pole location indicate that the system is lightly damped. The system-dynamics analysis is important in determining the MPPT bandwidth and designing a linear controller to regulate v_{pv}.

The case study in Section 5.1.2 has a nominal duty cycle of 64.9%, which represents the steady-state condition when the output voltage is nominal and the PV module is at STC. When the system enters the steady state, a small perturbation of 0.5% is applied periodically to the duty cycle in order to evaluate the step response of v_{pv}. For comparison,

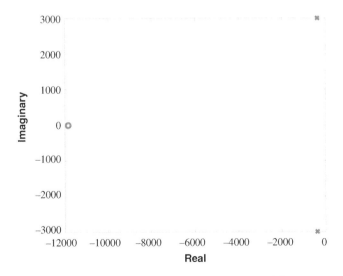

Figure 6.3 Location of poles and zero of the system model G_0.

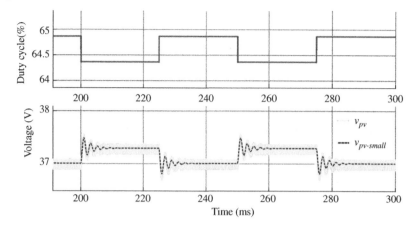

Figure 6.4 Dynamics comparison of simulation model and small-signal model when buck topology is used for the PVSC.

the small-signal model works in parallel with the simulation model that was described in Figures 5.6 and 5.7.

Figure 6.4 is a comparison of the output of the developed small-signal model and the simulation model. It should be noted that the small-signal models in (6.22) or (6.23) accept the small-signal perturbation of the duty cycle as the input, and output the small-signal value. A DC offset should be added to the small-signal model output to create an equal level for the comparison. The value can be derived from the steady-state condition. The output of the small-signal model has no switching ripples, but captures the critical dynamics during the transient period. Since the small-signal model is developed according to the MPP condition, the model matches very well at the nominal operating condition, which is around a voltage of 37 V. The negative gain is reflected by the voltage response, which is opposite to the variation of the duty cycle.

6.3.3 Full-bridge Transformer Isolated DC/DC as the PV-link Power Interface

A schematic of a full-bridge isolated DC/DC converter used for a PV-link power interface is shown in Figure 5.10. Even though the circuit is more complex than the buck topology, the dynamic modeling approach can be the same. The system dynamics can be discovered from the switching operation in CCM.

Either pair of switches is on-state:

$$L\frac{di_L}{dt} = Nv_{pv} - V_0 \tag{6.24a}$$

$$C_{in}\frac{dv_{pv}}{dt} = i_{pv} - Ni_L \tag{6.24b}$$

All switches are off-state:

$$L\frac{di_L}{dt} = -V_o \tag{6.25a}$$

$$C_{in}\frac{dv_{pv}}{dt} = i_{pv} \tag{6.25b}$$

Averaging is applied:

$$\frac{di_L}{dt} = \underbrace{\frac{1}{L}[2dNv_{pv} - V_o]}_{f(v_{pv}, d, i_L)} \tag{6.26a}$$

$$\frac{dv_{pv}}{dt} = \underbrace{\frac{1}{C_{in}}[i_{pv} - 2dNi_L]}_{g(v_{pv}, d, i_L)} \tag{6.26b}$$

where d is the switching duty cycle of each power switch.

Due to the nonlinear characteristics in (6.26), linearization is required to derive the small-signal model at the nominal operating condition. Based on the predefined steady state, the small-signal model can be derived using (6.10) and expressed in the state-space form:

$$\begin{bmatrix} \dfrac{d\tilde{i}_L}{dt} \\ \dfrac{d\tilde{v}_{pv}}{dt} \end{bmatrix} = \begin{bmatrix} 0 & \dfrac{2ND}{L} \\ -\dfrac{2ND}{C_{in}} & \dfrac{1}{R_{PV}C_{in}} \end{bmatrix} \begin{bmatrix} \tilde{i}_L \\ \tilde{v}_{pv} \end{bmatrix} + \begin{bmatrix} \dfrac{2NV_{PV}}{L} \\ -\dfrac{2NI_L}{C_{in}} \end{bmatrix} \tilde{d} \tag{6.27}$$

where the symbols of D, V_{PV}, and I_L represent the switching duty cycle, the PV terminal voltage, and the inductor current, which are considered to be constant at the steady state. The signals $\tilde{i}_L$ and $\tilde{v}_{pv}$ are the state variables that represent any small variations, and $\tilde{d}$ represents the control variable in the small-signal model.

To continue the case study in Section 5.1.3, a small-signal model is developed with the same system parameters and is expressed in the state-space form as

$$\begin{bmatrix} \dfrac{d\tilde{i}_L}{dt} \\ \dfrac{d\tilde{v}_{pv}}{dt} \end{bmatrix} = \begin{bmatrix} 0 & 3846 \\ -4740 & -765 \end{bmatrix} \begin{bmatrix} \tilde{i}_L \\ \tilde{v}_{pv} \end{bmatrix} + \begin{bmatrix} 438860 \\ -175510 \end{bmatrix} \tilde{d} \tag{6.28}$$

This can be transferred to a SISO transfer function, which represents the small-signal dynamics between $\tilde{v}_{pv}$ and the duty cycle, $\tilde{d}$:

$$\frac{\tilde{v}_{pv}(s)}{\tilde{d}(s)} = \frac{-(175510s + 208000)}{s^2 + 765s + 18228000} \tag{6.29}$$

This is a second-order system with two poles and one minimal-phase zero. The DC gain is negative. The undamped natural frequency is 3019 rad/s and the damping ratio is 0.13. The system dynamics are important in determining the MPPT bandwidth and designing a linear controller to regulate v_{pv}.

The case study in Section 5.1.3 has a nominal duty cycle 32.45%, which represents the steady-state condition when the output voltage is nominal and the PV module is at STC. When the system enters the steady state, a small perturbation of 0.5% is applied periodically to the duty cycle in order to evaluate the step response of v_{pv}. For comparison, the small-signal model works in parallel with the simulation model described in Figure 5.12.

Figure 6.5 compares the output of the small-signal model and the simulation model. Unlike the simulation model, the output of the small-signal model does not have

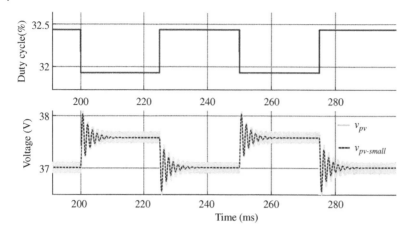

Figure 6.5 Dynamics comparison of simulation model and small-signal model.

switching ripples, but captures the critical dynamics during each transient state. Since the small-signal model is developed according to the MPP condition, the model matches very well at the nominal operating condition, which is around a voltage of 37 V. The negative gain is reflected by the voltage response, which is the opposite of the variation in the duty cycle. Therefore, the small-signal model is verified.

6.3.4 Boost Converter as the PV-link Power Interface

The design, analysis, and simulation of the boost topology were discussed in Section 5.1.4. Based on the schematic in Figure 5.16, the system dynamics can be derived.

Q on-state dynamics:

$$L\frac{di_L}{dt} = v_{pv} \tag{6.30a}$$

$$C_{in}\frac{dv_{pv}}{dt} = i_{pv} - i_L \tag{6.30}$$

Q off-state dynamics:

$$L\frac{di_L}{dt} = v_{pv} - V_o \tag{6.31a}$$

$$C_{in}\frac{dv_{pv}}{dt} = i_{pv} - i_L \tag{6.31b}$$

Averaging:

$$\frac{di_L}{dt} = \underbrace{\frac{1}{L}[v_{pv} - (1-d)V_o]}_{f(v_{pv},d,i_L)} \tag{6.32a}$$

$$\frac{dv_{pv}}{dt} = \underbrace{\frac{1}{C_{in}}[i_{pv} - i_L]}_{g(v_{pv},d,i_L)} \tag{6.32b}$$

where d is the switching duty cycle and the control variable.

The small-signal model can be derived using the standard linearization process, as expressed in (6.10). The small-signal model to represent the system dynamics is shown in state-space form in (6.33) and the transfer function is shown in (6.34):

$$
\begin{bmatrix} \dfrac{d\tilde{i}_L}{dt} \\ \dfrac{d\tilde{v}_{pv}}{dt} \end{bmatrix} = \begin{bmatrix} 0 & \dfrac{1}{L} \\ -\dfrac{1}{C_{in}} & \dfrac{1}{R_{PV}C_{in}} \end{bmatrix} \begin{bmatrix} \tilde{i}_L \\ \tilde{v}_{pv} \end{bmatrix} + \begin{bmatrix} \dfrac{V_O}{L} \\ 0 \end{bmatrix} \tilde{d}
\tag{6.33}
$$

$$
\frac{\tilde{v}_{pv}(s)}{\tilde{d}(s)} = \frac{-\dfrac{V_o}{LC_{in}}}{s^2 - \left(\dfrac{1}{R_{PV}C_{in}}\right)s + \dfrac{1}{LC_{in}}}
\tag{6.34}
$$

where V_{PV} and I_L represent the PV terminal voltage and the inductor current, respectively. These are considered to be constant in the steady state. The signals $\tilde{i}_L$ and $\tilde{v}_{pv}$ are the state variables, and $\tilde{d}$ represents the control variable in the small-signal model. This is a second-order system with two poles and no zero. It can be standardized in a general form, which is expressed in the transfer function:

$$
G_0(s) = \frac{K_0}{s^2 + 2\xi\omega_n s + \omega_n^2}
\tag{6.35}
$$

where the undamped natural frequency and damping factor are expressed as ω_n and ξ, respectively. To represent (6.34), the following coefficients can be derived:

$$
\omega_n = \frac{1}{\sqrt{LC_{in}}}
\tag{6.36a}
$$

$$
\xi = -\frac{\sqrt{L}}{2R_{pv}\sqrt{C_{in}}}
\tag{6.36b}
$$

$$
K_0 = -\frac{V_o}{LC_{in}}
\tag{6.36c}
$$

To continue the case study in Section 5.1.4, the small-signal model is developed with the same system parameters and expressed in state-space form as

$$
\begin{bmatrix} \dfrac{d\tilde{i}_L}{dt} \\ \dfrac{d\tilde{v}_{pv}}{dt} \end{bmatrix} = \begin{bmatrix} 0 & 5897 \\ -80000 & -16760 \end{bmatrix} \begin{bmatrix} \tilde{i}_L \\ \tilde{v}_{pv} \end{bmatrix} + \begin{bmatrix} 283061 \\ 0 \end{bmatrix} \tilde{d}
\tag{6.37}
$$

It can be transferred to a SISO transfer function, which represents the small-signal dynamics between $\tilde{v}_{pv}$ and the duty cycle, $\tilde{d}$:

$$
\frac{\tilde{v}_{pv}(s)}{\tilde{d}(s)} = \frac{-22644891961}{s^2 + 16760s + 471768583}
\tag{6.38}
$$

The undamped natural frequency is 21 720 rad/s and the damping ratio is 0.39. For the step response, the settling time can be estimated as 0.5 ms. The above information is useful in determining the MPPT bandwidth and designing a linear controller to regulate v_{pv}.

The case study in Section 5.1.4 has a nominal duty cycle of 22.9%, representing the steady-state condition when the output voltage is nominal and the PV module is at STC.

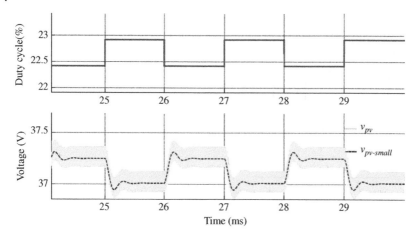

Figure 6.6 Dynamics comparison of simulation model and small-signal model when boost topology is used for PVSC.

When the system enters the steady state, a small perturbation of 0.5% is applied periodically to the duty cycle in order to evaluate the step response of v_{pv}. For comparison, the small-signal model works in parallel with the simulation model described in Figure 5.18.

Figure 6.6 compares the output of the small-signal model and the simulation model. Unlike the simulation model, the output of the small-signal model does exhibit switching ripples, but captures the critical dynamics during each transient state. Since the small-signal model is developed according to the MPP condition, the model matches very well at the nominal operating condition, which is around a voltage of 37 V.

6.3.5 Tapped-inductor Topology as the PV-link Power Interface

The design, analysis and simulation of the tapped-inductor topology were discussed in Section 5.1.5. Based on the schematic in Figure 5.21, the system's dynamic model can be derived accordingly.

Q on-state dynamics:

$$N_{112}L\frac{di_L}{dt} = v_{pv} \tag{6.39a}$$

$$N_{112} = \frac{N_1}{N_1 + N_2} \tag{6.39b}$$

$$C_{in}\frac{dv_{pv}}{dt} = i_{pv} - i_L \tag{6.39c}$$

Q off-state dynamics:

$$N_{112}L\frac{di_L}{dt} = N_{112}(v_{pv} - v_o) \tag{6.40a}$$

$$C_{in}\frac{dv_{pv}}{dt} = i_{pv} - i_L \tag{6.40b}$$

Over one switching cycle, the averaging process results in (6.41), which can be reorganized into (6.42):

$$N_{112}L\frac{di_L}{dt} = dv_{pv} + (1-d)N_{112}(v_{pv} - v_o) \tag{6.41a}$$

$$C_{in}\frac{dv_{pv}}{dt} = i_{pv} - i_L \tag{6.41b}$$

$$\frac{di_L}{dt} = \underbrace{\frac{N_{112}v_p v + dv_{pv} - dN_{112}v_{pv} + dN_{112}V_O}{N_{112}L}}_{f(v_{pv},d,i_L)} \tag{6.42a}$$

$$\underbrace{\frac{1}{C_{in}}[i_{pv} - i_L]}_{g(v_{pv},d,i_L)} \tag{6.42b}$$

where d is the switching duty cycle and the control variable. The small-signal model can be derived using the standard linearization process in (6.10). The small-signal model to represent the system dynamics in state-space form is

$$\begin{bmatrix} \dfrac{d\tilde{i}_L}{dt} \\ \dfrac{d\tilde{v}_{pv}}{dt} \end{bmatrix} = \begin{bmatrix} 0 & \dfrac{N_{112} + (1-N_{112})D_0}{N_{112}L} \\ -\dfrac{1}{C_{in}} & \dfrac{1}{R_{PV}C_{in}} \end{bmatrix} \begin{bmatrix} \tilde{i}_L \\ \tilde{v}_{pv} \end{bmatrix} + \begin{bmatrix} \dfrac{N_{112}V_O + (1-N_{112})V_{MPP}}{N_{112}L} \\ 0 \end{bmatrix} \tilde{d} \tag{6.43}$$

To continue the case study in Section 5.1.5, the small-signal model is developed with the same system parameters and expressed in state-space form as

$$\begin{bmatrix} \dfrac{d\tilde{i}_L}{dt} \\ \dfrac{d\tilde{v}_{pv}}{dt} \end{bmatrix} = \begin{bmatrix} 0 & 1231 \\ -80000 & -16771 \end{bmatrix} \begin{bmatrix} \tilde{i}_L \\ \tilde{v}_{pv} \end{bmatrix} + \begin{bmatrix} 167775 \\ 0 \end{bmatrix} \tilde{d} \tag{6.44}$$

This can be transferred to a SISO transfer function, which represents the small-signal dynamics between $\tilde{v}_{pv}$ and the duty cycle $\tilde{d}$:

$$\frac{\tilde{v}_{pv}(s)}{\tilde{d}(s)} = \frac{-13421997537}{s^2 + 16771s + 98515782} \tag{6.45}$$

This is a second-order system with two poles. The DC gain is negative. The undamped natural frequency is 9926 rad/s and the damping ratio is 0.84. For the step response, the settling time can be estimated as 0.5 ms. The above information is useful in determining the MPPT bandwidth and designing a linear controller to regulate v_{pv}.

The case study in Section 5.1.5 has a nominal duty cycle of 57.0% representing the steady-state condition when the output voltage is nominal and the PV module

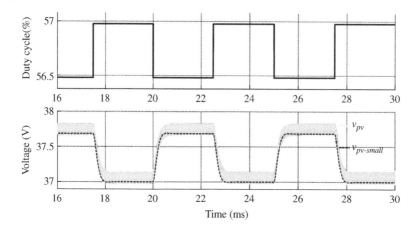

Figure 6.7 Dynamics comparison of simulation model and small-signal model when tapped-inductor topology is used for PVSC.

is at STC. When the system enters the steady state, a small perturbation of 0.5% is applied periodically to the duty cycle in order to evaluate the step response of v_{pv}. For comparison, the small-signal model works in parallel with the simulation model described in Figure 5.18.

Figure 6.7 compares the output of the small-signal model and the simulation result. Unlike the simulation model, the output of the small-signal model does not have switching ripples, but captures the critical dynamics during each transient state. Since the small-signal model is developed according to the MPP condition, the model matches very well at the nominal operating condition, which is around a voltage of 37 V.

6.3.6 Buck–boost Converter as the PV-link Power Interface

The design, analysis, and simulation of the buck–boost topology were discussed in Section 5.1.6. Based on the schematic in Figure 5.26, the system dynamic model can be derived:

Q on-state dynamics:

$$L\frac{di_L}{dt} = v_{pv} \tag{6.46a}$$

$$C_{in}\frac{dv_{pv}}{dt} = i_{pv} - i_L \tag{6.46b}$$

Q off-state dynamics:

$$L\frac{di_L}{dt} = -V_o \tag{6.47a}$$

$$C_{in}\frac{dv_{pv}}{dt} = i_{pv} \tag{6.47b}$$

Averaging:

$$\frac{di_L}{dt} = \underbrace{\frac{1}{L}[dv_{pv} - (1-d)V_o]}_{f(v_{pv},d,i_L)} \tag{6.48a}$$

$$\frac{dv_{pv}}{dt} = \underbrace{\frac{1}{C_{in}}[i_{pv} - di_L]}_{g(v_{pv},d,i_L)} \tag{6.48b}$$

where d is the switching duty cycle and the control variable.

The linearization process in (6.10) should be applied to derive the small-signal model at the nominal steady state. Based on the system dynamics in (6.48), the small-signal model for the buck–boost power interface can be expressed in state-space form as

$$\begin{bmatrix} \frac{d\tilde{i}_L}{dt} \\ \frac{d\tilde{v}_{pv}}{dt} \end{bmatrix} = \begin{bmatrix} 0 & \frac{D}{L} \\ -\frac{D}{C_{in}} & \frac{1}{R_{PV}C_{in}} \end{bmatrix} \begin{bmatrix} \tilde{i}_L \\ \tilde{v}_{pv} \end{bmatrix} + \begin{bmatrix} \frac{V_{PV} + V_O}{L} \\ -\frac{I_L}{C_{in}} \end{bmatrix} \tilde{d} \tag{6.49}$$

To continue the case study in Section 5.1.6, a small-signal model is developed with the same system parameters and is expressed in state-space form as

$$\begin{bmatrix} \frac{d\tilde{i}_L}{dt} \\ \frac{d\tilde{v}_{pv}}{dt} \end{bmatrix} = \begin{bmatrix} 0 & 1351 \\ -1249 & -531 \end{bmatrix} \begin{bmatrix} \tilde{i}_L \\ \tilde{v}_{pv} \end{bmatrix} + \begin{bmatrix} 200038 \\ -38919 \end{bmatrix} \tilde{d} \tag{6.50}$$

This can be transferred to a SISO transfer function, which represents the small-signal dynamics between $\tilde{v}_{pv}$ and the duty cycle $\tilde{d}$:

$$\frac{\tilde{v}_{pv}(s)}{\tilde{d}(s)} = \frac{-(38919s + 124931277)}{s^2 + 530s + 1687112} \tag{6.51}$$

This is a second-order system with two poles and one minimal-phase zero. The DC gain is negative. The undamped natural frequency is 1299 rad/s and the damping ratio is 0.20. For the step response, the settling time is about 15 ms. The above information is useful in determining the MPPT bandwidth and designing a linear controller to regulate v_{pv}.

The case study in Section 5.1.6 has a nominal duty cycle of 49.3%, representing the steady-state condition when the output voltage is nominal and the PV module is under STC. When the system enters the steady state, a small perturbation of 0.5% is applied periodically to the duty cycle in order to evaluate the step response of v_{pv}. For comparison, the small-signal model works in parallel with the simulation model that was described in Figure 5.27.

Figure 6.8 compares the output of the small-signal model and the simulation result. Since the small-signal model is developed according to the MPP condition, the model matches very well at the nominal operating condition, which is around a voltage of 37 V.

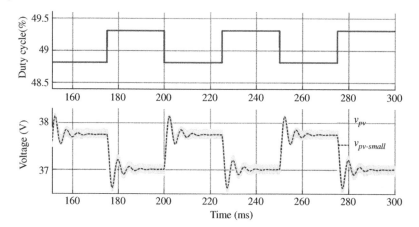

Figure 6.8 Dynamics comparison of simulation model and small-signal model when buck–boost topology is used for PVSC.

6.3.7 Flyback Converter as the PV-link power Interface

When a flyback topology is used as the PVSC, the system schematic is as shown in Figure 5.30d. A slight modification can be applied to the small-signal model of the buck–boost converter to form the model for the flyback topology. The winding turn ratio, n, should be included in the model:

$$\begin{bmatrix} \dfrac{d\tilde{i}_L}{dt} \\ \dfrac{d\tilde{v}_{pv}}{dt} \end{bmatrix} = \begin{bmatrix} 0 & \dfrac{D}{L} \\ -\dfrac{D}{C_{in}} & \dfrac{1}{R_{PV}C_{in}} \end{bmatrix} \begin{bmatrix} \tilde{i}_L \\ \tilde{v}_{pv} \end{bmatrix} + \begin{bmatrix} \dfrac{V_{PV} + V_O/n}{L} \\ -\dfrac{I_L}{C_{in}} \end{bmatrix} \tilde{d} \tag{6.52}$$

It should be noted that i_L is an internal signal of the flyback transformer that cannot be directly measured. However, it can still be used as the state variable for the dynamic model construction.

Based on the design example in Section 5.1.7, the small-signal model can be derived as (6.53) and (6.54):

$$\begin{bmatrix} \dfrac{d\tilde{i}_L}{dt} \\ \dfrac{d\tilde{v}_{pv}}{dt} \end{bmatrix} = \begin{bmatrix} 0 & 1351 \\ -1318 & -545 \end{bmatrix} \begin{bmatrix} \tilde{i}_L \\ \tilde{v}_{pv} \end{bmatrix} + \begin{bmatrix} 200034 \\ -41080 \end{bmatrix} \tilde{d} \tag{6.53}$$

$$\dfrac{\tilde{v}_{pv}(s)}{\tilde{d}(s)} = \dfrac{-(41080s + 133757931)}{s^2 + 545s + 1780840} \tag{6.54}$$

In this case, the DC gain is negative. The undamped natural frequency is 1335 rad/s and the damping ratio is 0.20. Figure 6.9 compares the output of the developed small-signal model and the simulation result. The simulation result shows that the small-signal model captures the critical system dynamics, but ignores the switching ripple in the PVSC.

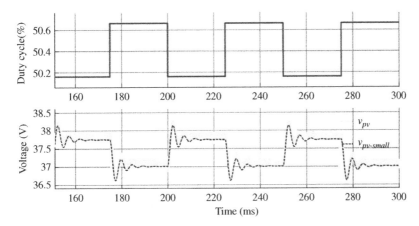

Figure 6.9 Dynamics comparison of simulation model and small-signal model when flyback topology is used for PVSC.

6.4 Dynamics of DC Bus Voltage Interfaced with Dual Active Bridge

As discussed in Sections 1.11 and 5.2, the battery buffer is an important component in DC microgrids and standalone systems, balancing the differences between power generation and load demand. Since dynamic analysis and feedback control play a critical role in real-world power electronics applications, it is important to find an effective way to characterize the system dynamics for regulating the charge and discharge cycles of batteries. The control objectives include

- current regulation in the battery bulk charge cycle
- voltage regulation in the absorption charge cycle
- voltage regulation of the DC bus in DC microgrids
- seamless transition between charge and discharge.

The battery power interface should be able to regulate the DC-link voltage in a DC microgrid through regulated power, either injection or extraction, from the DC bus. The DC bus can be modeled as a capacitor, C_{DC}, in parallel with an equivalent load resistor, R_L. When a dual active bridge (DAB) is used as the power interface, the system circuit is as shown in Figure 6.10. The inductance, L, is not considered for the dynamic model since it forms an impedance effect and interacts with the high-frequency switching. The dynamic effect of the inductance can be separated from the dominant frequency for the overall system. The dynamics of the DC-link voltage are dominated by the RC circuit including C_{DC} and R_L. This can be expressed as

$$C_{DC}\frac{dv_{dc}}{dt} = i_{dc} - i_o \qquad (6.55)$$

which can be further derived to

$$C_{DC}\frac{dv_{dc}}{dt} = \frac{P_{bat}}{V_{DC}} - \frac{v_{dc}}{R_L} \qquad (6.56)$$

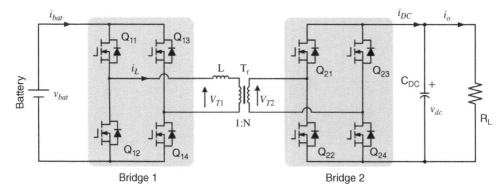

Figure 6.10 Bidirectional DC/DC circuit with dual active bridge topology for output voltage regulation.

where v_{dc} is the voltage variable and V_{DC} represents its steady state value (Syed et al. 2015). Other symbols are defined in Figure 6.10. Following (5.56) and (6.56):

$$\frac{dv_{dc}}{dt} = \frac{V_{bat}\varphi(\pi - \varphi)}{\omega L \pi N C_{DC}} - \frac{v_{dc}}{R_L C_{DC}} \tag{6.57}$$

where V_{bat} is the steady-state value of the battery voltage. There is nonlinearity in (6.57) due to the φ^2 term. It can be linearized by following the general form in (6.7). Therefore, the small-signal model can be expressed as

$$\frac{d\tilde{v}_{dc}}{dt} = \frac{V_{bat}}{\omega L \pi N C_{DC}}\left(1 - \frac{2\varphi}{\pi}\right)\tilde{\varphi} - \frac{\tilde{v}_{dc}}{R_L C_{DC}} \tag{6.58}$$

where $\tilde{\varphi}$ and $\tilde{v}_{dc}$ represent the small-signal variants of the phase shift and the output voltage, respectively. The model is based on the steady-state condition, which is determined by the phase shift φ and the battery voltage V_{bat}. The model can be transformed to the s-domain, as

$$\frac{\tilde{v}_{dc}(s)}{\tilde{\varphi}(s)} = \frac{\dfrac{R_L V_{bat}}{\omega L \pi N}\left(1 - \dfrac{2\varphi}{\pi}\right)}{R_L C_{DC}s + 1} \tag{6.59}$$

which can also be expressed as a standard first-order transfer function:

$$G_0(s) = \frac{K_0}{\tau_0 s + 1} \tag{6.60}$$

where $K_0 = \frac{R_L V_{bat}}{\omega L \pi N}\left(1 - \frac{2\varphi}{\pi}\right)$ and $\tau_0 = R_L C_{DC}$.

Variation of the system parameters and operating conditions changes the dynamics, as summarized in Table 6.1. The parameters C_{DC}, L, N, and f_{sw} are determined during the design stage of the DAB. They are considered to be constant during system operation. The system dynamics is mainly changed by the variation of φ and R_L. The small-signal model is based on the steady-state values for the operating condition. As shown in (6.60), an increase of R_L slows down the system response through increasing the time constant τ_0. The value is also proportional to the static gain, K_0. The control variable, φ, only affects the static gain, K_0, in an opposite direction: an increase of φ reduces the value of K_0.

Table 6.1 Parameter impact on system dynamics.

Parameter	Change	Magnitude (K_0)	Time constant (τ_0)
C_{DC}	Increasing	No effect	Increasing
L	Increasing	Decreasing	No effect
N	Increasing	Decreasing	No effect
f_{sw}	Increasing	Decreasing	No effect
φ	Increasing	Decreasing	No effect
R_L	Increasing	Increasing	Increasing

Based on the case study in Section 5.2, the specification of the DAB system is summarized in Table 5.1. When the system is operated at the nominal condition, for which the power is 750 W and the phase angle is 45°, the equivalent load resistance can be calculated as $R_L = 193\,\Omega$. For evaluation purposes, the capacitor across the DC bus is assigned as $C_{DC} = 1\,\mu F$. A higher value of the capacitance increases the value of the time constant and slows down the system dynamics.

Based on these parameters, the dynamic model is derived as $K_0 = 323$ and $\tau_0 = 193\,\mu s$. When the system enters the steady state, a small perturbation of ± 0.005 rad is applied periodically to the phase shift for evaluation purposes. For comparison, the small-signal model works in parallel with the simulation model that was described in Figure 5.36.

Figure 6.11 compares the output of the developed small-signal model and the simulation result using the switching model. The output of the small-signal model shows no information about the switching ripples, but captures the critical dynamics in response to step changes of the phase-angle shifting. The waveform shows the first-order response, which is predicted by the mathematical model, as expressed in (6.60). Since the small-signal model is developed according to the steady state, the model matches very well at the nominal operating condition, which is around the DC-link voltage of 380 V.

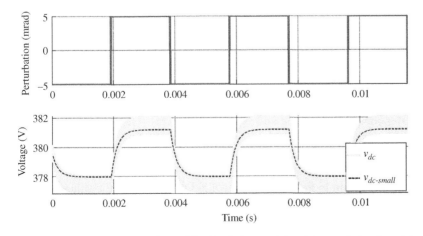

Figure 6.11 Dynamics comparison of the simulation model and the small-signal model.

6.5 Dynamics of DC Link for AC Grid Connection

As discussed in Section 5.3, the DC link is the port for DC-to-AC conversion, as shown in Figure 5.44. The dynamics of the DC link are described in (5.79) where the dynamics of v_{dc} are caused by the interaction of $i_{dc}(t)$, $i_{inv}(t)$, and the DC-link capacitor, C_{dc}. In the single-stage conversion system, the DC link is the same as the PV link. The modeling follows the same principle as for the two-stage conversion system. The dynamics of the DC link for DC/AC conversion is discussed in the following for single-phase and three-phase systems.

6.5.1 Single-phase Connection

In single-phase grid connections, a high capacitance is applied to the DC link to filter the double-line frequency ripple. The significant energy buffer makes for relatively slow dynamics in the DC-link voltage. It should be noted that the amplitude of $i_{dc}(t)$ is determined by the PV generator, which is affected by the status of the MPPT operation and instantaneous environmental conditions.

The dynamics of the DC-link voltage can be represented by the interaction of the input power (p_{dc}) and output power (p_{inv}) of the DC link, as expressed in (5.81). Thus, the energy equilibrium is

$$C_{dc}\frac{dv_{dc}}{dt} = \frac{p_{dc} - p_{grid}}{v_{dc}} \tag{6.61}$$

which gives

$$\frac{dv_{dc}}{dt} = \underbrace{\frac{1}{C_{dc}}\frac{2p_{dc} - v_{mag}i_{mag}}{2v_{dc}}}_{h(v_{dc},i_{mag})} \tag{6.62}$$

when the DC-to-AC conversion loss is neglected. The variables v_{mag} and i_{mag} represent the amplitude of the grid voltage and current, respectively.

A small-signal model can be derived for the steady-state condition in terms of the constant values of the DC power (P_{dc}) and the grid-voltage amplitude (V_{mag}). The small-signal model is represented by the variation of the DC-link voltage, $\tilde{v}_{dc}$, in response to the small-signal variation of the grid current, $\tilde{i}_{mag}$.

Following the linearization process:

$$\frac{d\tilde{v}_{dc}}{dt} = \frac{\partial h}{\partial v_{dc}}\bigg|_{SS}\tilde{v}_{dc} + \frac{\partial h}{\partial i_{mag}}\bigg|_{SS}\tilde{i}_{mag} \tag{6.63}$$

the small-signal model can be derived and expressed as

$$\frac{d\tilde{v}_{dc}}{dt} = -\frac{2P_{dc} - V_{mag}I_{mag}}{2C_{dc}V_{dc}^2}\tilde{v}_{dc} - \frac{V_{mag}}{2C_{dc}V_{dc}}\tilde{i}_{mag} \tag{6.64}$$

According to the discussion in Section 5.3.1, the difference between the values of $2P_{dc}$ and $V_{mag}I_{mag}$ can be considered to be zero in the steady state. Therefore, the small-signal model representing the variation of the DC-link voltage in response to an amplitude

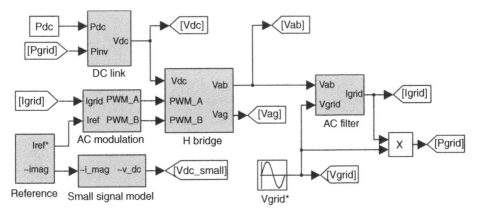

Figure 6.12 Simulink model for evaluating the small-signal model.

change of the grid injection current is expressed by

$$\frac{d\tilde{v}_{dc}}{dt} = -\frac{V_{mag}}{2C_{dc}V_{dc}}\tilde{i}_{mag} \tag{6.65}$$

The model shows the integral characteristics of the DC-link voltage in response to any small variation of the extracting current from its steady state. The static gain is a negative value, given by $-V_{mag}/(2C_{dc}V_{dc})$. This shows that any shift from equilibrium can lead the DC-link voltage to deviate at a rate that corresponds to the static gain of the small-signal model.

A case study is carried out to verify the modeling process. This is based on the system discussed in Section 5.4.1 and described in Table 5.5. In the steady state, the current values though the DC link are computed as $I_{dc} = 7.59$ A and $I_{mag} = 18.53$ A, as illustrated in Figure 5.53. The DC-link capacitance, C_{dc}, is rated as 24 mF. The Simulink models, including the DC link and the GSC, are shown in Figure 6.12. The blocks for the H bridge, AC modulation, AC filter, and DC link were introduced in Chapter 5 and illustrated in Figures 5.50, 5.52, 5.60, and 5.46, respectively. The small-signal model follows the expression in (6.65) and is modeled as

$$\frac{\tilde{v}_{dc}(s)}{\tilde{i}_{mag}(s)} = -\frac{16.95}{s} \tag{6.66}$$

Figure 6.13 shows the simulation result when a perturbation of 0.38 A is added to the control input, equivalent to 5% of the steady-state value, I_{mag}. The step change is implemented in the reference block. The increasing variation is noticeable in the waveform of i_{grid} at 60 ms. The DC-link voltage is expected to drop in response to the imbalance between the injection and extraction. The DC-link voltage response can be evaluated by comparing the outputs of the small-signal model and the large-signal simulation. When the waveforms are plotted together, they match in terms of the slope. The output of the mathematical model, indicated as $v_{dc-small}$, does not show any information about the double-line frequency ripples. The model captures the critical dynamics of the integral behavior. The Simulink model shows all the details, including the switching ripples and the double-line frequency ripple that appears in the waveform of v_{dc}. The small-signal

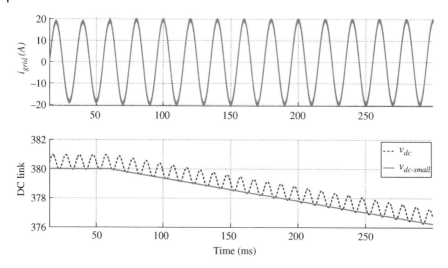

Figure 6.13 Dynamics comparison of the simulation model and the small-signal model.

model is developed at the steady-state equilibrium, and the model output deviates when the operation is away from the predefined steady state.

6.5.2 Three-phase Connection

A similar analysis can be applied to three-phase DC/AC conversion in connection with the DC link. In the steady state, the equilibrium of the current value is expressed as

$$P_{dc} - \frac{3V_{LN}I_{mag}}{\sqrt{2}} = 0 \tag{6.67}$$

Assuming the injected DC current is constant, when a small perturbation, $\tilde{i}_{mag}$ is introduced, the variation of the DC-link voltage in the small-signal is

$$C_{dc}\frac{d(v_{dc} + \tilde{v}_{dc})}{dt} = I_{dc} - \frac{3V_{LN}(I_{mag} + \tilde{i}_{mag})}{\sqrt{2}V_{dc}} \tag{6.68}$$

Linearization leads to the small-signal model:

$$\frac{\tilde{v}_{dc}(s)}{\tilde{i}_{mag}(s)} = \left(-\frac{3V_{LN}}{\sqrt{2}V_{dc}C_{dc}}\right)\frac{1}{s} \tag{6.69}$$

A case study is carried out to verify the modeling process, based on the system discussed in Section 5.3.2 and specified in Table 5.4. In the steady state, the current values though the DC link are computed as $I_{dc} = 5.15$ A and $I_{mag} = 6.18$ A, as illustrated in Figure 5.63. The DC-link capacitance, C_{dc}, is 2.7 mF. The Simulink models can be built with blocks for the CSI, AC modulation, AC filter, and DC link, as introduced in Section 5.3.2. The small-signal model follows the expression in (6.69) and is identified as

$$\frac{\tilde{v}_{dc}(s)}{\tilde{i}_{mag}(s)} = -\frac{305.13}{s} \tag{6.70}$$

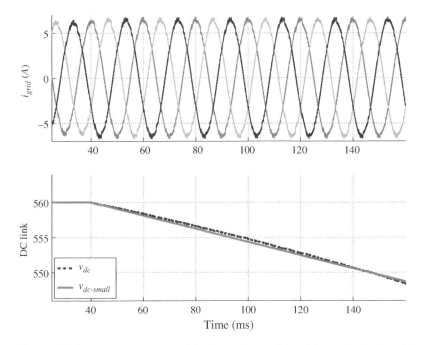

Figure 6.14 Dynamics comparison of the simulation model and the small-signal model.

Figure 6.14 shows the simulation results when a perturbation of 0.31 A is added to the control input, equivalent to 5% of the steady state value, I_{mag}. The increasing variation is noticeable in the waveform of i_{grid} at 40 ms. The DC-link voltage response can be evaluated by comparing the output of the small-signal model with the large-signal simulation. The DC-link voltage is expected to drop in response to the imbalance of injection and extraction. When the waveforms are plotted together, they match in terms of slope of the signals, $v_{dc-small}$ and v_{dc}. The output of the mathematical model captures the critical dynamics regarding the integration of the current deviation. Furthermore, the waveform, v_{dc}, does not show as much ripple as single-phase conversion, even though the DC-link capacitance is nine times lower. This shows that the balanced operation of three phases is very effective at minimizing the use of the DC-link capacitance. Since the small-signal model is developed at the steady state point, a model output deviation appears when the operation is away from the predefined steady state.

6.6 Summary

In this chapter, the mathematical models were developed to represent the dynamics of the PV link and the DC link. It starts with the state-space averaging approach to remove the nonlinear effect of on/off switching that is found in most power converters. Standard linearization is therefore introduced to develop a piecewise linear model – the small-signal model – in order to represent the dynamic response at a certain steady state. The nonlinearity in the PV output characteristics is also linearized and represented by either the dynamic conductance or resistance. According to the analysis, the increasing magnitude of the dynamic resistance, R_{pv}, reduces the value of the damping factor ξ in the mathematical model. The system becomes lightly damped, which can cause oscillation or even instability when the operating point enters the current source region since the value of R_{pv} becomes high. Therefore, a constraint is required to maintain the operating point within the zones of the power and voltage source.

For dynamic modeling of the PV link, the analysis and development are based on the case study with buck, boost, full-bridge transformer isolated, buck–boost, and flyback converters used for the PVSCs. Besides the value of R_{pv}, the ratio of inductance, L, and the capacitance, C_{in} influence the damping performance. The design should avoid very lightly damped systems. The values of L and C_{in} represent the undamped natural frequency, which is one indicator of the dynamic speed. This becomes a tradeoff because the fast response requires low values of of L and C_{in}, but high values are preferred for low ripple levels in voltage and current. The modeling process and verification are also demonstrated, based on the design examples from Section 5.1. The modeling process is important in synthesizing controllers for PV-link voltage regulation. The simulation plots verify the analysis and dynamic expectations.

In the case of battery power interfaces, the dynamic model is explored for dual active bridge topology. It is shown that the topology is suitable for use as the power interface for DC microgrids or systems with DC links. As far as the voltage dynamics of the DC link are concerned, the system exhibits a first-order transfer function according to the small-signal model.

In grid-connected PV systems, the voltage variation of the DC link responds to the interaction between the injected current and extracted current, corresponding to the capacitance of the DC link. The DC-link voltage dynamics exhibit integral behavior, which is expressed in the small-signal model. The output characteristics are proved by plotting the model output against the large-signal simulation result.

In summary, it is an important step to verify the developed mathematical model by comparing the output to the large-signal simulation. It is important for the small-signal model to capture the key system dynamics, but it can ignore switching ripples in both voltage and current signals.

Problems

6.1 Duplicate the simulation result in the design cases in this chapter. It is important to be familiar with the principles of state-space averaging, linearization, dynamic analysis, and verification of the small-signal model. The design examples also cover various PV-side converters and show the dynamics at the DC link for interconnection with single-phase and three-phase grids.

6.2 Based on the system design of the PVSCs in the problem section of Chapter 5:
a) Derive the mathematical models using the techniques of state-space averaging and linearization.
b) Verify the derived mathematical models with the corresponding simulation models by plotting together all the important waveforms.

References

Middlebrook R and Cuk S 1976 A general unified approach to modelling switching-converter power stages *Power Electronics Specialists Conference, 1976 IEEE*, pp. 18–34.

Syed I, Xiao W and Zhang P 2015 Modeling and affine parameterization for dual active bridge DC-DC converters. *Electric Power Components and Systems* **43**(6), 665–673.

Xiao W and Zhang P 2013 Photovoltaic voltage regulation by affine parameterization. *International Journal of Green Energy* **10**(3), 302–320.

Xiao W, Dunford WG, Palmer PR and Capel A 2007 Regulation of photovoltaic voltage. *Industrial Electronics, IEEE Transactions on* **54**(3), 1365–1374.

7

Voltage Regulation

As discussed in Chapter 5, AC current should be regulated to satisfy the power quality standards required for grid interconnection. Voltage regulation is also important in PV power systems for maximum power point tracking (MPPT) and to balance the power flow between the generation and grid injection (El Moursi et al. 2013). This chapter focuses on the voltage regulation through control loops for grid-connected PV systems.

By studying the I–V characteristics of PV cell output, the voltage and the current at MPP can represent the highest power level. Thus MPPT can be achieved by regulating either the PV output voltage or current to the optimal value in order to produce the highest power. However, the MPP refers only to the steady state at STC. The location of the MPP always changes, affected by the level of solar irradiance and the cell temperature. The question becomes: "Which is more suitable to be the controlling variable for MPPT: voltage or current?"

Xiao et al. (2007) show that the regulation of the PV terminal voltage is better than the current regulation. The irradiance can change suddenly, as a result of broken clouds or other unpredictable conditions, as illustrated in Figure 1.7. This results in the PV output current following a wide operating range in order to track the MPP. However, Figure 4.8 shows that changing radiation levels only slightly affect the voltage at the MPP. Even though the variation of cell temperatures shifts the voltage at MPP significantly, the dynamics is generally slow and predictable compared to the effect of changes of irradiance. With changes of solar irradiance, the PV output voltage at the MPP, which is commonly 70–82% of the open-circuit voltage, is more predictable than the output current, which varies over a significant range (Xiao et al. 2007). This is another advantage of using the voltage to represent the MPP in PV power systems. The optimal operating point becomes predictable and constrained when the PV output voltage is regulated. Therefore, it is voltage regulation at the PV link that is discussed in the following sections. The discussion is based on the dynamic models developed in Section 6.3.

7.1 Structure of Voltage Regulation in Grid-connected PV Systems

In the two-stage interfacing topology, as shown in Figure 7.1, two voltage-regulation loops are present. The PV-side converter and grid-side converter are indicated by PVSC and GSC in the diagram. The first control loop regulates the PV terminal voltage to follow the MPP as determined by the MPPT algorithm. The MPP can be represented by

Photovoltaic Power System: Modeling, Design, and Control, First Edition. Weidong Xiao.
© 2017 John Wiley & Sons Ltd. Published 2017 by John Wiley & Sons Ltd.
Companion Website: www.wiley.com/go/xiao/pvpower

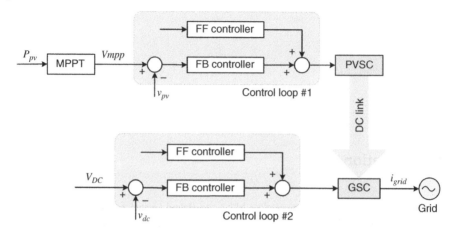

Figure 7.1 Control diagrams of two-stage conversion system for grid interconnection. FF, feedforward; FB, feedback.

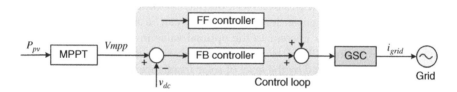

Figure 7.2 Control diagrams of single-stage conversion system for grid interconnection.

the voltage at the PV link, V_{mpp}. The second control loop regulates the DC-link voltage (v_{dc}) and determines the amount of active power to be injected into the grid. The power balance between the PV power generation and the active power extraction varies the DC-link voltage, v_{dc}.

In the single-stage interface, as shown in Figure 7.2, the DC link is merged with the PV link, the voltage of which should be regulated to follow the MPP. In both configurations, the injected current, i_{grid}, should be regulated to ensure high power quality at the point of grid connection.

FB and FF in the control diagrams denote feedback and feedforward, respectively. The FB controller is fundamental to form a feedback control loop with an error-correction function. The feedforward controller is commonly considered an additional function. The two-degree-of-freedom approach of the feedforward path provides flexibility and therefore improves control performance.

In this chapter, the linear control techniques that can be applied for voltage regulation are introduced. Analysis for stability and robustness is discussed in the following section. Design and implementation of voltage regulation for the PV and DC links are described, based on the converter topologies that were discussed in Chapter 5 and the corresponding mathematical models in Chapter 6. Digital control technologies for practical implementation are introduced at the end of the chapter.

7.2 Affine Parameterization

Affine parameterization is an approach to designing linear controllers, sometimes referred to as Q-parameterization or Youla parameterization. The approach was proposed by Youla et al. (1976). The design starts with an open-loop analysis that targets the proper balance of internal stability and control performance. A transformation is applied to derive the controller for a closed-loop implementation. This has several advantages over conventional controller syntheses in terms of ease of understanding, ease of use, and straightforward stability and performance analysis (Xiao and Zhang 2013). The design approach is outlined in Figure 7.3, which shows the important transition from the open-loop design to the closed-loop implementation.

Stability is defined as bounded-input, bounded-output (BIBO) in control systems. The transfer function, G_0, represents the nominal model of the controlled system. The open-loop configuration has a series form, which is straightforward for stability analysis. In the series form, a transfer function Q(s) is introduced:

$$\frac{Y(s)}{R(s)} = Q(s)G_0(s) \tag{7.1}$$

If both Q(s) and $G_0(s)$ are BIBO, the overall system is internally stable (Goodwin et al. 2001). However, the series form does not correct if the output deviates from the reference signal, R. Deviations can be caused by factors such as model uncertainty and disturbance.

As shown in Figure 7.3, the closed-loop implementation indicates the correction that will reduce or eliminate the error between the reference, R, and the plant output, Y. The transfer function of the closed-loop system is

$$\frac{Y(s)}{R(s)} = \frac{C(s)G_0(s)}{1 + C(s)G_0(s)} \tag{7.2}$$

which has a rational form. The stability analysis is not as straightforward as for the series form in (7.1). Special tools, such as Bode and Nyquist plots, are commonly required for the stability and robustness analysis. The affine parameterization has advantages over both representations. The system is specified by the open-loop transfer functions using Q(s). The implementation is based on a closed loop, which provides corrections

Figure 7.3 Illustration of Q-parameterization.

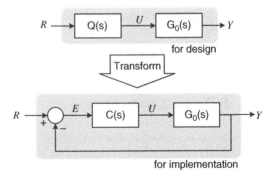

for design

Transform

for implementation

against output deviations. The transforming function between Q(s) and C(s) can be derived according to the equivalence between (7.1) and (7.2) and is expressed as

$$C(s) = \frac{Q(s)}{1 - Q(s)G_0(s)} \tag{7.3}$$

For affine parameterization, the controller synthesis follows the following procedure:

1) A closed-loop transfer function $F_Q(s)$, is defined for the performance expected with consideration of the dynamic characteristics of $G_0(s)$.
2) The function Q(s) can be derived from

$$Q(s) = F_Q(s)G_0(s)^{-1} \tag{7.4}$$

   where $G_0(s)^{-1}$ is the inverse form of $G_0(s)$.
3) The function C(s) is derived from (7.3). The format of C(s) does not follow any fixed structure, but can be constructed from the design of $F_Q(s)$ and the plant model of $G_0(s)$. The feasibility of the derived C(s) should be studied and verified before practical implementation.
4) The function $C(s)G_0(s)$ can be analyzed by Bode or Nyquist plots to illustrate the relative stability in terms of phase margin, gain margin, and sensitivity peak. The model uncertainty and non-ideal factors, such as predictable time delays, can be taken into account in this step.
5) If the system robustness and relative stability are sufficient, the controller function, C(s), is confirmed and ready for the closed-loop implementation. If not, the design should go back to the first step to revise the expected closed-loop function.

Constraints should be applied for the Q-parameterization as follows:

- The nominal transfer function of $G_0(s)$ should be stable and invertible. If it includes the non-minimal phase zero, it is not invertible. A pre-process should be applied to separate the non-minimal phase sector from the minimal phase sector.
- The order of $F_Q(s)$ should be specified by applying the transformation in (7.4), so that the function Q(s) is always proper.
- A feasibility study should be performed to evaluate if C(s) can be physically implemented.
- If the system performance is not satisfactory, the controller should be retuned to guarantee robustness and stability. Model uncertainty should always be considered for practical system designs. It should be noted that all small-signal models represent only a specific operating condition. The system operation is generally wider than the specified condition.
- If the system performance cannot be satisfied by controller tuning, the process should restart from the beginning.

7.3 PID-type Controllers

The majority of control systems in industry are based on the PID-type controllers. The name of PID refers to the fixed controller structure, which is formed from proportional, integral, and derivative terms. A standard PID controller can be written in parallel form:

$$C_{PID}(s) = \frac{U(s)}{E(s)} = K_P + \frac{K_I}{s} + \frac{K_D s}{\tau_d s + 1} \tag{7.5}$$

Table 7.1 Variation of PID-based controllers.

Type	Equation	Description
P	$C_P = K_P$	P-type controller is simple and effective for design and implementation. High gain of K_P can minimize but cannot eliminate steady-state errors.
I	$C_I = \frac{K_I}{s}$	I-type controller introduces 90° phase lag, which causes a slow process. It might be used in noisy environment.
PI	$C_{PI} = K_P + \frac{K_I}{s}$	PI controller is very popular in industry since it can completely eliminate steady-state errors and satisfy most design requirements.
PD	$C_{PD} = K_P + \frac{K_D s}{\tau_d s + 1}$	PD controllers provide a phase lead that can be used to improve damping. It cannot eliminate steady-state error.
PID	$C_{PID} = K_P + \frac{K_I}{s} + \frac{K_D s}{\tau_d s + 1}$	A PID controller is the complete solution, including all terms. Tuning of PID controllers is difficult due to the need to determine four parameters.

where $U(s)$ and $E(s)$ are the output and input of the PID controller respectively. The gains of the proportional, integral, and derivative terms are denoted as K_P, K_I, and K_D respectively. A first-order filter with the parameter τ_d is usually applied to the derivative term since the derivation is very sensitive to high-frequency noise. Table 7.1 outlines the variants of PID-type controllers, including P, I, PI, and PD types.

Figure 7.4 illustrates the dynamics of P, I, and D terms in response to the step change of the error signal, $e(t)$. This is based on the example controller:

$$C_{PID}(s) = \frac{U(s)}{E(s)} = 1.5 + \frac{4}{s} + \frac{0.2s}{0.02s + 1} \tag{7.6}$$

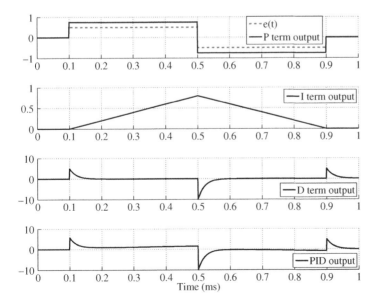

Figure 7.4 Actions of P, I, and D terms in the time domain.

The proportional term output is always proportional to the instantaneous error signal. The contribution from the integral term is proportional to both the magnitude of the error and its duration. The accumulation of the instantaneous error continues until the residual steady-state error is eliminated. Therefore, the integral term is important to eliminate the steady-state error in a control loop. The D term acts only in the transient time of $e(t)$. Based on the direction of the error, it produces an instantaneous control action to reduce the error.

Among the various PID-type controllers, PI is widely used since it has a relatively simple format and the steady-state error can be removed by the integral term. The P type is occasionally used because of its simplicity. It should be noted that the PD controller is also useful for practical applications. The PD controller, as shown in Table 7.1, is commonly coupled with a first-order low-pass filter for practical implementations. The transfer function can be derived to the new format:

$$C_{PD} = K_P \frac{(\tau_d + K_D/K_P)s + 1}{\tau_d s + 1} \tag{7.7}$$

Since the signs of K_P and K_D are the same, the value of $(\tau_d + K_D/K_P)$ is higher than that of τ_d, which is positive. Therefore, the PD controller is equivalent to a lead compensator, which always shows a positive phase angle. The Bode plot in Figure 7.5 shows the characteristics of a PD controller in the frequency domain for an example with $K_P = 4$, $K_D = 0.5$, and $\tau_d = 0.1$. When a PD controller is applied in a closed loop, the phase-lead feature can be used to improve the phase margin within a certain frequency region. Damping can also be improved when a PD controller is adopted. However, similar to

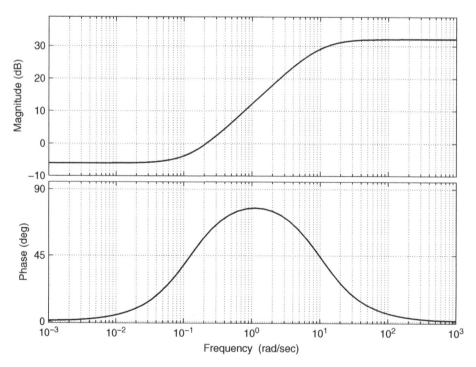

Figure 7.5 Bode diagram of typical PD controller.

P-type controllers, the drawback lies in the steady-state error, which cannot be completely removed.

7.4 Desired Performance in Closed Loop

The desired closed-loop transfer function $F_Q(s)$ should be specified according to the affine parameterization. For most control systems, $F_Q(s)$ can be expected to be either first-order or second-order form, as expressed in (7.8) and (7.9), respectively.

$$F_Q(s) = \frac{1}{\alpha s + 1} \tag{7.8}$$

$$F_Q(s) = \frac{\omega_{cl}^2}{s^2 + 2\xi_{cl}\omega_{cl}s + \omega_{cl}^2} \tag{7.9}$$

In the first-order transfer function, one parameter represents the system dynamics, which is the time constant, α. The settling time of the step response can be estimated to be 4α.

The second-order system can be identified from both the damping rate and the response speed. The damping factor in the expected closed-loop function (7.9) is shown as ξ_{cl}, which is the performance index corresponding to the oscillation scale in a step response. In the time domain, the oscillating level can also be measured by the percentage of overshoot (PO). The samples of the correlation between the PO and the damping factor are shown in Table 7.2. The damping factor, ξ_{cl}, is usually specified from 0.6 to 1 for closed-loop performance because these values balance the performance between the PO and the response speed.

The undamped natural frequency in the expected closed-loop function (7.9) is shown as ω_{cl}, which is the measurement of the oscillation frequency and an indicator of the response speed. The settling time of the step response can be estimated by (7.10) for the second-order transfer function in (7.9). Therefore, the closed-loop performance can be specified by its PO and settling time. The value of ξ_{cl} can be assigned from the PO specifications in Table 7.2. Meanwhile, the value of ω_{cl} can be determined from the expectation of the settling time according to

$$T_S \approx \frac{4}{\xi_{cl}\omega_{cl}} \tag{7.10}$$

When the expected closed-loop performance is defined by (7.9), the transfer function can be expressed in another equivalent form:

$$F_Q(s) = \frac{1}{\alpha_2 s^2 + \alpha_1 s + 1} \tag{7.11}$$

Table 7.2 Percentage of overshoot for damping factors in second-order systems.

Damping factor ξ_{cl}	0	0.1	0.2	0.3	0.4	0.5	0.6	0.7	0.8	0.9	1.0
PO (%)	100	72.9	52.9	37.2	25.4	16.3	9.5	4.6	1.5	0.2	0.0

where the parameters should be calculated from (7.12) to be equivalent to (7.9).

$$\alpha_1 = \frac{2\xi_{cl}}{\omega_{cl}} \tag{7.12a}$$

$$\alpha_2 = \frac{1}{\omega_{cl}^2} \tag{7.12b}$$

In affine parameterization, the selection of first or second order should be based on the relative model degree in the plant transfer function, $G_0(s)$, in order to make Q(s) proper. The relative order is evaluated by the difference between the number of poles and non-minimal-phase zeros. When the relative model degree of $G_0(s)$ is two, the closed-loop transfer function is selected as second order in the format of (7.9). When the relative model degree of $G_0(s)$ is one, the closed-loop transfer function is selected as first order in the format of (7.8). Both are shown in examples in the following sections.

7.5 Relative Stability

Absolute stability is defined as bounded-input, bounded-output (BIBO), which is essential for all control systems. As expressed in (7.1), the system is always BIBO stable if both Q(s) and $G_0(s)$ are stable transfer functions. The closed-loop stability can be proved since the implementation is transformed from the Q-parameterization, which is illustrated in Figure 7.3.

When the controller is derived, and meanwhile, the absolute stability is no longer relevant, it becomes important to evaluate the relative stability in terms of phase margin and gain margin. These criteria measure how much the closed-loop system is away from either instability or self-oscillation. The margins are typically measured by either a Bode diagram or a Nyquist plot of $C(s)G_0(s)$ in the frequency domain.

As an example, the transfer function of $G_0(s)$ is derived as (7.13) and the controller of $C(s)$ is developed as (7.14). The Bode diagram of $C(s)G_0(s)$ can be plotted to illustrate the phase margin and gain margin. The critical stability point refers to the conditions $|C(j\omega G_0(j\omega)| = 1$ and $\angle C(j\omega)G_0(j\omega) = -180°$, and reflects the continuous oscillation without damping that is the boundary between stability and instability.

$$G_0(s) = \frac{2000000}{s^2 + 1000s + 2000000} \tag{7.13}$$

$$C(s) = 0.1 + \frac{400}{s} \tag{7.14}$$

Based on the example expressed in (7.13) and (7.14), the Bode diagram can be plotted, as shown in Figure 7.6. At the $-180°$ phase crossing, the gain of $|C(j\omega_{cg}G(j\omega_{cg}|$ is measured as 0.3 where $\omega_{cg} = 1.6 \times 10^3$. In the Bode diagram, the gain margin in decibels is derived from (7.15). The value of G_{m-dB} is 10.5 dB, as shown in Figure 7.6.

$$G_{m-dB} = 20log_{10}\left(\frac{1}{|C(j\omega_{cg}G(j\omega_{cg}|}\right) \tag{7.15}$$

The phase margin is measured when the loop gain of $|C(j\omega_c)G_0(j\omega_c)|$ is unity, which is 0 dB in decimal representation. The angular difference from $-180°$ is the phase margin. In the example system, the phase lag is 97.2° at a frequency of 431.7 rad/s, and there is a unity gain. Therefore, the phase margin is calculated as 82.8°, as shown in Figure 7.6.

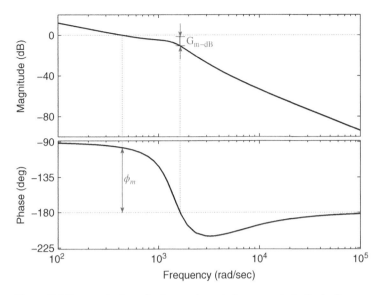

Figure 7.6 Demonstration of gain margin and phase margin by Bode diagram.

Figure 7.7 Demonstration of gain margin and phase margin using Nyquist plot.

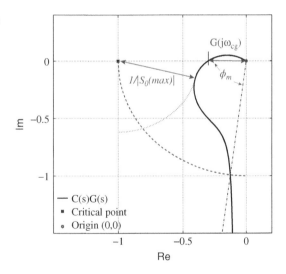

The Nyquist plot is another useful tool to evaluate the relative stability. Figure 7.7 plots $C(s)G_O(s)$ in the frequency domain, for the same model. A critical point is marked at the location $(-1, 0)$, which is the condition of $|C(j\omega G_o(j\omega)| = 1$ and $\angle C(j\omega)G_o(j\omega) = -180°$. At the $-180°$ phase crossing, the gain of $|C(j\omega_{cg}G(j\omega_{cg}|$ is measured as 0.3, corresponding to the gain margin of 10.5 dB in the Bode diagram. The crossing of the unity circuit indicates the condition of $|C(j\omega G_o(j\omega)| = 1$. The phase margin (ϕ_m) can also be measured as 82.8° in the Nyquist plot, as shown in Figure 7.7. The frequency scale is not directly indicated on the standard Nyquist plot, which is a disadvantage for stability margin illustration compared to the Bode diagram.

7.6 Robustness

A closed-loop system with consideration of the input disturbance (D_i), output disturbance (D_o), and measurement noise (D_m) is illustrated as Figure 7.8. The system output including the source of disturbance and measurement noise can be expressed as

$$Y = \frac{C(s)G_0(s)}{1 + C(s)G_0(s)}R - \frac{C(s)G_0(s)}{1 + C(s)G_0(s)}D_m + \frac{G_0(s)}{1 + C(s)G_0(s)}D_i$$
$$+ \frac{1}{1 + C(s)G_0(s)}D_o \tag{7.16}$$

In a closed-loop system, the sensitivity function is defined and expressed as

$$S_0(s) = \frac{1}{1 + C(s)G_0(s)} \tag{7.17}$$

Based on the previous example, the sensitivity function is illustrated by a Bode diagram, as shown in Figure 7.9. The function shows high-pass features: the magnitude is low in the low-frequency band, but is high in the high-frequency band. It is important to limit the magnitude peak of $S_0(s)$ to achieve robust system control. The sensitivity peak is defined as the maximum magnitude in $S_0(s)$:

$$S_0(max) = max[S_0(j\omega)] \tag{7.18}$$

The peak gain is measured as 1.6 and calculated as 4.2 dB, as shown in Figure 7.9. The peak can also be found in the Nyquist plot, as shown in Figure 7.7. The sensitivity function is plotted as a circle tangential to the polar plot. The radius that is measured as 0.62 indicates the reciprocal of the nominal sensitivity peak of 1.6. The radius of the sensitivity circle is expected to be larger for better robustness.

In general, the Nyquist plot can give information about the phase margin, gain margin, and the sensitivity peak, as shown in Figure 7.7. The information can also be illustrated by two Bode diagrams for $C(s)G_0(s)$ and $S_0(s)$, as illustrated in Figures 7.6 and 7.9. It is important to evaluate the robustness of the control loop before controller implementation. As a rule of thumb, the phase margin is expected to be more than 60°. The magnitude of the sensitivity peak should be lower than 2 to guarantee robust operation.

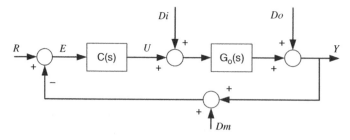

Figure 7.8 Demonstration of gain margin and phase margin by Nyquist plot.

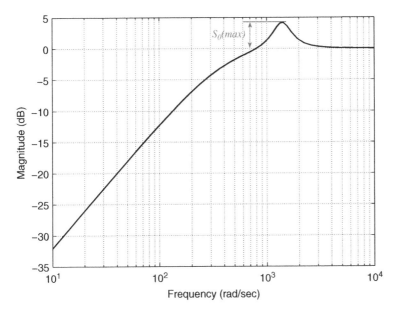

Figure 7.9 Demonstration of sensitivity peak by Bode diagram.

7.7 Feedforward Control

Feedforward control can be deployed to improve system performance when the system model and the behavior of the controlled variable are well understood. It is important to understand the roles of feedback and feedforward in a typical control system.

Table 7.3 gives a side-by-side comparison of the feedback and feedforward control approaches. The feedforward term is considered as the supplement of a feedback loop that aims for either improved command following or disturbance rejection. A typical hybrid control system including both a feedback control loop and a reference

Table 7.3 Feedback versus feedforward.

Feedback control approach	Feedforward control approach
Feedback implementation is based on closed-loop structure.	Feedforward implementation is based on open-loop structure in a direct form.
Feedback system can be designed without sufficient knowledge of the plant model.	The stability evaluation is simple due to the direct form and sufficient system knowledge.
Self-correction is always available in closed-loop systems to minimize the error between command signals and controlled variables.	Due to lack of the self-correction, feedforward is seldom used in an independent manner.
Closed-loop stability and robustness should be always evaluated in the system design and synthesis.	System design requires system model and good knowledge of the system.

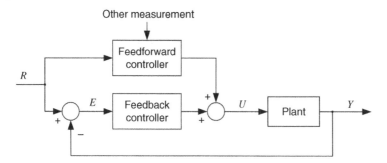

Figure 7.10 Hybrid control system using both feedback and feedforward controllers.

feedforward term is shown in Figure 7.10. The closed loop is designed to compensate for errors in the system model or expected disturbances. The feedforward component can make the system reach the steady state rapidly and reject disturbances effectively. Additional measurements might be necessary to apply feedforward control effectively.

For most PV-side converters and grid-side converters in PV power systems, the power interface is regulated by the switching duty cycle. The switching duty cycle can be estimated from the voltage-conversion ratio in the steady state. The reference for the feedforward term can be determined from the expected duty cycle based on the measurement of the input and output voltages.

7.8 Voltage Regulation in PV Links

Small-signal models for various PV-link converters were derived in Section 6.3. Six transfer functions were derived and set out in (6.23), (6.29), (6.38), (6.45), (6.51), and (6.54). They showed negative gains in the transfer functions, which indicates that the PV-link voltage runs in the opposite direction to the control variable. Therefore, the six models can be classified into two general formats:

$$G_0(s) = \frac{K_0}{s^2 + 2\xi\omega_n s + \omega_n^2} \tag{7.19}$$

$$G_0(s) = \frac{K_0(\beta s + 1)}{s^2 + 2\xi\omega_n s + \omega_n^2} \tag{7.20}$$

7.8.1 Boost Converter for PV Links

When the boost converter is used as the PVSC, the modeling process is as described in Section 6.3.4. An example model was derived and shown in (6.38), having a general second-order format, the same as (7.19). Following affine parameterization, the desired closed-loop transfer function can be defined as in (7.9). The function Q(s) can be derived from (7.4) and expressed as:

$$Q(s) = \frac{s^2 + 2\xi\omega_n s + \omega_n^2}{K_0(\alpha_2 s^2 + \alpha_1 s + 1)} \tag{7.21}$$

The function Q(s) is stable, since the poles are the same as those in the predefined transfer function, $F_Q(s)$. The feedback controller can be derived from (7.3) as

$$C(s) = \frac{s^2 + 2\xi\omega_n s + \omega_n^2}{K_0 s(\alpha_2 s + \alpha_1)} \tag{7.22}$$

The controller shown in (7.22) can be expressed in PID format as

$$C(s) = K_p + \frac{K_i}{s} + \frac{K_d}{\tau_d s + 1} \tag{7.23}$$

where, the PID parameters can be derived from (7.22) as

$$\tau_d = \frac{\alpha_2}{\alpha_1} \tag{7.24a}$$

$$K_i = \frac{\omega_n^2}{K_0 \alpha_1} \tag{7.24b}$$

$$K_p = \frac{2\xi\omega_n \alpha_1 - \omega_n^2 \alpha_2}{K_0 \alpha_1^2} \tag{7.24c}$$

$$K_d = \frac{\alpha_1^2 - 2\xi\omega_n \alpha_1 \alpha_2 + \omega_n^2 \alpha_2^2}{K_0 \alpha_1^3} \tag{7.24d}$$

In the example discussed in Sections 5.1.4 and 6.3.4, the small-signal model was derived, as shown in (6.38). Referring to the normalized form in (7.20), the parameters can be derived as $\xi = 0.39$, $\omega_n = 21720\,\text{rad/s}$, and $K_0 = -2.2645 \times 10^{10}$.

To improve the closed-loop performance over the original setup, the desired damping factor in the closed-loop transfer function is chosen as 0.7. The desired undamped natural frequency is assigned to be four times the value in ω_n, which is calculated as $86\,881\,\text{rad/s}$. Following the transfer function in (7.12), the desired closed-loop transfer function is specified as $\alpha_2 = 1.324 \times 10^{-10}$ and $\alpha_1 = 1.6114 \times 10^{-5}$. Following (7.22), the controller can be parameterized as

$$C(s) = -\frac{s^2 + 16760s + 471768583}{s(3s + 364900)} \tag{7.25}$$

which can be transformed to PID format as

$$C(s) = -0.0353 - \frac{1293}{s} - \frac{2.4502 \times 10^{-6}}{8.2214 \times 10^{-6} s + 1} s \tag{7.26}$$

where the PID parameters are $K_p = -0.0353$, $K_i = -1293$, $K_d = -2.4502 \times 10^{-6}$, and $\tau_d = 8.2214 \times 10^{-6}$. It should be noted that the signs of the controller constants K_p, K_i, and K_d should be evaluated when the controller is converted to PID format. The negative gains of the controller correspond to the negative value of the static gain in the plant model, which refers to the drop of the PV-link voltage when the duty cycle is increased.

After design, the relative stability and system robustness should be evaluated. From the controller expressed in (7.23) and the plant transfer function (6.38), which was derived in Chapter 6, the gain margin can be measured as ∞, the phase margin as $65.2°$, and the sensitivity peak as 1.27 or 2.11 dB. The information can be illustrated in a

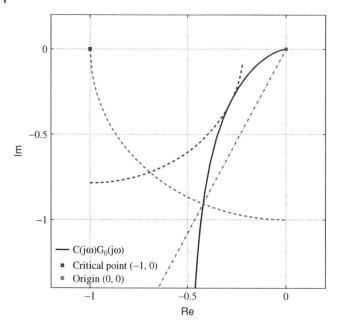

Figure 7.11 Demonstration of gain margin, phase margin, and sensitivity peak by Nyquist plot.

Nyquist plot, as shown in Figure 7.11. The values of the stability margins and sensitivity peak ensure robust voltage regulation.

It should be noted that the above stability analysis is only valid at the nominal operating condition. The controller is developed according to the small-signal model, which is linearized from a nonlinear model. To guarantee stable and robust control, a comprehensive study accommodating more potential operating conditions and representing more environmental variations can always be performed.

The closed-loop performance can be evaluated in the time domain using the simulation model that was developed in Section 5.1.4. The voltage reference is changed from 37.0 V to 37.5 V periodically, as shown in Figure 7.12. The simulated waveform shows the signals of v_{pv} and p_{pv}, and the switching duty cycle, which is the control variable. The result shows that the PV terminal voltage follows the reference variation. The PV power output deviates from the MPP (288.3 W) when the PV voltage is regulated away from the optimal operating voltage, which is represented by $V_{MPP} = 37.0$ V. Ripples appears in the waveforms because the simulation is based on the switching model that was introduced in Section 5.1.4.

The closed-loop performance can also be compared to the open-loop response without the feedback controller. The open-loop response in response to the step change in duty cycle is shown in Figure 6.6. Due to the light damping, the step change causes significant overshoots and oscillation. The settling time is estimated to be 0.4 ms. The percentage overshoot is 26.9%. With the feedback controller, the settling time is reduced to 70 μs. The overshoot is about 4.7%, as illustrated in Figure 7.12. The simulation proves the effectiveness of the controller design and the closed-loop implementation in improving the dynamic response.

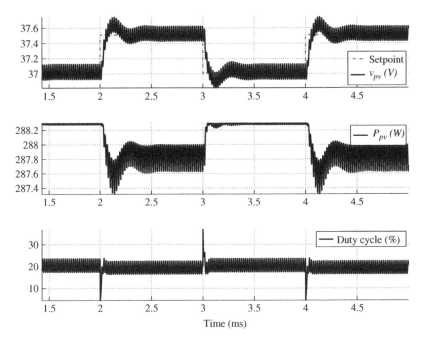

Figure 7.12 Waveform of PV voltage regulation using boost converter.

7.8.2 Tapped-inductor Topology for PV Links

The modeling process when the tapped inductor topology is used is described in Section 6.3.5. The small-signal model was derived and shown in (6.45). The second-order transfer function has the same format as (7.19). The controller synthesis process is the same as for the boost converter since both share the same format of the system model. Using the affine parameterization, the desired closed-loop transfer function can be defined as (7.12). The function Q(s) can be derived from (7.4) and expressed as (7.21).

The example discussed in Section 6.3.5 is used, where the small-signal model is expressed in (6.44) and (6.45). The model can be represented in standard format as (7.19), with the parameters as $\xi = 0.84$, $\omega_n = 9925$ rad/s, and $K_0 = -1.3422 \times 10^{10}$.

Based on the plant dynamics, the desired damping factor in the closed-loop transfer function is chosen as 0.85. The desired undamped natural frequency is assigned to be four times ω_n, which is calculated as 39 702 rad/s. Following the transfer function in (7.12), the desired closed-loop transfer function is specified as $\alpha_2 = 6.344 \times 10^{-10}$ and $\alpha_1 = 4.030 \times 10^{-5}$. Using the function in (7.22), the controller can be designed as

$$C(s) = -\frac{s^2 + 16770s + 98515782}{s(8.515s + 574716)} \tag{7.27}$$

which can be transformed to PID format by the transformation expressed in (7.23) and (7.24). The parameters K_p, K_i, and K_d are negative, as expressed in (7.28), which validates the PID transformation:

$$C(s) = -0.0266 - \frac{171.42}{s} - \frac{1.3452 \times 10^{-6}}{1.4816 \times 10^{-5}s + 1}s \tag{7.28}$$

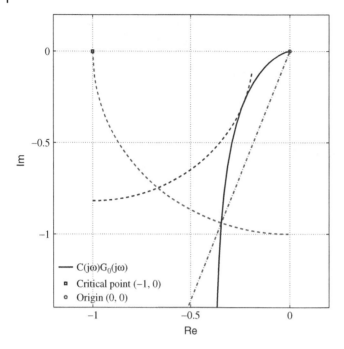

Figure 7.13 Demonstration of gain margin, phase margin, and sensitivity peak by Nyquist plot.

Based on the controller expressed in (7.27) and the plant transfer function (6.45), as derived in Section 6.3.5, the gain margin is measured as ∞, the phase margin as 71.8°, and the sensitivity peak is 1.20 or 1.60 dB. The information can be illustrated in the Nyquist plot, as shown in Figure 7.13. The values of the stability margins and sensitivity peak ensure robust voltage regulation.

The closed-loop performance can also be evaluated in the time-domain simulation using the model developed in Section 5.1.5. The voltage setpoint changes from 37.0 V to 37.5 V periodically, as shown in Figure 7.14. The simulated waveform shows the signals of v_{pv} and p_{pv}, and the switching duty cycle, which is the control variable. The result shows that the PV terminal voltage follows the reference variation. Ripples appear in the signals because the simulation model is based on the switching details. The PV power output deviates from the MPP (288.3 W) when the PV voltage is regulated away from the optimal operating voltage, which is represented by $V_{MPP} = 37.0$ V.

The closed-loop performance can also be compared to the open-loop response when the feedback controller has not been implemented. The open-loop response in response to the step change in duty cycle is shown in Figure 6.7. The settling time is estimated to be 0.4 ms. The percentage overshoot is measured at 0.7%. With the feedback controller, the settling time is reduced to 0.1 ms. The overshoot is measured as 0.6%, as illustrated in Figure 7.14. The simulation proves the effectiveness of the controller design and the closed-loop implementation.

7.8.3 Buck Converter as the PV-link Power Interface

The modeling process when the buck topology is used as the PVSC has been described in Section 6.3.2. An example model is derived and shown in (6.23), which shows two

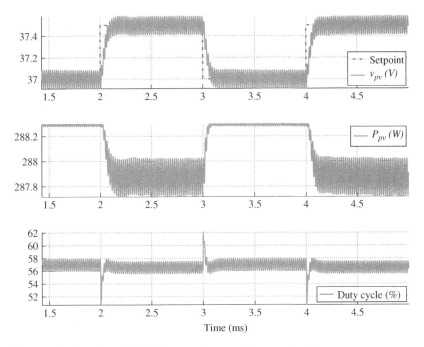

Figure 7.14 Waveform of PV voltage regulation using tapped inductor converter.

stable poles and one minimal phase zero. The format of the transfer function is the same as the general format in (7.20). The relative degree of the model is one because one minimal phase zero is present. Following the affine parameterization, the desired closed-loop transfer function should be defined in order to make the transfer function Q(s) proper. Therefore, the first-order transfer function in (7.8) is adopted. The function Q(s) can be derived using (7.4) and expressed as

$$Q(s) = \frac{s^2 + 2\xi\omega_n s + \omega_n^2}{K_0(\beta s + 1)(\alpha s + 1)} \tag{7.29}$$

The function Q(s) is stable, since both β and α are positive. The feedback controller can be derived from (7.3) as

$$C(s) = \frac{s^2 + 2\xi\omega_n s + \omega_n^2}{K_0 s(\alpha\beta s + \alpha + \beta)} \tag{7.30}$$

The controller shown in (7.30) can also be expressed in the parallel PID format, as shown in (7.23), and the parameters can be derived as

$$\tau_d = \frac{\alpha\beta}{\alpha + \beta} \tag{7.31a}$$

$$K_i = \frac{\omega_n^2}{K_0(\alpha + \beta)} \tag{7.31b}$$

$$K_p = \frac{2\xi\omega_n(\alpha + \beta) - \omega_n^2\alpha\beta}{K_0(\alpha + \beta)^2} \tag{7.31c}$$

$$K_d = \frac{(\alpha + \beta)^2 - 2\xi\omega_n(\alpha + \beta)\alpha\beta + \omega_n^2\alpha^2\beta^2}{K_0(\alpha + \beta)^3} \tag{7.31d}$$

The small-signal model for the example discussed in Sections 5.1.2 and 6.3.2 was derived and shown in (6.23). Referring to the normalized form in (7.20), the parameters can be derived as $\xi = 0.13$, $\omega_n = 3019\,\text{rad/s}$, $\beta = 8.4 \times 10^{-5}$, and $K_0 = 5.2 \times 10^8$.

The value of α can be assigned from

$$\alpha = \frac{1}{N_F \xi \omega_n} \tag{7.32}$$

where the closed-loop dynamics can be expected to be faster than the open-loop of the plant. In this case, the value of N_F is assigned as 8. Therefore, the controller can be derived from (7.30) as

$$C(s) = -\frac{s^2 + 765s + 9114000}{14.33s^2 + 213722s} \tag{7.33}$$

The controller transfer function can be transformed to the parallel PID format as

$$C(s) = -7.2170 \times 10^{-4} - \frac{42.6437}{s} - \frac{4.6306 \times 10^{-6}}{6.7058 \times 10^{-5}s + 1}s \tag{7.34}$$

Under the nominal condition, stability analysis for the phase and gain margins are unnecessary since the nominal transfer function of the closed-loop system is the first-order function in (7.8). It always shows $-90°$ in phase across all frequencies. The gain margin is ∞. For a first-order system, neither oscillation nor overshoot is expected in the step response. However, the derivation of the small-signal model is based on the assumption of one steady-state operating condition. When the operating deviates, the plant model changes, so the closed-loop transfer function is no longer the same as the nominal model. Therefore, oscillation in terms of overshoot or undershoot is also expected from each step change. A comprehensive stability study accommodating more potential operating conditions and representing environmental variations can always be conducted.

The closed-loop performance can be demonstrated in the time-domain simulation using the model developed in Section 5.1.2. The voltage setpoint changes from 37.0 V to 37.5 V periodically, as shown in Figure 7.15. The simulated waveform in Figure 7.15 shows the signals of v_{pv} and p_{pv}, and the switching duty cycle, which is the control variable. The result shows the PV terminal voltage follows the reference variation. The closed-loop performance can also be compared to the open-loop response without the feedback controller. The open-loop response in response to the step change in duty cycle is shown in Figure 6.4. Due to the light damping, the step change shows significant overshoots and oscillation. The settling time is estimated to be 9.6 ms. The percentage of overshoot is 69%. With the feedback controller, the settling time is reduced to 1.7 ms. The step response shows minor over- and undershoots, as illustrated in Figure 7.15, because the simulation model is based on the practical operation of the DC/DC buck converter. The simulation proves the effectiveness of the controller design and the closed-loop implementation in improving the dynamic response.

7.8.4 Buck–boost Converter as the PV-link Power Interface

The buck–boost topology was introduced in Section 5.1.6 and then mathematically modeled in Section 6.3.6. The small-signal model was derived as (6.49), which shows two poles and one minimal phase zero. The format of the transfer function can be

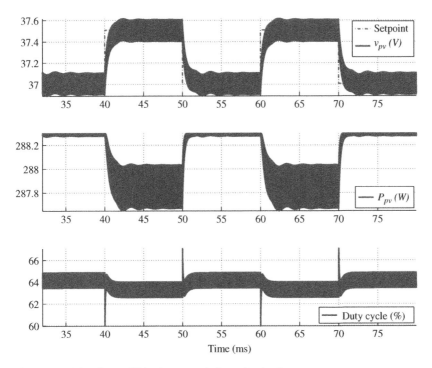

Figure 7.15 Waveform of PV voltage regulation using buck converter.

expressed by the general format in (7.20). According to the affine parameterization, the desired closed-loop transfer function should be defined in the standard form in (7.8) since the relative degree of the plant model is one. The function Q(s) can be derived using (7.4) and expressed as (7.29). The feedback controller can be derived as (7.30) using the transformation in (7.3).

The small-signal model discussed in Section 6.3.6 is expressed in (6.50) and (6.51). Referring to the normalized form in (7.20), the parameters of the plant model can be derived as $\xi = 0.20$, $\omega_n = 1299 \, \text{rad/s}$, $\beta = 1.56 \times 10^{-4}$, and $K_0 = -2.50 \times 10^8$.

The value of α can be assigned from (7.32), where the closed-loop dynamics are expected to be faster than the open-loop of the plant. In this case, the value of N_F is assigned to be 8. Therefore, the controller can be derived from (7.30) as

$$C(s) = -\frac{s^2 + 531s + 1687112}{18.34s^2 + 156625s} \tag{7.35}$$

The controller transfer function can be transformed to the parallel PID format as

$$C(s) = -0.0021 - \frac{10.77}{s} - \frac{6.14 \times 10^{-6}}{1.17 \times 10^{-4}s + 1}s \tag{7.36}$$

It is unnecessary to evaluate the relative stability according to the nominal small-signal model. The transfer function to represent the close-loop system becomes a first-order function, as expressed in (7.8), which is always stable and robust. It should be noted that the condition of stability and robustness is only valid at the nominal operating condition.

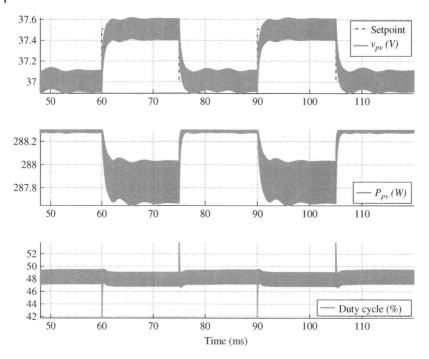

Figure 7.16 Waveform of PV voltage regulation using buck–boost converter.

A comprehensive study of stability and robustness is always recommended to accommodate different potential operating conditions in practical circumstances.

The closed-loop performance can be demonstrated in the time-domain simulation using the model developed in Section 5.1.6. The voltage setpoint changes from 37.0 V to 37.5 V periodically, as shown in Figure 7.16. The simulated waveform shows the signals of v_{pv} and p_{pv}, and the switching duty cycle, which is the control variable. The result shows that the PV terminal voltage follows the reference variation.

The closed-loop performance can also be compared to the open-loop response without the feedback controller. The open-loop response in response to the step change in duty cycle is shown in Figure 6.4. Due to the light damping, the step change shows significant overshoots and oscillation. The improvement of the voltage regulation is noticeable in Figure 7.16 in terms of fast response and reduced overshooting. Minor oscillations are noticeable after each step change, which represents the practical operation of the buck–boost converter because the operating point is away from the condition for which the small-signal model was developed.

7.8.5 Flyback Converter as the PV-link Converter

The flyback topology was introduced in Section 5.1.7 and then mathematically modeled in Section 6.3.7. The small-signal model was derived as (6.52), which shows two poles and one minimal phase zero. The format of the transfer function can be expressed by the general format in (7.20). According to affine parameterization, the desired closed-loop transfer function should be defined as (7.8) since the relative degree of the plant model is one. The function Q(s) can be derived using (7.4) and expressed as (7.29). The feedback controller can be derived as (7.30) by following the transformation in (7.3).

The small-signal model for the example discussed in Section 6.3.7 is expressed in (6.53) and (6.54). Referring to the normalized form in (7.20), the parameters of the plant model can be derived as $\xi = 0.20$, $\omega_n = 1335\,\text{rad/s}$, $\beta = 1.56 \times 10^{-4}$, and $K_0 = -2.64 \times 10^8$.

The value of α can be assigned from (7.32), where the closed-loop dynamics can be expected to be faster than the open-loop of the plant. In this case, the value of N_F is assigned as 8. Therefore, the controller can be derived from (7.30) as

$$C(s) = -\frac{s^2 + 545s + 1780840}{18.84s^2 + 162011s} \tag{7.37}$$

The controller transfer function can be transformed to the parallel PID format as

$$C(s) = -0.0021 - \frac{10.99}{s} - \frac{5.93 \times 10^{-6}}{1.16 \times 10^{-4}s + 1}s \tag{7.38}$$

The closed-loop performance can be demonstrated in the time-domain simulation using the model developed in Section 5.1.7. The voltage setpoint changes from 37.0 V to 37.5 V periodically, as shown in Figure 7.17. The simulated waveform show the signals of v_{pv} and p_{pv}, and the switching duty cycle, which is the control variable. The result shows the PV terminal voltage follows the reference variation.

The closed-loop performance can also be compared to the open-loop response without the feedback controller. The open-loop response in response to the step change in duty cycle is shown in Figure 6.9. Due to the light damping, the step change shows

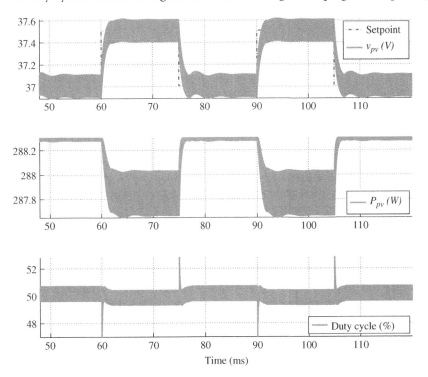

Figure 7.17 Waveform of PV voltage regulation using buck–boost converter.

significant overshoots and oscillation. The improvement of the voltage regulation is noticeable in Figure 7.17 in terms of fast response and reduced overshoot. Minor oscillations appear after each step change, representing the practical operation of the buck–boost converter.

7.9 Bus Voltage Regulation for DC Microgrids

When the dual active bridge (DAB) is used as the battery power interface for a DC microgrid, it is necessary to regulate the DC bus voltage. The variation of the DC bus voltage results from the imbalance between power injection and extraction.

As discussed in Section 6.4, a piecewise-linear model is derived for the system and shows the first-order dynamics, as shown in (6.60). The affine parameterization can be used to synthesize the control loop for voltage regulation. When the desired closed-loop transfer function can be defined by (7.8), the function Q(s) can be derived from (6.60) and expressed as

$$Q(s) = \frac{\tau_0 s + 1}{K_0(\alpha s + 1)} \tag{7.39}$$

Following (7.3), the feedback controller can be derived as (7.40), which is a PI controller.

$$C(s) = \frac{\tau_0 s + 1}{K_0 \alpha s} = \underbrace{\frac{\tau_0}{K_0 \alpha}}_{K_P} + \underbrace{\frac{1}{K_0 \alpha}}_{K_I} \frac{1}{s} \tag{7.40}$$

The small-signal model for the example introduced in Section 5.2 was derived in Section 6.4, with $K_0 = 323$ and $\tau_0 = 193$ μs. To make the closed-loop performance faster than the original plant, the α is assigned a value of one-tenth of the value of τ_0. Following (7.40), the controller can be synthesized as

$$C(s) = 0.031 + \frac{161}{s} \tag{7.41}$$

The Bode and Nyquist plots are unnecessary for the reevaluation since the function $C(s)G_0(s)$ in the nominal condition is assigned to the function $1/\alpha s$, showing that the phase margin is 90° and the gain margin is ∞. As mentioned in the previous sections, additional stability analysis for more operating conditions is always recommended in order to represent the system's nonlinearity.

A test case is built to simulate the voltage regulation performance in response to load changes. The load resistance changes from 193 Ω to 385 Ω and back to 193 Ω. The significant change of the load causes the transition seen in the waveforms. The feedback loop is able to control the DC bus voltage in the steady state at a constant level, 380 V. Since the power is regulated by the phase-shift angle, it varies to a different level in the steady state to respond to the load change, as shown in Figure 7.18. The voltage regulation in a DAB is different from PWM for PV-side converters, where the duty cycle directly links to the steady-state conversion ratio of the input and output voltages. The phase-shift angle in the DAB correlates to the power level instead of the voltage-conversion ratio.

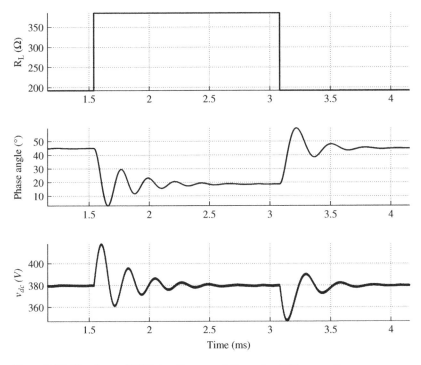

Figure 7.18 Waveform of DC-link voltage regulation using DAB converter.

7.10 DC-link Voltage Regulation for AC Grid Interconnections

In the two-stage conversion system, as shown in Figure 7.1, the DC-link voltage should be regulated to balance the power injection and extraction. The small-signal modeling was developed in Section 6.5 and showed the dynamics of the DC link to have integral characteristics.

The control diagram for the DC-link voltage regulation is shown in Figure 7.19 and includes both the feedback and feedforward controllers, indicated as C_{fb-dc} and C_{ff-dc}, respectively.

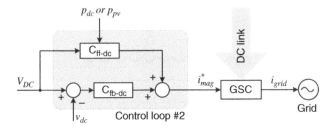

Figure 7.19 Diagram of DC-link voltage regulation with DC/AC grid-connected conversion.

7.10.1 Single-phase Grid Interconnection

As discussed earlier, feedforward control can be powerful when added to the conventional feedback control loop. The setpoint referring to the amplitude of the AC current is directed by the feedforward form, which is constructed with knowledge of the steady-state behavior. The injected power or current to the DC link can be measured directly or estimated from the measurement at the PV link. The information can be used to estimate how much current should be injected into the single-phase AC grid. When the value of P_{dc} is known, the steady-state amplitude of the AC current, I_{mag}, can be computed from (5.91), which was derived in Section 5.3.1. The steady-state equilibrium can be used to implement the direct-form computation for the feedforward control.

Under the ideal condition in the simulation, the feedforward controller alone can regulate the DC-link voltage regardless of the variation of the injected current from the PV power source. Based on the case study discussed in Section 6.5.1, the feedforward controller can be derived from (5.91) and implemented by

$$I^*_{mag} = \left(\frac{2}{V_{mag}} \right) P_{pv} \tag{7.42}$$

where V_{mag} and I_{mag} represent the amplitude of the grid voltage and grid current and the PV power p_{pv} is assumed to be equivalent to the DC power injection to the DC link.

The simulation result is illustrated in Figure 7.20, which indicates a step-down change of the injected DC power, p_{dc} at 365 ms.

The amplitude of the injected AC current, i_{grid}, is reduced by half immediately, which follows the command signal, I^*_{mag}. The command value is determined by the feedforward

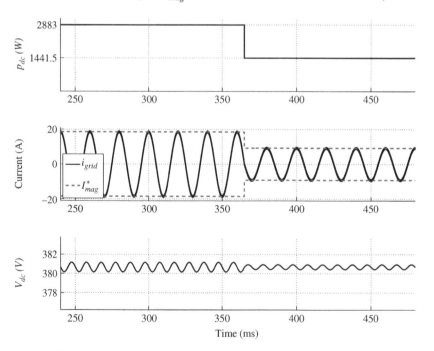

Figure 7.20 Performance of DC voltage regulation for single-phase grid interconnection using only feedforward controller.

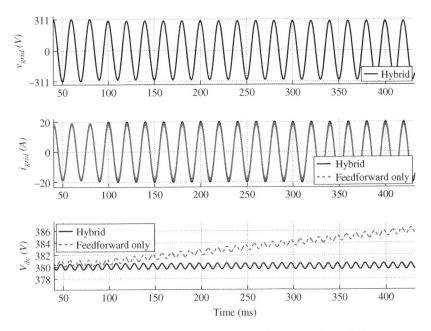

Figure 7.21 Performance of DC-link voltage regulation for single-phase grid interconnection using hybrid controller with both feedforward and feedback.

controller, as expressed in (7.42), in response to the variation. This results in the DC-link voltage being maintained at the same level regardless of the variance. This proves the effectiveness of the feedforward approach for DC-link voltage regulation. However, the mathematical model will never be accurate for a real-world PV power system. Therefore, it is common to combine the feedforward controller with a feedback controller, which formulates a closed loop to correct any error that results from unexpected and non-ideal disturbance and uncertainty.

The controller design for voltage regulation follows the case described in Section 6.5.1. The small-signal model, as expressed in (6.66), has an integral function. The feedback controller can by synthesized by using a P-type controller with $K_p = -10$. The implementation of the hybrid control approach is shown in Figure 7.19.

Figure 7.21 illustrates the effectiveness of the hybrid approach using both feedback and feedforward controls. A disturbance is introduced at 85 ms: the grid voltage dips by 5% to an RMS value of 209 V, which results in a reduction in the power extraction from the DC link, which can be seen in the waveform. Action should be taken to clear the imbalance between the injected and extracted power at the DC link.

Since the feedforward controller does not take the information, it maintains the same amplitude for grid current regulation by following the estimation of the injected DC power. When it is used alone, the DC-link voltage, v_{dc} starts to deviate from the nominal value due to the unbalanced power, because the feedforward controller is only based on pre-established knowledge and does not support correction of non-ideal factors or disturbances. With the hybrid controller, the feedback controller detects such differences and increases the reference value for grid-current regulation. Therefore, the DC-link voltage is maintained at the rated value regardless of the disturbance, as illustrated in Figure 7.21. The simulation demonstrates the effectiveness of the design approach, including the feedforward, feedback, and hybrid solutions.

The DC-link voltage contains the noticeable ripple appearing at the double-line frequency, as shown in both Figure 7.20 and Figure 7.21. This is caused by the significant difference between the constant DC injection and the fluctuating extraction from the single-phase DC/AC conversion, which has been explained in Section 5.3.1. The feedback signal that is sensed at the DC link is coupled with the double-line frequency ripples. Low-pass filter is not ideal for implementation due to the relative low frequency of the ripple voltage. A Low-pass filter can introduce significant phase lag into the control loop and reduce the system dynamic speed and stability margins. When the voltage setpoint is a constant value, V_{DC}, as shown in Figure 7.19, the ripples present at the error signal and the controller output, which eventually affects the grid current regulation and causes distortions. The issue can be minimized by introducing another feedforward form into the control system.

The peak-to-peak value of the DC-link ripple voltage can be determined from

$$\Delta V_{dc} = \frac{P_{dc}}{\omega_b V_{dc} C_{dc}} \tag{7.43}$$

according to (5.90), as derived in Section 5.3.1. Therefore, the DC-link voltage including the double-line frequency ripples is expressed as

$$v_{dc} = V_{DC} + \frac{\Delta V_{dc}}{2} \sin(2\omega_b t) \tag{7.44}$$

The signal can be added into the feedforward form to cancel the ripple that is sensed from the feedback signal.

The controller can be reconstructed by implementing one feedback controller with two feedforward paths, as shown in Figure 7.22. One is computed by (7.42), showing the steady-state relation between the DC power and the amplitude of the grid current.

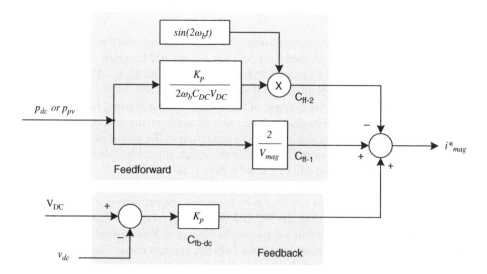

Figure 7.22 Diagram of DC-link voltage regulation implemented with both feedforward and feedback controllers.

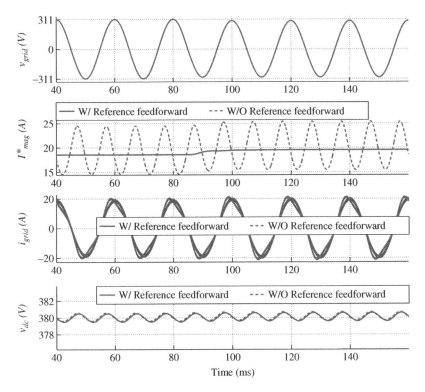

Figure 7.23 Performance of DC voltage regulation for single-phase grid interconnection using hybrid with both feedforward and feedback controllers.

Another direct form is constructed according to the ripple voltage, which can be computed by (7.43) and (7.44). Both feedforward terms use the measured DC power, either p_{dc} or p_{pv}, and known parameters including the AC grid frequency (ω_b), the amplitude of the grid voltage (V_{mag}), the DC-link capacitance (C_{DC}), and the DC-link reference voltage (V_{DC}). Synchronization with the grid signal is required to construct the feedforward term for the ripple cancellation. The feedback loop is formed by the proportional gain, K_p, as discussed in the previous section.

Figure 7.23 shows the simulation results for the hybrid controller implementation with one feedback controller and two feedforward forms. Without the implementation of the ripple-cancellation form, a ripple appears in the command signal, I^*_{mag}, which results in the distortion of the grid current, i_{grid}, as shown in Figure 7.23. This results from the feedback loop since the proportional gain ($K_p = -10$) amplifies the error and sends the ripple into the command.

When the complete controller shown in Figure 7.22 is implemented, the ripple is canceled in the command signal, I^*_{mag}, by the reference feedforward term. Distortion is noticeable in the waveform of the grid current, i_{grid}, when the feedforward term is not implemented to correct the double-line frequency ripple. The power quality of the injected current is maintained at a high level when the second feedforward term is implemented. For demonstration purposes, a disturbance is introduced at 85 ms: the grid voltage dips to the RMS value of 209 V. The variation results in a reduction in the power extraction from the DC link. The feedback controller increases its output, bringing the power balance back to the steady state and maintaining the same level of DC-link voltage.

It should be noted that the DC link is the same as the PV link in the single stage conversion system, as shown in Figure 5.44. The control system design can therefore be the same as described above.

7.10.2 Three-phase Grid Interconnection

When the injected power to the DC link, p_{dc}, is measured or estimated from the PV generation power, p_{pv}, the amplitude of the command signal for three-phase AC current regulation can be computed. When the system loss is ignored, an energy equilibrium can be created, which is expressed in as

$$P_{dc} = \frac{3V_{LN} I_{mag}}{\sqrt{2}} \tag{7.45}$$

$$I_{mag} = \frac{\sqrt{2}}{3V_{LN}} P_{dc} \tag{7.46}$$

where V_{LN} represents the line-to-neutral voltage, I_{mag} is the amplitude of the line current, and P_{dc} is the steady-state value of the injected DC power.

The feedforward term can be established accordingly, as illustrated in Figure 7.24, in which the feedback controller is also present. Any model uncertainty and disturbance can be corrected by the feedback control loop. The controller output is denoted as i^*_{mag}, which is the command signal to represent the amplitude of the three-phase current. It should be noted that the balanced three-phase DC to AC conversion does not introduce significant low-frequency voltage ripples appearing at the DC link. The dedicated feedforward term for the double-line frequency ripple cancellation, as shown in Figure 7.22, might not be applicable for the balanced three-phase cases. When either a single-line or a double-line fault happens, the three phases are no longer balanced and double-line frequency ripple appears at the DC link, which should be mitigated for fault ride-through.

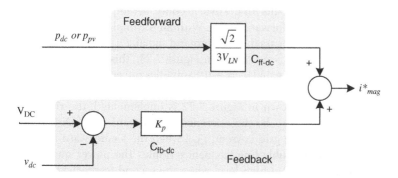

Figure 7.24 Diagram of DC-link voltage regulation implemented with both feedforward and feedback controllers.

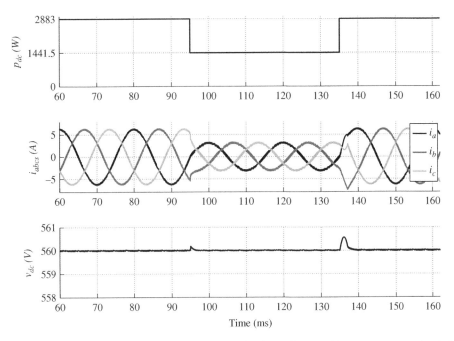

Figure 7.25 Performance of DC voltage regulation for three-phase grid interconnection using hybrid with both feedforward and feedback control.

The controller synthesis follows the same example and transfer function as derived in Section 6.5 and expressed in (6.70). Due to the integral characteristics of the transfer function, the feedback controller is P-type with $K_p = -10$. With both feedback and feedforward control, the simulation result is as shown in Figure 7.25. The DC-link voltage should be regulated at the rated value, 560 V.

The simulation includes a disturbance when the injected DC power changes from the rated level to half that level at 95 ms and back to the rated power again at 135 ms. It simulates a sudden irradiance variation common in PV power systems. No significant overshoot is noticeable in the DC-link voltage even though a significant power variation happens at the transient moment. The highest overshoot is measured as 0.6 V, about 0.1% of the rated DC-link voltage, when the step-up occurs at 135 ms. The three-phase current is regulated to accommodate this significant variation. The simulation demonstrates the effectiveness of the design approach of including feedforward, feedback, hysteresis, and hybrid control.

7.11 Sensor, Transducer, and Signal Conditioning

Sensors and transducers are essential in PV power systems, and can be considered as the "eyes" of the control systems. The voltage and current signals are usually measured. Additional information, such as solar irradiance and temperature, can be sensed to improve control system performance. Ideal measuring devices – including sensors, transducers, and signal conditioners – have the following features:

- The magnitude of output impedance is as low as possible to output strong signals against noise and signal distortion.
- The magnitude of input impedance is as high as possible to avoid any distortion to the measured signal.
- The bandwidth is as high as possible to capture all essential dynamics.
- The transducer output is linear over a large range.
- All parameters are time invariant.

A typical measurement for voltage is illustrated in Figure 7.26; the voltage signal v_i is measured by the sensor. The signal is transferred through the signal-conditioning circuit. The output signal v_o is presented at the input port of an analog-to-digital (A/D) converter for digital control implementation. The cascaded connection of the sensor and signal conditioner distorts the measured signal because of the input and output impedance, as expressed in (7.47):

$$v_o(j\omega) = \left[\frac{Z_{i2}(j\omega)}{Z_{o1}(j\omega) + Z_{i2}(j\omega)}\right]\left[\frac{Z_{i3}(j\omega)}{Z_{o2}(j\omega) + Z_{i3}(j\omega)}\right] v_i(j\omega) \tag{7.47}$$

Only if $|Z_{i2}(j\omega)| \gg |Z_{o1}(j\omega)|$ and $|Z_{i3}(j\omega)| \gg |Z_{o2}(j\omega)|$, can the distortion can be minimized.

Example circuits for measuring DC and AC voltage are shown in Figure 7.27. For DC voltage measurement, the resistor dividers, shown as R_1 and R_2, can scale the PV output voltage down to low levels, which fits the range of the sensing circuit. The capacitor is added as a low-pass filter to remove high-frequency noise, as illustrated in Figure 7.27a. Considering the unity gain and high-frequency bandwidth of the voltage follower, the transfer function in the s-domain can be expressed as

$$\frac{v_o(s)}{v_i(s)} = \frac{R_1}{R_1 R_2 Cs + R_1 + R_2} \tag{7.48}$$

which shows the first-order low-pass feature and the scaling factor of $R_1/(R_1 + R_2)$. It should be noted the resistance of R_2 can be accumulated in multiple resistors to achieve high scaling factors when v_i is significantly higher than v_o. The voltage follower is formed by the operational amplifier (labeled Op-amp), which has high input impedance and low output impedance. The output signal of the op-amp, v_o, can be transmitted to the sampler and A/D conversion unit for either digital control or data logging.

The AC voltage measurement circuit is shown in Figure 7.27b. Since the majority of A/D converters accept only low-voltage DC signals at the input port, the signal conditioner makes the AC/DC conversion. Voltage dividers are required to lower the AC

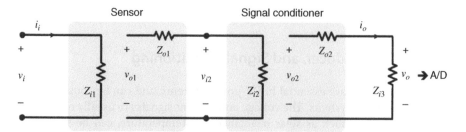

Figure 7.26 Cascade connection of sensor and signal conditioner.

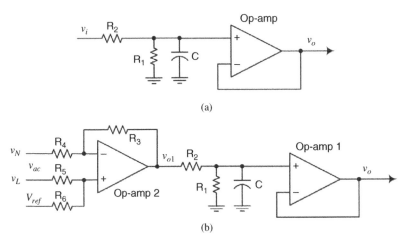

(a)

(b)

Figure 7.27 Example of voltage measurement circuits: (a) DC voltage measurement and signal conditioning; (b) AC voltage measurement and signal conditioning.

voltage to low levels. These dividers are formed by the resistor networks labeled R_1–R_6. To form a differential amplifier, it is usual to make $R_3 = R_6$ and $R_4 = R_5$. When a high scale-down factor is required, more resistors can be used, forming a resistor network for the function of voltage dividing. An offset voltage (V_{ref}) is applied at the op-amp input port to shift the AC signal to DC; this can be expressed as

$$v_{o1} = V_{ref} + \frac{R_3}{R_4} v_{ac} \tag{7.49}$$

The function of the measurement circuit can also be illustrated by the signal comparison between the input (v_{ac}) and the output (v_o), as shown in Figure 7.28. It can be seen that the waveform of v_o is DC, with and offset voltage, V_{ref}. The voltage follower is also used to lower the output impedance to minimize noise contamination in the measured

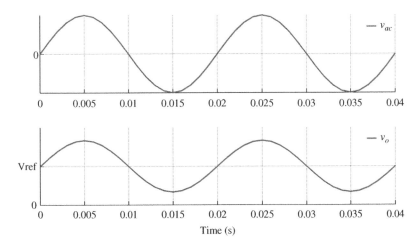

Time (s)

Figure 7.28 Waveform in AC voltage measurement.

signal during transmission. The capacitor is also included in the voltage divider circuit in order to attenuate high-frequency noise, as shown in Figure 7.27b. Thus, the transfer function to represent the input and output of the filter is expressed as

$$\frac{v_o(s)}{v_{o1}(s)} = \frac{R_1}{R_1 R_2 Cs + R_1 + R_2} \tag{7.50}$$

7.12 Anti-windup

All physical systems show a certain level of constraint; the input and output should be limited to their top or bottom scale. For the example of switching-mode power converters, the duty cycle of the PWM ranges from 0 to 100%. Furthermore, the values of voltage and current can reach certain limits. Therefore, limiters are commonly implemented to constrain the control variable before the plant to avoid misplacements. However, controllers are mainly synthesized from mathematical models, which are based on ideal operating conditions.

Figure 7.29 illustrates a PID-controlled system with integration of a limiter. The PID controller output can reach the saturation limit if the error value between the reference and the plant output is significant. Before the error is completely eliminated, the integration term in PI or PID controller continuously accumulates the error regardless of the saturation. Depending on the time period of the existing error, the integral term can raise the controller output up to a high value. The limiter constrains the controller output by saturation. The correction happens until the sign of the error changes, but the correction takes a long time because of the continuous integration. Therefore, the phenomenon is also called "integral windup" or "integrator windup." In general, the windup effect results in sluggish operation, which causes significant over- or undershoot in the time-domain response. A significant value of the error can result from many factors, such as sudden setpoint changes, significant disturbances, and improper controller design.

Following the example discussed in Section 7.8.1, the windup can be illustrated by the setpoint change. The periodic perturbation to the reference increases from 0.5 V to 3 V, as shown in Figure 7.30. At the moment of the step change in the setpoint, the error is so significant that the duty cycle demanded from the controller is between −79% and 115%. The value is out of the limit, which is defined as 0–90%, since the boost converter is short-circuited if the PWM duty cycle reaches 100%. The over-limit eventually

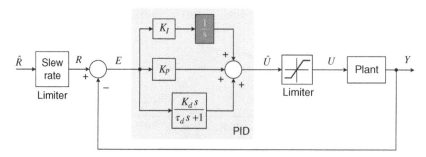

Figure 7.29 Feedback system with limiter.

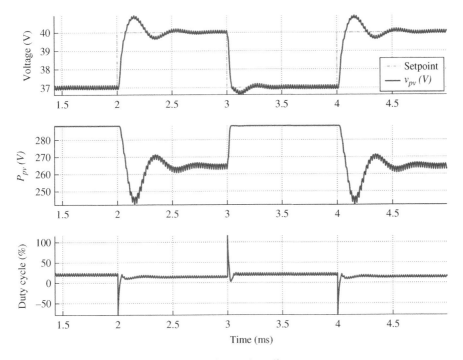

Figure 7.30 Voltage regulation illustrating the windup effect.

saturates the PWM generator and results in the windup phenomenon, which manifests as significant overshoot and slow recovery. The overshoot voltage is measured as 0.91 V, which is equivalent to 30.0% of the relative value. The settling time can be measured as 0.42 s. Due to the MPP deviation, the PV output power is significantly reduced during the transient time.

The method to avoid the windup effect is called as the "anti-windup scheme." As shown in Figure 7.29, a slew-rate limiter is used to prevent windup. When a controller is designed, the closed-loop transfer function is known. The response time to the setpoint change becomes predictable, which is based on the closed-loop dynamic analysis. The slew rate limiter can be designed accordingly, which is based on the concept of feed-forward control and two degrees of freedom. This is considered as an effective way to eliminate the windup side effect that is caused by any sudden change of the setpoint.

Based on the same case, the slew rate is implemented as 10 V/s to limit the changing rate of the reference voltage. Figure 7.31 demonstrates the simulation result, which can be compared with the case without any anti-windup implementation. The overshoot voltage is measured as 0.2 V, which is equivalent to 6.7% in the relative value. The settling time is monitored as 0.3 s. The control performance is significantly improved by comparison with the previous case study, as shown in Figure 7.30. The duty cycle is well maintained between 11% to 24%, which does not cause any saturation in the limiter. The MPP deviation is noticeably reduced during the transient time in comparison with the previous case study, as shown in Figure 7.30. The study shows the slew rate limit for the reference is an effective method to eliminate the windup that is caused by the sudden and significant change of the setpoint.

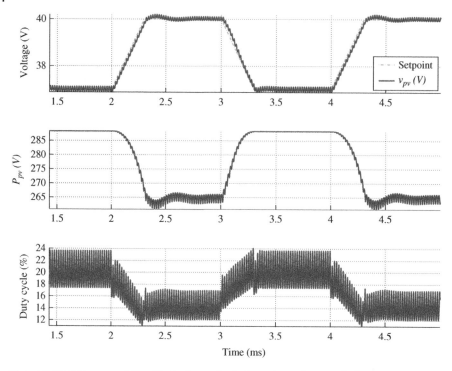

Figure 7.31 Voltage regulation illustrating anti-windup using a slew-rate limiter.

However, the slew-rate limiter cannot stop windup caused by unpredictable disturbances. For example, the solar irradiance can change dramatically and cause disturbances in the voltage regulation. Figure 7.32 shows a simulated case in which the irradiance varies periodically between 1000 W/m² and 200 W/m². The disturbance is out of the limit of 0–90% and is therefore large enough to cause saturation of the controller output.

The windup effect manifests in significant undershoots and overshoots in the v_{pv} signal during each transient time. The variation of the PV output voltage affects the open-circuit voltage and causes the output power to reset to zero during the transient period. The windup effect dramatically changes the operating point and causes significant power loss.

The fundamental principle of anti-windup is that the PID or PI control should change when the actuator hits the limit and cannot follow the controller command. One commonly used anti-windup mechanism is based on conditional integration, which constrains the integration output during actuator saturation. This is also known as the "clamping method." The input for the integral path is set to zero when windup is detected. The implementation of conditional integration is straightforward in digital control systems. Figure 7.33 shows the flowchart for detection of windup and introduction of anti-windup action. All variables used in the flowchart refer to the labels in Figure 7.29.

The saturation is determined when the input ($\hat{U}$) and output (U) of the limiter are different. The sign of the error signal (E) is used to evaluate the status. When the signs of

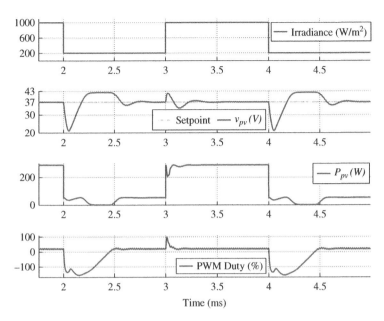

Figure 7.32 Voltage regulation illustrating windup effect caused by sudden irradiance variations.

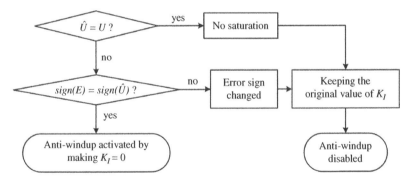

Figure 7.33 Detection and conditional integration for anti-windup action.

E and $\hat{U}$ are the same, the windup status can be determined. Therefore, the input to the integration path is set to zero: $K_I = 0$. When the signs of E and $\hat{U}$ are different, the PID controller should return to normal operation of the integral term. Conditional integration is also uses in PI controllers. For anti-windup implementation, the PID controller can be divided into a PD controller in parallel with an I-type controller.

A case study is now set out to prove the effectiveness of the clamping method. The operating conditions are the same as the previous example, with a disturbance caused by a step change of solar irradiance from 1000 W/m^2 to 200 W/m^2. The previous case study showed the windup effect in Figure 7.32. Figure 7.34 shows the effect when conditional integration is used as an anti-windup scheme. There is no dramatic voltage variation caused by the saturation effect. The PV output voltage no longer hits the open-circuit voltage and is maintained in the operating range. As a result, power loss

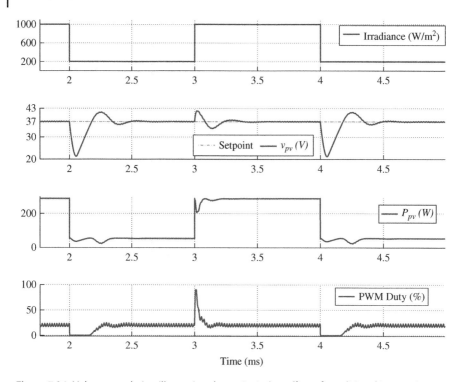

Figure 7.34 Voltage regulation illustrating the anti-windup effect of conditional integration.

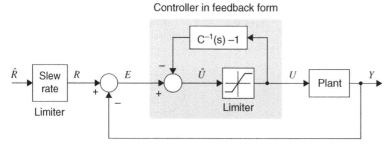

Figure 7.35 Illustration of feedback form of biproper controller for anti-windup.

is minimized even though a significant disturbance is caused by the solar irradiance change. The case study shows that the clamping method is an effective way to prevent the windup that is caused by disturbances.

Another anti-windup scheme was proposed by Goodwin et al. (2001). The implementation is not limited to the parallel-form implementation of the PID or PI controller; it is applicable to any controller type that is stable, biproper, and minimum phase. When a controller is synthesized and shown as $C(s)$, it can be inverted and implemented as the feedback form for anti-windup, as illustrated in Figure 7.35. The limiter is integrated inside the feedback form of the controller.

When the limiter takes no action, the value of $\hat{U}$ is equal to U. The controller transfer function is expressed as (7.51), which is equivalent to the original controller, $C(s)$. This indicates that no anti-windup action is taken.

$$\frac{U(s)}{E(s)} = \frac{1}{1 + [C(s)^{-1} - 1]} \tag{7.51}$$

The controller output can be expressed as

$$\hat{U}(s) = E(s) - [C^{-1}(s) - 1]U(s) \tag{7.52}$$

When saturation occurs, the value of $\hat{U}$ is unequal to U since the limiter is in action. The value of U is constrained by the saturation value, U_{sat}. This can be expressed as (7.53) or (7.54).

$$\hat{U}(s) = [E(s) - C^{-1}(s)U_{sat}] + U_{sat} \tag{7.53}$$

$$\hat{U}(s) = C^{-1}(s)(\hat{U}_{sat} - U_{sat}) + U_{sat} \tag{7.54}$$

The static gains of $C(s)$ and C^{-1} are positive in typical control systems. U_{sat} can represent either the upper limit or the lower limit of the constraint. When $\hat{U} > U_{sat}$, the controller output hits the upper limit. According to (7.54), the condition is maintained considering the positive value of the static gain of $C^{-1}(s)(\hat{U}_{sat} - U_{sat})$. When $\hat{U} < U_{sat}$, the controller output reaches the lower limit. The saturation is also maintained with consideration of the negative value of the static gain of $C^{-1}(s)(\hat{U}_{sat} - U_{sat})$. When the controller output is not saturated, the controller is back to the standard form in (7.51), which is equivalent to the controller function, $C(s)$.

It is important to note that the feedback form of the biproper controller, as shown in Figure 7.35, can only be applied for the case in which $C(s)$ shows a positive static gain. However, the static gain of $C(s)$ can be negative when it is derived for PV-link voltage regulation, as shown in (7.25). The sign should corrected before the anti-windup implementation. Therefore, the implementation of voltage regulation in PV links is as shown in Figure 7.36.

An example shows the effectiveness of the anti-windup approach. The operating condition is the same as the previous example of a step change of solar irradiance from 1000 W/m² to 200 W/m². The controller is implemented in the feedback form, as shown in Figure 7.36, since the static gain of $C(s)$ is negative for PV-link voltage regulation. Figure 7.37 shows the effect when the feedback form of biproper controller is applied as an anti-windup scheme. It exhibits similar performance to the conditional

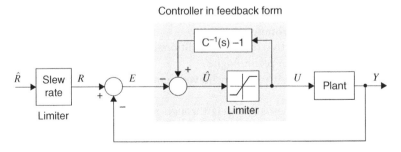

Figure 7.36 Illustration of feedback form of biproper controller for anti-windup with a negative static gain.

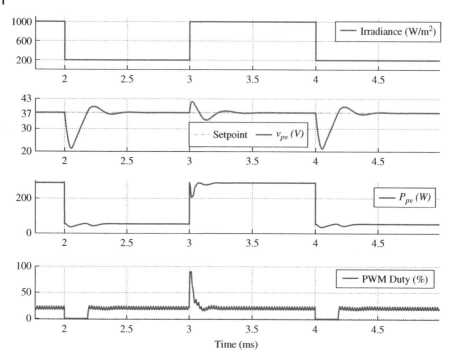

Figure 7.37 Voltage regulation illustrating the anti-windup effect of the feedback form of the controller.

integration scheme, as shown in Figure 7.34. Windup is prevented, and there is no dramatic voltage variation caused by the saturation effect. The PV output voltage does not reach the open-circuit voltage, as it does if no anti-windup mechanism is implemented.

7.13 Digital Control

Digital control techniques use software programming. They have the following advantages:

- Digital controllers are less susceptible to noise and parameter variation due to aging and changes in the operating environment.
- Digital control is more cost effective than the traditional analog control approach thanks to the latest microcontroller technologies.
- High accuracy and control bandwidth can be achieved thanks to the advanced microcontroller technology.
- Control routines can be maintained accurately through programming. One digital controller is capable of conducting the operations of multiple control loops.
- Digital control can handle complex control laws, controller tuning, and redesign.
- Digital controllers are capable of data storage, data communication, and can incorporate advanced human machine interfaces, such as graphical and digital displays.

- Component counts can be reduced to avoid the complications of analog control approaches.
- Overall power consumption can be reduced thanks to the low component count and advances in microcontroller technology.

A control system for PV power applications is commonly constructed in a hybrid format. A typical control system diagram is shown in Figure 7.38. This has three different signal formats: analog, discrete, and digital. The explanation of each signal format is found in Table 7.4.

The control loop is formulated by analog, discrete, and digital signals. A control system can be designed in the continuous time domain and implemented in a hybrid system using the digital controller. The following procedure should be applied to construct a hybrid system with digital control.

1) Modeling should be initialized to derive the nominal plant model in the continuous time domain, such as the nominal transfer function, $G_0(s)$, which shows the fundamental information of the damping and response speed.
2) System review should always be performed to identify all constraints including disturbance, noise source, time delay, non-minimal phase zeros, damping levels, and model uncertainty.
3) The analog controller can be synthesized by various approaches, for example affine parameterization.
4) Frequency-domain analysis should be performed to evaluate the stability margins and sensitivity peak using either a Bode diagram or a Nyquist plot. All system constraints should be considered.

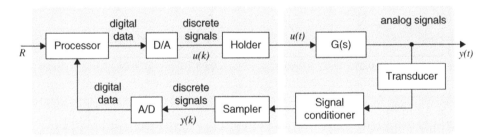

Figure 7.38 Hybrid system with digital controller.

Table 7.4 Category of signals.

Signal format	Alternative name	Description
Analog signal	Continuous-time signal	Continuous in time
		Measured signals of voltage and current
Discrete signal	Sampled signal	Pulse amplitude-modulated
		Generated by sampler and holder
Digital data	Digital signal	Coded numerical data

5) The digital redesign procedure includes the selection of the sampling frequency and the controller transformation from s-domain to z-domain. The controller is eventually implemented in the discrete-time format.
6) Time delays caused by the digital approach should be considered to see if the phase margin is still sufficient.
7) Evaluation can be based on time-domain simulation and then experimental testing.

The analog signal that is sensed by the transducer is continuous in time. Signal conditioning is usually required to manipulate the measured signal for the next stage of processing. It provides the functions of noise filtering, signal scaling, and impedance matching of the input and output terminals. The sampler produces the discrete signal from the analog signals. The analog-to-digital (A/D) and digital-to-analog (D/A) units are the interfaces between the discrete signals and digital data.

Figure 7.39 illustrates the voltage signal at 220 V RMS and 50 Hz. This can be transformed into two different discrete low-voltage DC signals for measurement; the majority of A/D converters accept only DC signals within an upper limit of 5 V or less. One type of transformation includes a DC offset to shift the AC signal to DC. The scaling factor is 20:1, as indicated in Signal 1 in Figure 7.39. The other signal captures only the amplitude of the AC signal, with a scale factor of 10:1. The resolution is better than Signal 1. The direction of zero crossing should be detected to differentiate the negative and positive cycles. The signal is shown as Signal 2 in Figure 7.39. The AC to DC conversion and scaling can be achieved with the signal-conditioning circuit shown in Figures 7.27 and 7.38.

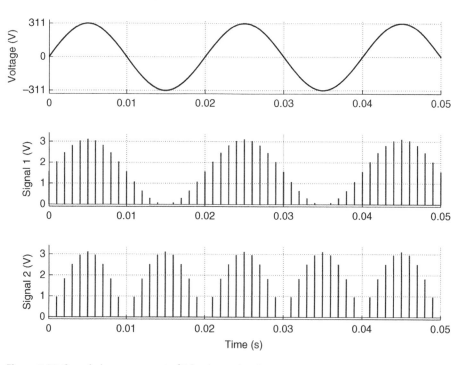

Figure 7.39 Sampled measurement of AC voltage signal.

As shown in Figure 7.38, the command signal (R) can be either analog from outside or a digital input embedded inside the processor's memory. The controller function is executed and programmed inside the processor with memory support. The holder is required to maintain the discrete signal and convert it into continuous-time signals. Even though high-order holds are available, the common type is the zero-order holder (ZOH), which holds each sample value for one sample interval. In the s-domain, the ZOH can be expressed as the simple form of

$$G_{ZOH} = \frac{1 - e^{-sT_s}}{s} \tag{7.55}$$

where T_s is the sampling time interval. It should be noted that the sampler, A/D converter, processor, D/A converter, and holder can be integrated in a microcontroller, which significantly reduces the component count of the overall system. Advanced transducers can be integrated with digital processors, signal conditioning, samplers, and A/D converters, and are able to output digital signals. Data communication is required to interface intelligent transducers with control processors.

The sampling time period is controlled by the timer. The overflow interrupt of the timer triggers the digital control process, as shown in the flowchart of Figure 7.40. The processor first reads out the measured variables from the data buffer of the A/D unit. The A/D unit should be reactivated for the next sampling and A/D conversion. Data calibration is needed to relate the measured variable to the digital number. If any abnormality is found in the measured variable, protection or some other emergent process should be triggered. Otherwise, the normal process to run the programmed control algorithm is performed. The controller output is updated and then sent out to the plant through the holder. For the power interface in PV power systems, the controller output signal

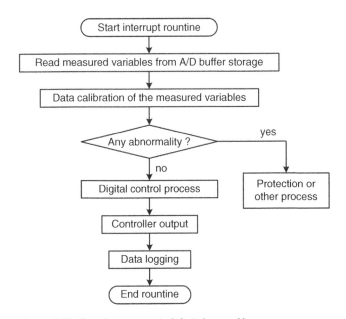

Figure 7.40 Flowchart in a typical digital control loop.

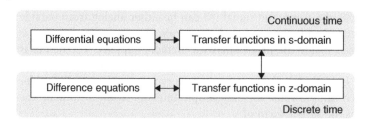

Figure 7.41 Dynamic system representation and transformation in discrete and continuous time.

is mostly represented by the duty cycle of the PWM or sometimes phase-shift angles. Data logging can be conducted to save the measured variable as historic data.

7.13.1 Continuous Time and Discrete Time

A dynamic system can be represented and analyzed in either continuous time or discrete time. Figure 7.41 illustrates the system representation and transformation. In continuous time, the differential equations and transfer functions in the s-domain can be transformed to discrete time through a z-transformation. The sampling rate should always be included for the transformation.

The sampling frequency in digitally controlled systems is the flow rate of digital values. It is limited by the bandwidth of the sensing components and the processing speed of the digital controller. Nyquist–Shannon sampling theory defines the minimal frequency of the sampling rate that allows the sampled signal sequence to capture all the important dynamic information from the continuous-time signal. The sampling rate should be at least twice the maximum component frequency of the analog system. The Nyquist frequency is half the value of the sampling frequency, which is the upper band limit in digital control system dynamics. In general, a high sampling frequency is desirable for the best representation of the analog system. However, constraints should be applied because of the physical limits or the coefficient resolution due to the z-transformation.

7.13.2 Digital Redesign

The approach of converting the analog controller into a digital controller is termed "digital redesign." An analog controller is first designed by synthesis methods, such as affine parameterization. The stability margins and robustness should be evaluated before the transformation from analog to digital. Two common methods are used to approximate a digital controller that fulfills the function of an analog controller.

The bilinear method, which is also called a "Tustin transformation," can be used to transform from the s-domain to the z-domain:

$$s \approx \frac{2}{T_s}\left(\frac{1-z^{-1}}{1+z^{-1}}\right) \tag{7.56}$$

where T_s refers to the sampling period. For example, a PD controller, as expressed in (7.7), can be converted to a digital controller in the z-domain and expressed as

$$C_{PD}(z) = \frac{(K_P T_s + 2K_P \tau_d + 2K_D) + (K_P T_s - 2K_P \tau_d - 2K_D)z^{-1}}{(2\tau_d + T_s) + (T_s - 2\tau_d)z^{-1}} \tag{7.57}$$

For the digital controller implementation, it can be converted into a difference equation as

$$u(k) = -a_1 u(k-1) + b_0 e(k) + b_1 e(k-1) \tag{7.58}$$

where,

$$a_1 = \frac{T_s - 2\tau_d}{2\tau_d + T_s} \tag{7.59a}$$

$$b_0 = \frac{K_P T_s + 2K_P \tau_d + 2K_D}{2\tau_d + T_s} \tag{7.59b}$$

$$b_1 = \frac{K_P T_s - 2K_P \tau_d - 2K_D}{2\tau_d + T_s} \tag{7.59c}$$

The symbols u and e represent the controller output and the error value in the control loop. The index k is applied to represent the operating sequence in discrete time.

Another approximation is based on how the pole, zero, and gain match between the analog controller and its digital counterpart. It is called the "matched pole-zero method." The approximation for both pole and zero match is

$$z_i \approx e^{s_i T_s} \tag{7.60}$$

where z_i represents the zero or pole in the z-domain and s_i is the zero or pole in the s-domain.

The pole and zero of the PD controller in the s-domain can be derived as (7.61a) and (7.61b) respectively.

$$p_c = -\frac{1}{\tau_d} \tag{7.61a}$$

$$z_c = -\frac{K_P}{K_P \tau_d + K_D} \tag{7.61b}$$

The analog controller is then transformed to a digital controller by the match of pole and zero:

$$C_{PD}(z) = K_{PZ} \frac{1 - e^{z_c T_s} z^{-1}}{1 - e^{p_c T_s} z^{-1}} \tag{7.62}$$

The gain can be back-calculated by (7.63) to match the static gain in the s-domain.

$$K_{PZ} = \frac{Kp(1 - e^{p_c T_s})}{1 - e^{z_c T_s}} \tag{7.63}$$

The analog controller in (7.25) can be transformed to a digital controller using either method. The sampling frequency is assigned as 40 kHz. The Matlab function, "c2d," is used for the transformation. The transfer functions based on the Tustin approximation and the matched pole-zero method are expressed in (7.64) and (7.65), respectively.

$$C_{Tustin}(z) = \frac{-0.1697 + 0.2450z^{-1} - 0.1143z^{-2}}{1 - 0.7935z^{-1} - 0.2065z^{-2}} \tag{7.64}$$

$$C_{match}(z) = \frac{-0.1310 + 0.1863z^{-1} - 0.0861z^{-2}}{1 - 1.048z^{-1} - 0.0478z^{-2}} \tag{7.65}$$

The Bode diagrams of the three controllers are shown in Figure 7.42. The Bode plots for the digital controllers stop at the Nyquist frequency, which is 20 kHz, equivalent to

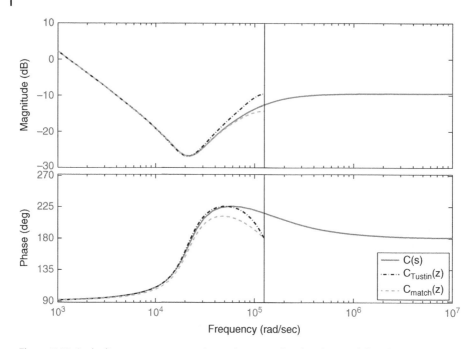

Figure 7.42 Bode diagrams to compare the analog control with redesigned digital counterparts.

125 660 rad/s. This indicates that the Nyquist frequency is the upper bound of the frequency in digital systems. However, the analog controller, shown as $C(s)$, does not have a frequency limit. The Bode diagram also shows the good match in the low-frequency region and the deviation when the frequency is close to the Nyquist frequency.

Because of the implementation of anti-windup, the standard PID controller can be expressed with a separate integral term:

$$C_{PID}(s) = \underbrace{\frac{(K_P\tau_d + K_D)s + K_P}{\tau_d s + 1}}_{C_{PD}} + \underbrace{\frac{K_I}{s}}_{C_I} \tag{7.66}$$

The digital transformation can be divided into two parts: the PD term and the I term. The digitalization of the PD controller has been demonstrated in (7.57). The matched pole-zero method cannot be directly used for the transformation of the I term. It can be approximated by the bilinear method as (7.67) or the Euler method, which is expressed in (7.68).

$$C_I(z) = \frac{K_I T_s (1 + z^{-1})}{2(1 - z^{-1})} \tag{7.67}$$

$$C_I(z) = \frac{K_I T_s}{1 - z^{-1}} \tag{7.68}$$

The PID controller is therefore represented as a combination of two controllers: $P_{PD}(z) + P_I(z)$.

Based on the analog controller in (7.25), the transformation is conducted for $P_{PD}(z)$ and $P_I(z)$ separately. The sampling frequency is 40 kHz and the bilinear method is

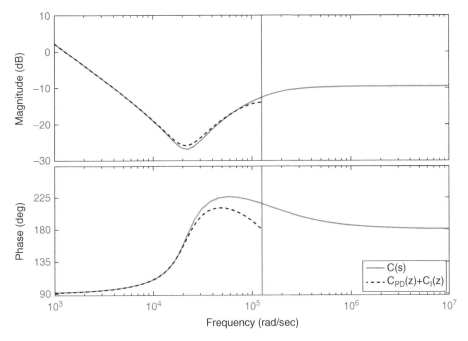

Figure 7.43 Bode diagrams to compare analog control with the redesigned digital counterpart.

applied. The controllers are derived as:

$$C_{PD}(z) = \frac{-0.1221 + 0.8847z^{-1}}{1 - 0.0478z^{-1}} \tag{7.69}$$

$$C_I(z) = \frac{-0.01616 - 0.01616z^{-1}}{1 - z^{-1}} \tag{7.70}$$

When saturation is detected, the input to the integral function, as shown in (7.70), can be reset to zero to prevent any windup effect. The approximation can be verified by Bode plots, as shown in Figure 7.43, where $C(s)$ indicates the analog controller, and the digital formation is shown as $C_{PD(z)} + C_I(z)$. The plots show the frequency characteristics of the analog controller (7.25), and the digital approximation matches up to the Nyquist frequency, which verifies the transformation.

7.13.3 Time Delay due to Digital Conversion and Processing

In the hybrid control loop, as shown Figure 7.38, additional dead time or time delay is introduced by the sampler, A/D and D/A units, holder, and the digital controller process. For microcontrollers, one A/D unit is integrated with the sampler, and this is commonly shared by the N channels that are distributed by the analog multiplexer, as shown in Figure 7.44. The input voltage for A/D conversion is maintained by the capacitor. Time delays can be expected from the A/D conversion time and capacitor charging time. The inputs of the multiplexer are connected to signal-conditioning circuits; these were discussed in Section 7.11. The low output impedance of the signal conditioners is ideal to achieve fast sampling.

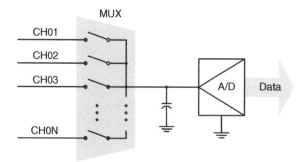

Figure 7.44 Typical sensing circuit including the A/D converter and the analog multiplexer for multiple channels.

When the controlled variable is sampled and converted to digital data, the control algorithm is operated by software and this consumes a certain amount of time. The controller output – the PWM duty cycle in controlling power interfaces – can only be updated by the next PWM cycle. Time delays can be expected in the output stage, sensors, and the signal conditioning. The total time delay can be estimated to be several sampling time periods, depending on the programming and digital implementation. The time delay introduces a phase lag and lowers the phase margin of a closed-loop system. The deduction of the phase margin can be calculated as

$$\phi_D = \frac{T_d \times \omega_{CP} \times 360°}{2\pi} \tag{7.71}$$

where T_d and ω_{CP} are the delay and the cross-frequency when the phase margin is measured.

A case study is based on the example discussed in Section 7.8.1. The analog controller, as shown in (7.25), was originally designed without consideration of any time delay. The phase margin is 65.2° with the associated frequency (ω_{CP}) of 56 315 rad/s. For digital control, the time delay is estimated as 50 μs if the sampling frequency is 40 kHz and two

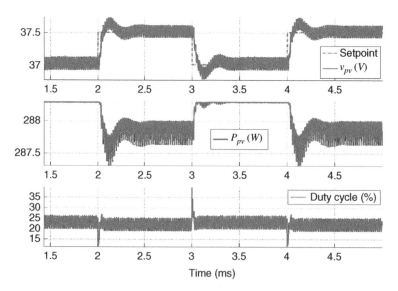

Figure 7.45 Waveform of voltage regulation by digital controller.

sampling time periods is the total delay time. The deduced phase margin is calculated from (7.71) as 161.3°. This will result in unstable operation of the closed loop. When the sampling frequency is increased to 400 kHz, the phase margin becomes 49° since the phase margin deduction is calculated as 16°. Figure 7.45 illustrates the voltage regulation to follow the step change in the reference between 37 V and 37.5 V. The time delay of 5 µs is included in the closed loop and the simulation. In practical systems, the non-ideal factor should be always considered in the control loop synthesis along with consideration of the tradeoff between the control performance and system robustness.

In general, increasing the sampling frequency can reduce the delay time, but requires fast operation in digital controllers. If the 400 kHz sampling frequency is impractical for implementation, the analog controller should be redesigned or re-tuned with a more conservative approach. The original design should provide more phase margin against the deduction that is caused by the digital control approach.

7.14 Summary

This chapter focuses on model-based approaches for controller synthesis in PV power systems. The fundamental knowledge of control systems is introduced first. This is based on linear control theory, including the subjects of affine parameterization, PID-type controllers, stability margins, and robustness performance. The control of grid-connected PV systems is divided into two parts: voltage regulation for the PV link and for the DC link. Bus voltage regulation for DC microgrids when the dual active bridge is used for the battery power interface is considered.

Affine parameterization for development of linear controllers for regulating the PV-link voltage is introduced. The chapter covers controller synthesis for the common topologies used for the PV-side converter: boost, buck, tapped inductor, buck–boost, and flyback. The controller synthesis for the full-bridge transformer isolated DC/DC converter is neglected since the analysis is the same as for the non-isolated buck converter.

In a case study, the relative stability is evaluated and based on the nominal plant model, which is a small-signal representation. It has good stability margins since affine parameterization is used for the closed-loop design. For practical implementations, a comprehensive evaluation in terms of phase margin, gain margin, and sensitivity peak can be conducted to cover the wide range of operating conditions away from the nominal condition. Even though the controllers are developed by following small-signal models, the evaluation of the closed-loop operation is based on time-domain simulations, using the models that were developed in Chapter 5. The simulation models can incorporate nonlinearity, including high-frequency switching operations.

Feedforward controllers are used in the voltage regulation of the DC link and are effective for fast response to disturbances caused by active power variations. For the single-phase grid interconnection, the feedforward controller is also used to mitigate the double-line frequency ripple that is coupled with the DC-link voltage. The ripple is fed back to the controller and affects the reference signal for grid-current regulation.

Sensors, transducers, and signal-conditioning circuits are briefly included in the discussion. The system design constraints are presented, along with consideration of the system bandwidth and integral windup in real-world applications. Even though windup is not widely discussed in relation to PV power systems, it can degrade control performance and reduce PV power harvesting. The chapter also introduces anti-windup solutions. The slew-rate limiter of the setpoint is effective in eliminating the windup that is caused by sudden changes of the reference value. Conditional integration is effective in eliminating windup with several causes, such as disturbances due to significant irradiance variations. However, it requires the integral path to be distinguished from the controller format. The feedback form of a biproper controller is a general and effective anti-windup solution when the controller is stable, non-minimal, and biproper.

Finally, the digital control approach based on the technique of digital redesign is briefly discussed. A complete design approach, including the modeling process, dynamics analysis, analog controller synthesis, digital redesign, system revisit, and evaluation is presented, step by step, for readers to follow. For power electronic applications, a time delay is introduced by the steps of the digital process: sampling, A/D conversion, computation, zero-order holder, and D/A conversion. This delay can dramatically reduce stability margins and cause oscillations and even instability. Special attention should be given to digital redesign, with consideration of the time delays caused by digital processing.

Problems

7.1 Duplicate the simulation results for the design cases presented in this chapter. The process is valuable in becoming familiar with the principles of dynamic analysis, stability evaluation, and controller synthesis, simulation, and verification. The design examples also cover the control approach for various PV-side converters and the voltage regulation of DC links for interconnection to single-phase and three-phase grids.

7.2 Based on the system design and modeling of the PVSCs in the problem section of Chapter 6:
   a) Identify the mathematical model involved in terms of damping factor, undamped natural frequency, and static gain value.
   b) Use affine parameterization to design a suitable controller for the closed loop.
   c) Use a Bode diagram and a Nyquist plot to identify the phase margin, gain margin, and sensitivity peak.
   d) Use time-domain simulation to simulate the system response to variation of the setpoint.
   e) Use time-domain simulation to simulate the system response to a significant variation of the solar irradiance.
   f) Evaluate whether windup occurs during the transient for a setpoint change or a sudden irradiance variation.
   g) Use anti-windup to improve the system response.

References

El Moursi MS, Xiao W and Kirtley Jr JL 2013 Fault ride through capability for grid interfacing large scale PV power plants. *IET Generation, Transmission & Distribution* **7**(9), 1027–1036.

Goodwin G, Graebe S and Salgado M 2001 *Control System Design*. Prentice Hall.

Xiao W and Zhang P 2013 Photovoltaic voltage regulation by affine parameterization. *International Journal of Green Energy* **10**(3), 302–320.

Xiao W, Dunford WG, Palmer PR and Capel A 2007 Regulation of photovoltaic voltage. *Industrial Electronics, IEEE Transactions on* **54**(3), 1365–1374.

Youla D, Bongiorno J and Jabr H 1976 Modern Wiener–Hopf design of optimal controllers Part I: The single-input-output case. *IEEE Transactions on Automatic Control* **21**(1), 3–13.

8

Maximum Power Point Tracking

The output of PV cells is limited by the current, voltage, and power. In a steady state of solar irradiance and cell temperature, there is a single operating point where the output of the voltage and current results in the maximum power output. The maximum power point (MPP) is occasionally called the "peak power point" (PPP) or "optimal operating point" (OOP) in the literature (Xiao et al. 2006). When the power is plotted against voltage (a P–V plot), the peak power point can be easily recognized, as shown in Figure 1.9.

In most PV power systems, a control algorithm called maximum power point tracking (MPPT) is used to take full advantage of the available solar energy (Xiao et al. 2011). The algorithm is occasionally called "peak power tracking." The term MPPT is used in this book. The real-time operation is to control the PV-side power interface so that the operating characteristics of the load and the PV array always match each other at the MPP. The control objective is to maximize the power output for highest solar energy harvesting at any given instant.

This chapter introduces the MPPT techniques and provides background knowledge about recent developments. Simulations are used to demonstrate the operations that are widely used for practical implementation.

8.1 Background

The principle behind MPPT is the conductance or resistance match between the PV generator output and the load condition, as illustrated in Figure 8.1. In grid-connected systems, the load condition is not as straightforward as in standalone systems, but it can be equivalent to the active power level extracted from the PV generation. The plot shows that the MPP of the PV cell is located at 0.54 V and 8.2 A. When a load rated at 15.2 S in conductance or 66 mΩ in resistance is connected, the maximum power generation is achieved, which is 4.43 W. Any deviation from the ideal load resistance of 66 mΩ, such as an applied load resistance of 0.1 Ω or 0.05 Ω makes the PV cell output deviate from the MPP and produce less power, as illustrated in Figure 8.1.

The majority of loads require either constant voltage or current. When they are directly coupled with the PV generator, the load impedance cannot always be adjusted for the purpose of MPPT. However, when a power interface is connected between the load and PV generator, the load conductance or the impedance at the PV generator terminals can be varied through power conversion. The evolution is demonstrated in Figure 8.2, where a power interface is used between the PV generator and the

Photovoltaic Power System: Modeling, Design, and Control, First Edition. Weidong Xiao.
© 2017 John Wiley & Sons Ltd. Published 2017 by John Wiley & Sons Ltd.
Companion Website: www.wiley.com/go/xiao/pvpower

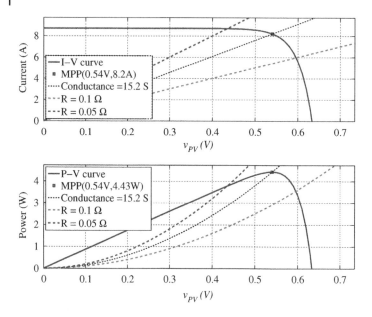

Figure 8.1 Conductance match for maximum power point tracking: top, based on I–V curve; bottom, based on P–V curve.

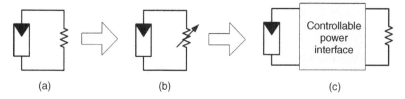

Figure 8.2 Evolution of maximum power point tracking: (a) direct resistor match; (b) variable load match; (c) controllable power interface.

resistive load. The power converter can change the input I–V characteristics from the input to the output and give the equivalent resistance to match the MPP at the PV output terminal. This is the fundamental approach in the latest MPPT technologies.

As discussed in Section 4.1.3, the solar irradiance and cell temperature affect the MPP significantly, as illustrated in Figures 4.8 and 4.9. Dynamic tracking is required to follow unpredictable changes of environmental conditions, particularly the solar irradiance. The cell temperature usually varies with slower dynamics than the irradiance.

The diversity of materials used makes the I–V characteristics different from one PV to another, as illustrated in Figure 1.10. The fill factor (FF) is one way to describe the PV output characteristics; this was introduced in Section 1.6. It is well known that the FF values of crystalline-based PV cells are higher than those of thin-film devices. This fact should be considered to optimally specify the MPPT parameters for different PV output characteristics.

Furthermore, switching-mode power converters are widely used as the power interfaces for PV power generation. As demonstrated in Section 5.1, switching ripples at the PV output terminals make dynamic tracking difficult and reduce tracking efficiency

Table 8.1 Loss analysis of switching-ripple voltage at the PV output terminals.

Percentage voltage ripple (%)	1.0	2.0	5.0	10.0
Percentage power loss (%)	0.03	0.11	0.70	2.99

(Xiao 2007). The power losses caused by ripple voltage in one specific crystalline-based PV module are summarized in Table 8.1. The values can be used as a reference to design the PV-side converter and the control algorithm that will prevent either significant power losses or MPPT malfunctions due to the appearance of voltage ripple. The measurement of voltage and current should be carefully designed since the value difference between the upper and lower peaks in opposite directions can result in significant errors in the measured data.

An MPPT algorithm must generally measure or estimate the PV power output and its variations. For the highest power output, the tracking function is to determine the optimal reference, which can be used as the command or reference to control the power interface. The direct command for switching-mode power converters is either the PWM duty cycle or the phase-shift angle. The reference for the MPPT operation that represents the PV output characteristics can be either the PV-link voltage or current.

The PV-link voltage is considered the most effective reference for MPPT (Xiao et al. 2007b). A feedback loop is required to regulate the PV-link voltage to follow the reference value for the highest power output. An implementation was investigated in Section 7.8 and shown as the block diagrams in Figures 7.1 and 7.2.

It is important to design a proper MPPT algorithm for the power interface in order to achieve the most effective solar energy harvesting. Various techniques have been proposed for the MPPT algorithm (Xiao 2007; Xiao et al. 2011):

- linear approximation
- heuristic search
- extreme value search
- sliding mode
- extremum-seeking control
- real-time identification
- particle swarm optimization (PSO)
- dividing rectangles (DIRECT) algorithm
- intelligent control.

The linear approximation method tries to derive a fixed percentage between the MPP and other measurable signals, such as the open-circuit voltage and short-circuit current. The linear approximation method generally leads to a simple and inexpensive implementation. They are also designed to avoid the problems caused by trial-and-error approaches, commonly referred to as "heuristic search." Some studies have indicated that the optimal operating voltage of a PV module is always very close to a fixed percentage of the open-circuit voltage. MPPT can be achieved by using the open-circuit voltage to predict the optimal operating condition. Similarly, studies have also shown that the optimal operating current can be predicted as a fixed percentage of the short-circuit current. However, the method of linear approximation will not be discussed further since it has been proven to be inaccurate due to the variation of PV materials and aging and temperature effects (Xiao 2007).

8.2 Heuristic Search

For MPPT, heuristic search is one of the most well-known algorithms thanks to its simplicity and effectiveness. The heuristic search method normally refers to the hill climbing (HC) algorithm, which is an optimization technique. The idea can be simply understood as follows: the top of a hill can be identified if the direction of movement is always up. The algorithm can be used for MPPT since the P–V curve of PV output is hill-shaped. For MPPT, an iteration is applied to change the system operating condition and then sense if the change produces increasing power output (up the hill). If so, the next increment is made in the same direction. In principle, repeating this operation will eventually lead to a maximum, which is the MPP.

The search can also be based on either the P–I curve or the curve of power versus switching duty cycle (P–D), since switching-mode converters are widely used as the PV power interface (Xiao and Dunford 2004b). The switching duty cycle is the control variable used for the majority of PV-side converters, as discussed in Section 5.1. The implementation of the HC algorithm for MPPT is illustrated in Figure 8.3.

The symbol x is the control reference, which can be either the voltage or current of the PV link, or the switching duty cycle if the P–D curve is used for the tracking. The symbol Δx is a constant reflecting the incremental value. The subscripts "new" and "old" are used to represent the recent and historical values of the PV output power and the control variable x. At the end of each MPPT cycle, the latest reference, x_{new}, is output for control.

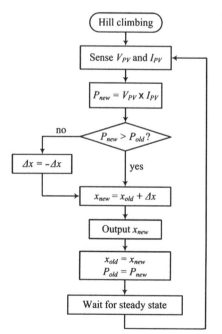

Figure 8.3 Hill climbing algorithm for maximum power point tracking.

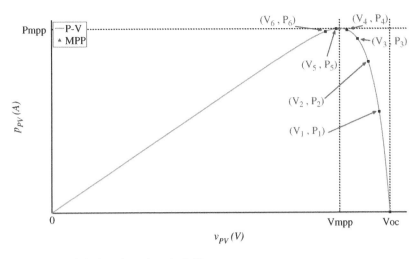

Figure 8.4 Hill climbing based on the P–V curve.

Special attention is given to the "wait for steady state" step at the end of the MPPT loop. When a perturbation is applied, the controller should wait until the system responds and enters a steady state and then make the next measurement. This is important to avoid any misleading information during transient stages, which would cause a malfunction of the MPPT algorithm. The waiting time can be specified in the tracking frequency. The value should be determined according to the system dynamics. The system-dynamics analysis for PV-link voltage regulation was described in Chapters 6 and 7. The tracking time should always be longer than the settling time of the voltage-regulation loop so as to guarantee a stable MPPT operation.

Figure 8.4 illustrates a normalized P–V curve used to demonstrate the HC operation. The initial operation point is commonly referred to the open-circuit condition, V_{oc}, which shows no power output. When the HC algorithm starts, the new operating point (V_1, P_1) is recognized by the MPPT algorithm. The next direction of movement is decided by the corresponding change of power level. Since the new operating point (V_1, P_1) makes the PV modules output more power than the previous one, a relocation to (V_2, P_2) is made by continuing in the same direction. This continues until the movement from (V_5, P_5) to (V_6, P_6), when the MPPT algorithm senses a reduction of the output power by comparing the values of P_5 and P_6. The next direction for the operating point is to move back to (V_5, P_5), and then (V_4, P_4). This process continues until the operating point moves backwards and forwards around the MPP (V_5, P_5). The illustration shows that the HC-based MPPT is capable of finding the local MPP and regulating the system around it.

In the example in Section 5.1.4, the boost topology is used as the PV-side converter (PVSC). At STC, the MPP is the operating condition when the switching duty cycle is 22.9%. The variation of the duty cycle can change the PV power output, as illustrated in Figure 8.5. The hill-shaped P–D curve suits the operation of HC-based MPPT. Following the MPPT procedure, as set out in Figure 8.3, the continuous perturbation of the duty cycle with observation of the power-change direction can identify the MPP, which is at (22.9%, P_{mpp}). The simulation is based on the switching-mode power converter, and

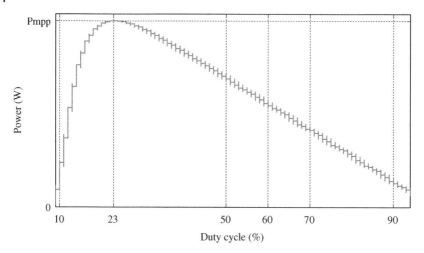

Figure 8.5 Hill climbing based on the P–D curve.

the overshoot and undershoot are noticeable in the recorded curves, caused by the step changes of the duty cycle.

In the steady state, the tracking accuracy of HC-based MPPT is determined by the step size, Δx. The smaller the step size, the higher accuracy in the steady state, since the final oscillation is close to the true MPP, as marked in Figure 8.4. On the contrary, a high value of Δx indicates a fast tracking process, since fewer steps are required to reach the area near the MPP. Continuous perturbation and observation is generally a robust and effective way to locate the MPP in real-time, regardless of any variations of solar irradiance and cell temperature.

HC-based MPPT is commonly referred to as the method of perturbation and observation (P&O), and it is frequently discussed in the MPPT literature. It should be noted that the principle of P&O is the same as the HC algorithm; P&O is simply a dedicated term used for the MPPT algorithm in PV power systems. The HC algorithm is a generic optimal algorithm, with uses in fields other than MPPT. A perturbation is always required to run heuristic search algorithms, such as HC or P&O. The two terms are equivalent. In the following sections, only HC-based MPPT is referred to, so as to avoid confusion.

A Simulink model can be built to simulate HC-based MPPT, as shown in Figure 8.6. It can be implemented by following the flowchart in Figure 8.3. The latest measured power level *Pnew* is compared with the previous measurement *Pold* to determine the sign of the next perturbation. The perturbation amplitude is implemented in the unit delay block, indicated by *delta-x*. The sign of *delta-x* is always determined by the comparison between *Pnew* and *Pold*. The MPPT output is shown as *Xnew* and can be either increased or decreased from the previous value, *Xold*. The tracking frequency is implemented in the three unit delay blocks using the settling of the sample time. Special attention is given that the sampling time of the unit delay blocks matches the MPPT tracking speed, which is specified by the system dynamic analysis. The sampling time for MPPT should not be confused with the sampling time for the overall system simulation.

Continuous oscillation around the MPP is an intrinsic problem of the standard HC-based MPPT algorithm (Xiao and Dunford 2004b). If the incremental step is low, it takes a long time to find the MPP initially or after a change in environmental conditions.

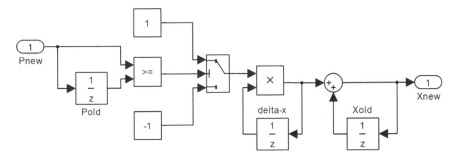

Figure 8.6 Simulink model of the hill-climbing-based maximum power point tracking.

However, if a high value of Δx is used, the power fluctuation is significant around the MPP and results in energy waste, as detailed in Table 8.1. The tradeoff between tracking accuracy and speed should always be considered when a fixed value of Δx is used for HC-based MPPT.

8.3 Extreme-value Searching

One development of MPPT is based on the extreme value theorem (EVT), according to which the extremum, either maximum or minimum, occurs at a critical point, as expressed in (8.1). The critical point x_0 is the local extremum found from $y(x_0)$, where y is a function of x. The critical point can be continuously tracked and updated to follow (8.1).

$$\left.\frac{dy(x)}{dx}\right|_{x=x_0} = 0 \tag{8.1}$$

For the PV output, the power reaches a local peak if the condition is satisfied by (8.2), which assumes that p_{pv} is a function of v_{pv}. Based on the EVT, an MPPT algorithm can be developed.

$$\left.\frac{dp_{pv}(v_{pv})}{dv_{pv}}\right|_{v_{pv}=V_{mpp}} = 0 \tag{8.2}$$

As discussed in Chapter 4, the function p_{pv} with respect to v_{pv} can be established. However, the coefficients of the function are constant only when the environmental conditions in terms of solar irradiance and cell temperature are steady. In the real world, the function is time-variant due to environment variations. Furthermore, the mathematical model coefficients of the PV cell are difficult to identify in real time.

One approach uses a numerical approach – the Euler method – to approximate the operation of the extreme value search. The approximation approach does not require a mathematical model to represent the PV output characteristics. The truncation error should be considered in the numeral differentiation, as pointed out by Xiao et al. (2007a). Therefore, the numerical differentiation can be expressed by (8.3) and (8.4), which are based on the forward Euler and backward Euler methods, respectively.

$$\left.\frac{dp_{pv}}{dv_{pv}}\right|_{V_{k-1}} = \frac{P_k - P_{k-1}}{\Delta V_k} + O(\Delta V^2) \tag{8.3}$$

where $\Delta V_k = V_k - V_{k-1}$.

$$\left.\frac{dp_{pv}}{dv_{pv}}\right|_{V_k} = \frac{P_k - P_{k-1}}{\Delta V_k} + O(\Delta V^2) \tag{8.4}$$

The expressions are represented in discrete time, where P_k and P_{k-1} are the adjacent records of the measured power, and V_k and V_{k-1} are sequential values of the measured voltage. The local truncation error for the Euler methods is equal to $O(\Delta V^2)$, which stands for the order of ΔV. The expressions in both (8.3) and (8.4) indicate first-order accuracy (Xiao et al. 2007a). The local truncation error is defined to represent how well the exact solution satisfies the numerical scheme. The capital O notation is used to characterize the residual term of a truncated infinite series in mathematics. According to (8.3) and (8.4), the MPP is considered to be tracked if the condition of (8.5) is satisfied when the value of the truncation error is insignificant and $P_k \approx P_{k-1}$.

$$\frac{P_k - P_{k-1}}{\Delta V_k} + O(\Delta V^2) = 0 \tag{8.5}$$

Another MPPT algorithm that is frequently discussed in literature, and which is based on the EVT, is the incremental conductance method (IncCond). Its mathematical expression is

$$\frac{dp_{pv}}{dv_{pv}} = \frac{d(v_{pv})i_{pv}}{dv_{pv}} = i_{pv} + v_{pv}\frac{di_{pv}}{dv_{pv}} = 0 \tag{8.6}$$

where the differentiation is changed from dp/dv to di/dv since $p = v \times i$. The numerical approximation based on the Euler method is

$$I_k + V_k\frac{(I_k - I_{k-1})}{\Delta V} + O(\Delta V^2) = I_k + V_k\frac{\Delta I_k}{\Delta V_k} + O(\Delta V^2) \approx 0 \tag{8.7}$$

The IncCond method is established according to the equilibrium and is expressed as

$$-\frac{I_k}{V_k} \approx \frac{\Delta I_k}{\Delta V_k} \tag{8.8}$$

The flowchart of the IncCond algorithm is illustrated in Figure 8.7. When the condition

$$-\frac{I_k}{V_k} < \frac{\Delta I_k}{\Delta V_k} \tag{8.9}$$

is met, the current operating point is expected to be on the left-hand side of the MPP in the I–V curve. An increase of v_{pv} should be executed in order to approach the MPP. This is equivalent to the condition, $dp/dv > 0$. On the other hand, the algorithm sends a command to decrease v_{pv} when the equivalent condition, $dp/dv < 0$ is detected. The perturbation that is caused by the reference change can be stopped if the equilibrium in (8.8) is satisfied according to the IncCond algorithm. This indicates that the MPP has been successfully located and that the perturbation should be stopped. This aims to eliminate a steady-state oscillation around the MPP, which is an intrinsic problem of using the HC-based MPPT algorithm.

However, there are still oscillations under stable environmental conditions when the IncCond algorithm is applied. The condition in (8.2) is seldom satisfied in practical

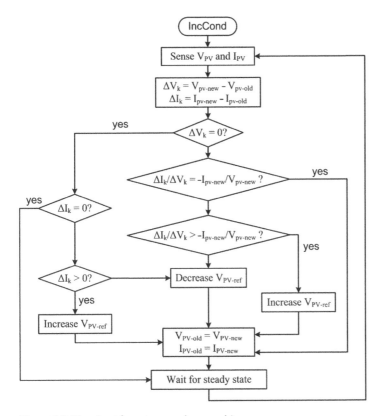

Figure 8.7 Flowchart for extreme value searching.

PV systems. The numerical approximation of the maximum power condition rarely matches the true MPP, which is represented in the continuous time by $dp_{pv}/dv_{pv} = 0$.

This issue, as explained by Xiao et al. (2007a), is caused by the numerical approximation. According to (8.3) and (8.4), the local truncation error of a numerical derivation is always present in Euler methods. Section 8.8 will introduce one approach for reducing the truncation error in order to improve the EVT-based MPPT algorithm.

8.4 Sampling Frequency and Perturbation Size

It should be noted that the MPPT implementation for both HC and EVT is based on digital control in discrete time. The flowcharts of the HC and IncCond, as shown in Figures 8.3 and 8.7, respectively, always update the control reference and indicate the waiting time for the system entering the next steady state. Therefore, one important parameter to execute the MPPT is the perturbation frequency, which is denoted as f_{mppt}. When the dynamic model is derived at the PV link, the value of f_{mppt} can be determined from the settling time of each step response. For example, if the P–D curve is used for the tracking, the dynamic model developed in Section 5.1 should be used to evaluate the settling time. If the P–V curve is applied, the system dynamics is considered using the voltage regulation loop, as described in Sections 6.3 and 7.8.

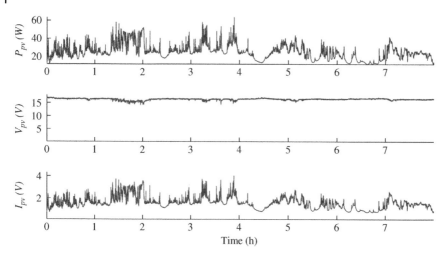

Figure 8.8 Waveforms of PV power (P_{pv}), voltage (V_{pv}), and current (I_{pv}) measured on 9 July 2006.

The perturbation size should be determined with consideration of the ripples that are caused by the switching-mode power interface. For practical applications, the signal-to-noise ratio (SNR) should be considered to avoid any misleading information for the MPPT operation. The noise and ripples in the measured signal can cause malfunctions of MPPT algorithms. The tradeoff of the tracking speed in transient states and the accuracy in steady states should also be considered.

Determining the optimal step size becomes even more difficult if weather variation patterns are considered. Figure 8.8 shows the waveform of the measured power, voltage, and current of a specific circumstance. The eight-hour test was performed in Vancouver, Canada. It is noticeable that the PV power level varies dramatically over a daily period and causes significantly fast dynamics for the MPPT algorithm. For the best energy harvesting under any specific weather conditions, the dynamic performance of MPPT should be emphasized more than the steady state. Large perturbations are expected, which allows for fast tracking of changes of the MPP.

Based on the same PV power system, Figure 8.9 shows a different example, which is a good sunny day. The PV generating power increases smoothly in the morning and decreases slowly in the afternoon. The environmental conditions become predictable. This is commonly considered as an ideal condition for energy harvesting, and the MPPT algorithm shows good performance in the steady state. Some dry areas, such as the Arabian Gulf countries, show similar patterns year round. Comparing the case studies shown in Figure 8.8 and 8.9, a single optimal value of the perturbation size is difficult to choose due to the tradeoff of the tracking speed and accuracy and the regional differences (Du et al. 2015).

8.5 Case Study

The buck converter used for the PV-side power interface was designed in Section 5.1.2, dynamically modeled in Section 6.3.2, and its control was described in Section 7.8.3. This example is based on the same system parameters, and will reveal the approaches

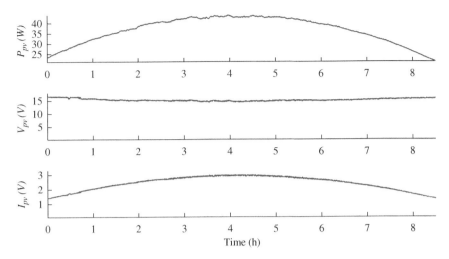

Figure 8.9 Waveforms of PV power (P_{pv}), voltage (V_{pv}), and current (I_{pv}) measured on 24 June 2006.

Table 8.2 Dynamic analysis of the buck converter used for the PV-side power interface.

Configuration	Overshoot	Settling	Description in	Illustration
Original plant for P–D MPPT	69%	9.6 ms	Section 6.3.2	Figure 6.4
Closed-loop for P–V MPPT	≈ 0	1.7 ms	Section 7.8.3	Figure 7.15

to system design, modeling, and control. According to the control diagram, as shown in Figure 7.1, an MPPT block needs to be implemented. Therefore, the system dynamics should be analyzed before the design of the MPPT. The important information about the original dynamics and the closed-loop performance are summarized in Table 8.2. For this case study, the MPPT operation is based on the HC algorithm.

Without the voltage regulation loop, HC-based MPPT can be operated with the P–D curve, which shows the MPP with respect to the switching duty cycle. The minimal perturbation time can be determined from the settling time, which is estimated as 9.6 ms. When the voltage regulation loop is implemented, the system response is improved, with a settling time of 1.7 ms. The minimal perturbation time, which is the response of P_{pv} to a change of v_{pv}, can be selected from the settling time. With a voltage regulation loop, HC-based MPPT can operate five times faster than without.

Figure 8.10 illustrates the overall simulation model, including the blocks for the PV array and DC/DC buck converter. The control functions are the PWM signal generator, the MPPT function, and the PID controller for PV-link voltage regulation. The MPPT algorithm evaluates the power variation and decides the reference for the PV-link voltage. The voltage regulation loop follows commands in order to stay at the MPP regardless of variations of irradiance and temperature.

The MPPT sampling frequency is 400 Hz according to the settling time when the voltage regulation loop is implemented. The simulation sampling frequency 50 MHz, of which one thousand samples are filled for each switching cycle. It should not be confused by the MPPT sampling frequency.

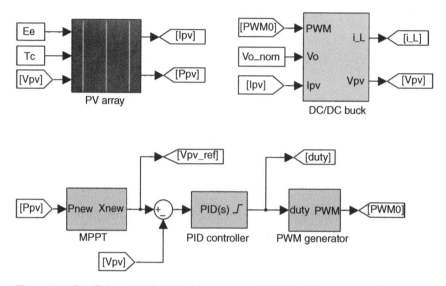

Figure 8.10 Simulink model of the buck converter used for PV-side power interface with MPPT.

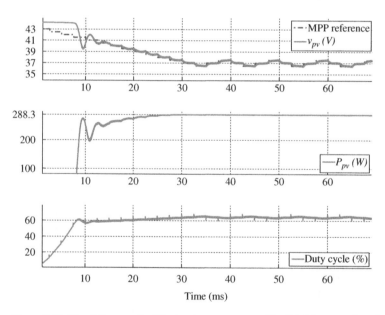

Figure 8.11 Simulation result of the buck converter used for PV-side power interface with MPPT.

The simulation result is shown in Figure 8.11, and includes the MPPT reference signal that is generated by the MPPT algorithm. The voltage regulation loop follows the commands to reach the MPP. After the transient stage, about 34 ms, the HC-based MPPT algorithm continues the perturbation process and maintains the operating point around the MPP. The peak power is 288.3 W. This is shown in the power waveform and agrees with the PV module specification, as shown in Figure 5.4. The switching duty cycle is the direct control variable for the DC/DC buck converter, and is also shown in Figure 8.11.

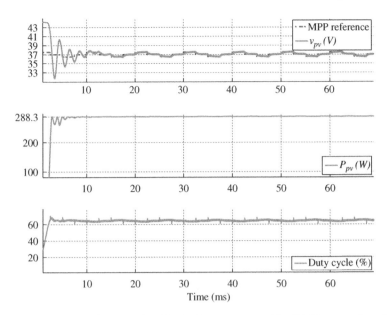

Figure 8.12 Simulation result of the buck converter used for PV-side power interface with MPPT.

The case study demonstrates the HC-based MPPT function and shows the effectiveness of the proposed design, modeling, control, and simulation approaches.

For the regulation for the PV output voltage, the initial position of the MPP can be estimated from the open-circuit voltage, as discussed at the beginning of Chapter 7. In this case, the initial value can be set to be 38 V, which is equivalent to 85% of the open-circuit voltage. The simulation result with the implementation of the initial value is illustrated in Figure 8.12. This shows the steady state is reached for the MPP at 16 ms, in contrast to the 34 ms in Figure 8.11. This shows that an effective way to avoid long tracking times at the initial stage and demonstrates the advantage of using the voltage control loop. The PV-link voltage is an useful reference for MPPT and other purposes, since it does not vary significantly with the irradiance.

8.6 Start-stop Mechanism for HC-based MPPT

One drawback of HC-based MPPT is the intrinsic oscillation at the steady state, which causes energy waste. Even though the incremental conductance method is based on the extremum search theorem, it cannot completely stop oscillations in the steady state due to the truncated error of the numerical differentiation, as described in Section 8.3. One proposed solution is the start-stop mechanism (Khan and Xiao 2016). A three-point oscillation pattern can be identified in the steady state when the MPP is found, as shown in Figures 8.11 and 8.12. A detailed view of three-point oscillation is given in Figure 8.13. This includes the waveform of v_{pv} in a time series together with the P–V curve. The HC-based algorithm directs the operating point alternately to the three points, a, b, and c, which causes the steady oscillation.

The simple idea proposed is to stop the active tracking when the three-point oscillation pattern is identified. Two operation modes are defined: one is called

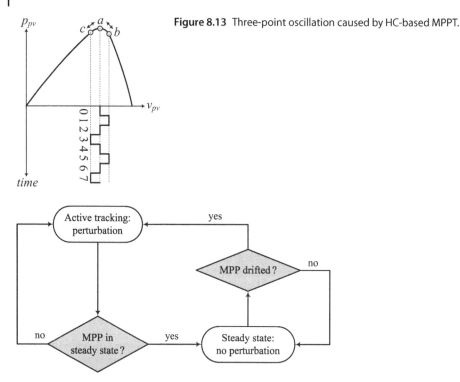

Figure 8.13 Three-point oscillation caused by HC-based MPPT.

Figure 8.14 Flowchart of the start-stop mechanism for hill-climbing operation.

"active tracking" and the other is the "steady state," as shown in Figure 8.14. The HC algorithm represents the active tracking mode, comparing the power variation and the perturbation direction. The steady-state operating mode stops the perturbation and maintains a constant reference for system regulation. PV-link voltage regulation is commonly used to maintain the steady state. The system status is continuously evaluated to detect if the MPP has changed, something that is commonly caused by a variation of environmental conditions. When this is confirmed, active tracking is reactivated until a new steady-state MPP is identified.

The oscillation pattern can be recognized when the perturbation is around the MPP. The perturbation direction changes every two cycles. This phenomenon can be expressed as

$$\Delta V_k = -\Delta V_{k-2} \tag{8.10}$$

where the perturbation step of the HC algorithm is defined as ΔV_k, where the k shows the data sequence in discrete time.

Figure 8.13 illustrates that the step sign is positive at the moment 1, but it is negative at moment 3. The sign is negative at moment 3, but it becomes positive at moment 5. The condition holds true repeatedly in the steady state and shows the three-point oscillation pattern. Therefore, the steady state around the MPP can be identified when three of four consecutive sampling cycles satisfy the condition of (8.10). For robust operation in a practical implementation, it is recommended to look at more than ten cycles before a stop decision is made.

When the perturbation is stopped, the system is controlled in the steady-state mode since the MPP has been found. The PV voltage is regulated to a constant value, which represents the MPP. The output power is continuously sensed and monitored during the steady-state operation. When the power level detected is different from the reference value, the MPP is considered to have drifted to a new value due to environmental variation. The active tracking via the HC algorithm should be activated again to find the new MPP. The start-stop mechanism formed in this way avoids the oscillation problem typical of the HC-based MPPT method.

To evaluate the effectiveness of the start-stop mechanism, a simulation based on the case discussed in Section 8.5 is conducted. A Simulink model is built to combine the start-stop mechanism with HC-based MPPT, as shown in Figure 8.15. Based on the measured power from the PV output, P&O can be conducted and compared with the historical value. The active tracking is implemented in the perturbation block, which is the standard format of the HC algorithm.

The start-stop mechanism estimates the condition and decides to pass the perturbation or stop it. When the steady state is detected, the perturbation signal is assigned to zero in order to stop the perturbation. Otherwise, it follows the output of the perturbation block. The signal shown as Xnew is the control output from the MPPT block, which is updated from the historic value, Xold. The signal represents the PV output voltage reference for the voltage feedback loop when the P–V curve is used for MPPT. The signal can also be the direct control signal, the switching duty cycle for the power interface, when the P–D curve, as illustrated in Figure 8.5, is applied for MPPT.

Figure 8.16 illustrates the Simulink implementation of the start-stop mechanism. Its inputs are the perturbation signal from the perturbation block and the latest power measurement. The status symbol "mode," commands active tracking if the value is 1. A value of 0 leads to the stop mode, which sets the signal Delta-x to zero. The three-point pattern is detected by evaluating (8.10). When the three-point pattern is clearly recognized over a number of cycles, the stop mode commences and the value of PV output power is recorded as the benchmark. The cycle limit is implemented inside the relay block; it is assigned a value of 11 in this case study.

When the system enters the stop mode, the benchmark value is continuously compared with the new measurement of the PV output power. When the difference becomes significant, the mechanism is reset to pass the perturbation and perform active HC tracking. In this case study, the threshold is assigned a value of 3 W. In general, the start-stop mechanism performs the function of evaluation and decision as illustrated in Figures 8.14 and 8.15.

The simulation results are illustrated in Figure 8.17, which includes the waveforms of v_{pv}, p_{pv}, and the switching duty cycle. When the three-point condition, as expressed

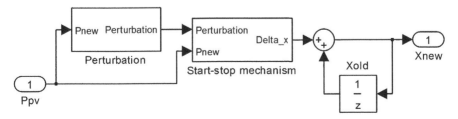

Figure 8.15 Simulink model for integration of start-stop mechanism with HC-based MPPT.

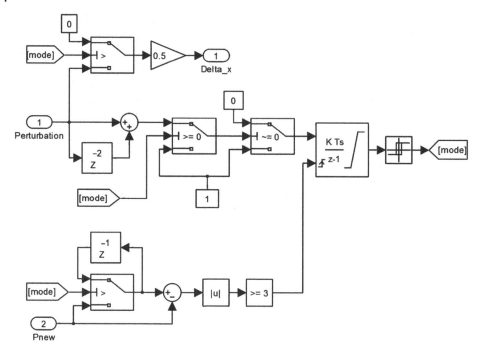

Figure 8.16 Simulink model of the start-stop mechanism.

in (8.10), is satisfied for 11 cycles consecutively, the perturbation is stopped from the MPPT block output, as shown at 33 ms. The voltage of the MPP is determined to be 37 V and is kept steady. A step change of the irradiance from 1000 W/m^2 to 500 W/m^2 is applied at 60 ms. The power variation triggers the active tracking mode to restart, the assumption being that the MPP has drifted to a new value. The active tracking locates the new MPP. After the condition of (8.10) is satisfied over 11 cycles, the perturbation operation is stopped again at 93 ms. The voltage of the new MPP is determined to be 36 V and is the system is then kept steady in this state. The start-stop mechanism has been proven to be a simple and effective way to avoid oscillations in the steady state and to improve the MPPT performance (Khan and Xiao 2016). The effectiveness and improvement can be demonstrated by comparing the simulation results in Figures 8.12 and 8.17.

8.7 Adaptive Step Size Based on the Steepest Descent

Another idea has been proposed to target the drawbacks of HC-based algorithms. This is called the adaptive HC method, and involves dynamic adjustment of the perturbation size (Xiao and Dunford 2004b). When there is an oscillation pattern during MPPT, it is ideal to make the perturbation size significant during the transient stage, but to assign it a low value when the steady state is reached for the MPP.

An optimization method can again be used to achieve the target. One algorithm is the steepest descent method, which is designed to find the nearest local minimum. The same principle can be applied to find the local maximum value – the MPP – when the gradient

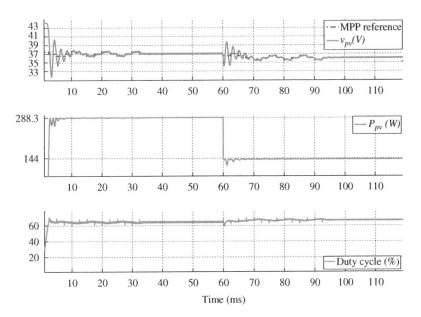

Figure 8.17 Simulation result of the buck converter used for PV-side power interface with the start-stop mechanism for MPPT.

direction is reversed (Xiao et al. 2007a). The method for MPPT can be derived as

$$v_{k+1} = v_k + \epsilon \left. \frac{dp}{dv} \right|_{v_k} \tag{8.11}$$

This is based on the function for PV output power in response to the voltage. The result of the differentiation shows both the tracking direction and the step size. Meanwhile, ϵ is the corrector. In theory, the implementation of the steepest descent allows fast tracking, accurate locating, and zero steady-state error for MPPT. To demonstrate the effectiveness of the approach, the ideal single-diode model (ISDM) is used, with the I–V characteristics expressed as in (4.8). The P–V relation of the PV cell under STC can be derived as

$$P_{pv} = I_{SCS} \times v_{pv} - I_{SS} \times v_{pv} \times e^{\left(\frac{v_{pv}}{V_{TCS}A_n} \right)} + I_{SS} \times v_{pv} \tag{8.12}$$

The derivative is

$$\frac{dP_{pv}}{dv_{pv}} = I_{SCS} + I_{SS} - I_{SS}e^{\left(\frac{v_{pv}}{V_{TCS}A_n} \right)} - \left(\frac{I_{SS}v_{pv}}{V_{TCS}A_n} \right) e^{\left(\frac{v_{pv}}{V_{TCS}A_n} \right)} \tag{8.13}$$

If a PV array is considered, the numbers of series and parallel connections, N_s and N_p, can be applied to the equation.

When the typical P–V curve of the PV output is studied, the derivative results in a relatively significant value when the operating point is far from the MPP. For the PV module in the case study, the derivative for each operating point can be derived from (8.13), as illustrated in Figure 8.18. When the derivative is computed at the MPP, the value approaches zero, according to the EVT. At an open-circuit voltage that is far from the MPP, the derivative has a value of −1.94, which indicates the direction to the MPP

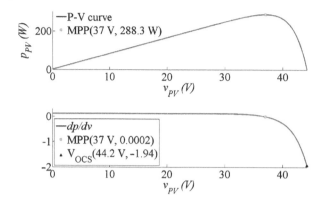

Figure 8.18 PV output: top, P–V curve; bottom, *dp/dv* curve.

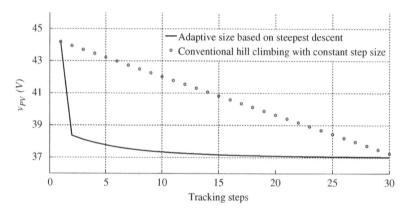

Figure 8.19 Maximum power point tracking based on steepest descent algorithm.

involves a reduction of the PV output voltage. When the derivative value is used as the reference for each tracking cycle, the step size can be automatically adjusted for fast tracking or a zero error in the steady state. The magnitude is calculated to be large when the operating point is far from the MPP and becomes lower and lower and ultimately zero when it reaches the MPP.

The simulation is illustrated in Figure 8.19, and shows the voltage of the MPP is 37 V, which matches the PV module output as shown in Figure 8.18. The correction factor, ϵ, which is expressed in (8.11), is chosen as 1, meaning that a large step is triggered by the steepest descent algorithm at the initial point. The step becomes smaller and smaller as the operating point approaches the MPP. In comparison with the conventional HC with a constant step size (0.5 V), the algorithm with an adaptive step size demonstrates the potential for fast tracking, and eventually eliminates steady-state errors when the MPP is located.

The discussion above is based on the ideal condition that all parameters are available for the PV output model, including solar irradiance and cell temperature. In reality, the power and voltage function is nonlinear and changes with environmental conditions, which are difficult to identify in real time. In most cases, the steepest descent algorithm cannot be built using (8.13) and a numerical approximation is used instead of a full mathematical model to determine the step direction and size (Xiao and Dunford 2004b). The tracking algorithm is therefore expressed by

$$v_{k+1} = v_k + \epsilon \left(\frac{P_k - P_{k-1}}{V_k - V_{k-1}} \right) \tag{8.14}$$

The adaptive step size is determined from ratio of the variation in power and the variation in the perturbation. However, this numerical approach always has a truncated error, as discussed in Section 8.3. An accurate approximation can be achieved by a reduction of the step size of the voltage, expressed as $\Delta V_k = V_k - V_{k-1}$. The voltage step is constrained to reach a certain lower limit when the noise and inaccuracy in measurement are considered in a practical implementation. Otherwise, the operation of (8.14) might fail, because of the inaccuracy when the SNR becomes insignificant.

The stability should be evaluated when numerical approaches, such as the Euler methods, are implemented in practical systems. Disturbances and measurement noise are two major causes of malfunctions. In terms of stability, numerical differentiation is more critical than numerical integration because numerical differentiation requires Lipschitz classes for stability (Rowland and Weisstein n.d.). Following (8.14), the numerical differentiation must satisfy the Lipschitz condition, which can be expressed as

$$|P_k - P_{k-1}| \leq C_{LIP} |V_k - V_{k-1}| \tag{8.15}$$

The coefficient, C_{LIP}, is a predefined constant. For MPPT, C_{LIP} can be defined as the maximum value in terms of dp/dv, which refers to the open-circuit condition. The magnitude can be determined from offline analysis of the P–V curve of the PV output.

When the PV module is selected, the output characteristics can be determined at STC. The design example is based on the PV module output illustrated in Figure 8.18. The coefficient, C_{LIP} should be assigned to a value of 1.94 since it represents the upper level of the values of dp/dv across the voltage range. For numerical stability, the Lipschitz condition should be computed and evaluated every MPPT cycle for numerical differentiation. This is effective in preventing any invalid numerical differentiation that would result in instability.

8.8 Centered Differentiation

Research has shown that MPPT based on the EVT is ideal in principle to achieve both fast tracking and zero steady-state errors. However, the approach is based on numerical differentiation, which shows truncation errors and is difficult to use. One alternative approach aims to minimize the truncated error in the numerical approximation and to show the advantage of extremum value searching for MPPT implementations (Xiao et al. 2007a). The methodology is based on centered differentiation.

In Figure 8.20, the MPP is located at the point, (V_k, P_k), which is centered between the two points, (V_{k-1}, P_{k-1}) and (V_{k+1}, P_{k+1}). The numerical differentiation can be based on the symmetric form, expressed as

$$\left. \frac{dp}{dv} \right|_{(V_k, P_k)} = \frac{P_{k+1} - P_{k-1}}{V_{k+1} - V_{k-1}} + O(\Delta V^3) \tag{8.16}$$

where $\Delta V = V_{k+1} - V_k$ or $V_k - V_{k-1}$ when the step is considered to be constant.

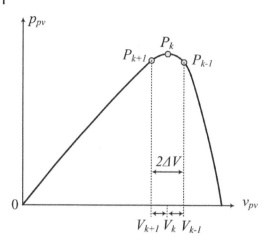

Figure 8.20 Centered differentiation based on the P–V curve.

When the MPP is reached, the conditions expressed in (8.17) and (8.18) hold true:

$$|P_{k+1} - P_{k-1}| < |P_k - P_{k-1}| \tag{8.17}$$

$$|P_{k+1} - P_{k-1}| < |P_{k+1} - P_k| \tag{8.18}$$

The value of the numerical differentiation by (8.16) is very close to zero when the MPP is centered. Therefore, this approach provides a more accurate approximation than the Euler-based methods since the local truncated error is significantly reduced. It is expressed as $O(\Delta V^3)$, which is of order 2. Unlike the forward or backward Euler methods, centered differentiation has no phase error when the approximation is analyzed in the frequency domain (Xiao et al. 2007a).

MPPT using centered differentiation is efficient when implemented with the steepest descent method and the start-stop mechanism (Xiao et al. 2007a). Figure 8.21 illustrates the flowchart to implement the algorithm.

Three measurements are needed for each computation of the centered differentiation, as expressed in

$$\left.\frac{dp}{dv}\right|_{(V_k, P_k)} \approx \frac{P_{k+1} - P_{k-1}}{2\Delta V} \tag{8.19}$$

Since the truncated error is significantly reduced, a stop decision can be made if the value of the numerical differentiation is so low that it can be ignored. When the steady state is reached, the PV output voltage is regulated to the value representing the MPP. Active tracking can be reinitialized when MPP drift is sensed by the algorithm introduced in Section 8.6. In the flowchart, the numerical differentiation is also evaluated by the Lipschitz condition, which was given in (8.15).

When the Lipschitz condition is not satisfied, the MPPT operation is switched to conventional HC operation with a fixed value of the perturbation size. Otherwise, centered differentiation is always used for active tracking, with the step size and direction determined with the steepest descent algorithm. The start-stop mechanism monitors the operating status and makes decisions on whether to start or stop active perturbation.

A comprehensive evaluation of the proposed algorithm has been reported by Xiao et al. (2013). Their study shows that the MPPT improvement is about 1% in

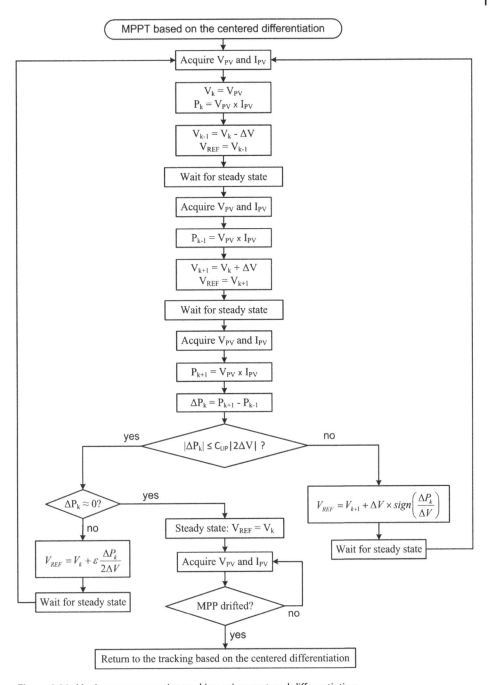

Figure 8.21 Maximum power point tracking using centered differentiation.

comparison with the conventional P&O algorithm. The long-term power generation is reported to increase by up to 4.5% when the MPPT algorithm shown in Figure 8.21 is applied. In practice, the algorithm is effective in accommodating changes in weather conditions and achieving high solar energy harvesting.

8.9 Real-time System Identification

For MPPT, the ideal is to directly determine the MPP of the PV array without any perturbation process. If the PV power output can be modeled and identified in real time, the variation of the MPP can be estimated and predicted to avoid the need for a search based on the trial and error. One study shows the potential for such an application. It uses a polynomial model and recursive least squares (Xiao et al. 2006).

8.9.1 Recursive Least Squares Method

A polynomial model for the I–V characteristics of PV cells was discussed in Section 4.7. This can model the outputs of both crystalline-based and thin-film cells. When the polynomial model has been established offline, the least squares (LS) method can be used to identify the variation of the model parameters in discrete time. An example is based on the Siemens ST-10 PV module. The PV output characteristics are simulated using the polynomial model and are plotted next to the measured data in Figures 4.28 and 4.29. The model parameters were identified by the LS method and are summarized in Tables 4.10 and 4.11.

Since the output characteristics of PV cells vary with changes in irradiance and temperature, real-time identification is required to track the parameter variations. When the model is identified, the local MPP can also be found using the mathematical model. In discrete time, the PV power output can be expressed as the polynomial function of the voltage in discrete time:

$$P_k = I_k V_k = b_{p0}(k)V_k + b_{p1}(k)V_k^2 + b_{p2}(k)V_k^3 + b_{p3}(k)V_k^4 \tag{8.20}$$

where $b_{p0}(k), b_{p1}(k), b_{p2}(k)$, and $b_{p3}(k)$ are the polynomial parameters. The sequential index k indicates the time-variant feature of the parameters. The polynomial coefficients are defined in a vector, $\theta(k)$:

$$\theta(k) = \begin{bmatrix} b_{p0}(k) \\ b_{p1}(k) \\ b_{p2}(k) \\ b_{p3}(k) \end{bmatrix} \tag{8.21}$$

The PV output voltage and its corresponding polynomials are defined in a vector $\phi(k)$:

$$\phi(k) = \begin{bmatrix} v_k \\ v_k^2 \\ v_k^3 \\ v_k^4 \end{bmatrix} \tag{8.22}$$

Therefore, the expression in (8.20) can be written in a simplified format:

$$P_k = \phi^T(k)\theta(k) \tag{8.23}$$

The LS algorithm can be performed when all data are collected for offline parameter estimation. The recursive least squares (RLS) algorithm is an adaptive method that is continuously and recursively operated in order to find the coefficients in real time. With RLS, the parameter vector can be estimated by following the recursive equations from (8.24) to (8.26):

$$M(k) = M(k)\phi(k) = \frac{W(k-1)\phi(k)}{\lambda + \phi^T(k)W(k-1)\phi(k)} \tag{8.24}$$

$$\hat{\theta}(k) = M(k)[P_k - \phi^T(k)\hat{\theta}(k-1)] \tag{8.25}$$

$$W(k) = \frac{[I - M(k)\phi^T(k)]W(k-1)}{\lambda} \tag{8.26}$$

$M(k)$ are the weighting factors that determine the direction of correction, $W(k)$ is a non-singular matrix that is specifically required for the RLS operation, and λ is the forgetting factor, which ranges from zero to unity. The forgetting factor is the parameter to weight the latest samples more than older ones for real-time identification.

The identification process can be illustrated in a flowchart, as shown in Figure 8.22. The recursive equations need to start with the initial condition, which is expressed as the matrix $W(0)$ and vector $\hat{\theta}(0)$. The PV output voltage and current are acquired in every sampling cycle. Following (8.24), the matrix $M(k)$ is updated for the correction. The forgetting factor λ is applied to distinguish recent data from older data. According to (8.25), the parameter vector $\theta(k)$ is updated every sampling cycle with the correction between the measured power, P_k, with an estimated value $\phi^T(k)\hat{\theta}(k-1)$. After the current cycle estimation, the historical data is updated for the next cycle. The cycle is repeated to identify the variation in the model parameters. The parameter vector, $\theta(k)$, is expected to converge and represent the system model regardless of any parameter variation.

Figure 8.22 System identification based on recursive least squares method.

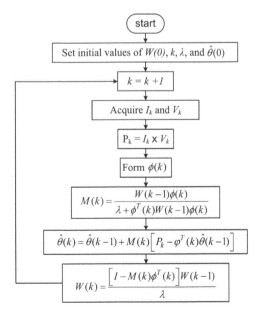

8.9.2 Newton–Raphson Method for MPP Determination

When the polynomial coefficients, $b_{p0}(k)$, $b_{p1}(k)$, $b_{p2}(k)$, and $b_{p3}(k)$ are identified in real time, the P–V characteristics are expressed by (8.20). The MPP can be found by satisfying the EVT, namely $dp/dv = 0$. Since the presented approach is model-based, a numerical approximation of the differentiation is no longer required, which is different from the EVT-based MPPT algorithms. For a fourth-order polynomial representing the P–V characteristics, the MPP can be located by satisfying

$$\frac{dp}{dv}\bigg|_{V_{MPP}} = b_{p1} + 2b_{p2}v + 3b_{p3}v^2 + 4b_{p4}v^3 = 0 \tag{8.27}$$

The function in (8.27) can be numerically solved using the Newton–Raphson method (NRM). An additional differentiation is also required, which is given by

$$\frac{d^2p}{dv^2} = 2b_{p2}v + 6b_{p3}v + 12b_{p4}v^2 \tag{8.28}$$

When the NRM is applied, the numerical solver is as illustrated in Figure 8.23. The derivatives in (8.27) and (8.28) are computed and lead to convergence and the local maxima. When the voltage value is identified and considered to be steady-state, the algorithm stops and outputs the result. The output value is considered as the MPP reference for the voltage regulation loop. The RLS and NRM algorithms should be implemented in the MPPT block to adjust the operating point for maximum PV power production.

8.9.3 Forgetting Factor

Exponential forgetting of data can be used to identify the time-varying parameters of PV panels in response to changes of solar irradiance and cell temperature. The parameter, λ, is the forgetting factor that determines the forgetting speed. When a constant forgetting factor is used with the RLS method, the lower its value, the faster the algorithm converges. However, a low value of λ may lead the matrix $W(k)$ to blow up and result in unstable estimations (An and Yao 1988). A tuning process is required to reach

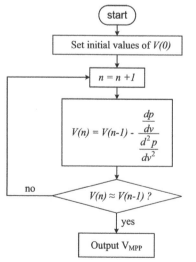

Figure 8.23 Model-based determination of the MPP using the Newton–Raphson method.

an optimal value of λ that balances the convergence speed and stability. It is desirable to make the value of λ adaptive to the system dynamics. The forgetting factor is adjusted according to the prediction error as

$$\lambda(k) = 1 - K_\lambda \varepsilon^2(k) \tag{8.29}$$

$$\varepsilon(k) = P_k - \hat{P}_k \tag{8.30}$$

where K_λ is a parameter chosen according to the system dynamics, and ε represents the error between the prediction and measurement of the output power.

The PV-link voltage is usually the variable to be tracked and regulated to follow the optimal operating condition. Temperature dominates the change of the optimal voltage to represent the MPP. Generally, the thermal dynamics are slow in comparison with other operating conditions; a sudden increase or decrease of temperature seldom occurs. Therefore, the forgetting factor needs to be tuned according to the dynamics caused by variation in temperatures.

8.10 Extremum Seeking

Extremum-seeking (ES) is an optimization approach to identify the local minima or maxima for nonlinear plant (Ariyur and Krstic 2003). A mathematical model of the non-linear plant is not required for use of the ES method. Local convergence allows it to be used for various areas (Liu and Krstic 2012). The implementation of the deterministic ES method is shown in Figure 8.24. The plant is represented by a black box, indicating the unknown and time-variant nature of the function, $p_{pv} = f(v_{pv})$.

A sinusoidal excitation signal, $A_p \sin(\omega t)$, should be added to the control variable in order to probe the nonlinearity and identify its gradient. A high-pass filter is applied to distinguish the excitation from the plant output from other components. Multiplication with the same perturbation signal, $\sin(\omega t)$, produces the estimate of the derivative, $f'(v_{pv})$. By considering the P–V curve, a negative value of the current time derivative indicates a position to the right of the MPP, and a positive value indicates the left-hand side. The integrator generates a value of v_{pv} leading in the direction that will make $f'(v_{pv}) = 0$. As shown in Figure 8.24, the ES-based tracker accepts the PV output power as the input and updates the PV-link voltage as the reference. Eventually, the algorithm drives the estimated voltage v_{pv} to the nearest maximum of the PV output power, p_{pv}. This basically matches the objective of MPPT. The synthesis of ES-based MPPT includes four parameters, which are summarized and explained in Table 8.3. The parameters are defined in the diagram for the ES algorithm shown in Figure 8.24.

Figure 8.24 Block diagram for extremum-seeking control.

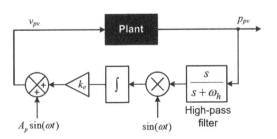

Table 8.3 Parameters and synthesis of ES scheme.

Parameter	Description and synthesis
A_p	The value of A_p is decided by the tradeoff of the residual error and the tracking speed.
k_e	A gain parameter that mainly determines the tracking speed. The higher the value, the faster the speed expected.
ω	The perturbation frequency, which must be distinguished from the plant dynamics and noise.
ω_h	Parameter to determine the cut-off frequency in order to eliminate the DC component in the output signal.

To demonstrate the technology, the polynomial model of the PV module is used, as developed in Section 4.7. The P–V characteristics of the ST-10 PV module are expressed as the fourth-order polynomial in (4.76), and the MPP can be found by solving

$$\frac{dp_{pv}}{dv_{pv}} = b_{p1} + 2b_{p2}v_{pv} + 3b_{p3}v_{pv}^2 + 4b_{p4}v_{pv}^3 = 0 \tag{8.31}$$

Based on the parameters shown in Table 4.11, the MPP is located at the point (15.51 V, 9.96 W), as shown on the P–V curve. For demonstration purposes, although the polynomial model is implemented inside a black box, as shown in Figure 8.24, it is unknown for the tracking algorithm. ES-based MPPT should be able to find the MPP without preknowledge of the system model.

Figure 8.25 shows the simulation model for ES tracking. The noise signal is intentionally coupled with the model output to simulate practical conditions. It reflects the ripples that are caused by the switching operation of the power converters or other high-frequency noise. To demonstrate the ES operation, Table 8.4 describes three different cases for which the different parameters are specified. The ES operation is expected to identify the voltage value of 15.51 V, which corresponds to the MPP.

Figure 8.26 shows the simulation result in the first case study. The MPP is located within 7 s by the ES algorithm. The peak-to-peak ripple of the PV output voltage is 2 V, corresponding to the assigned value of A_p.

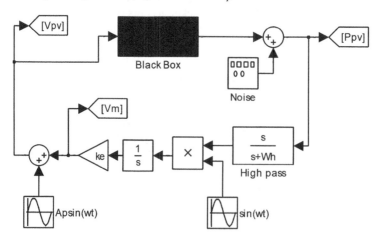

Figure 8.25 Simulink model for the extremum seeking scheme.

Table 8.4 ES parameters in different case studies.

Parameter	A_p	k_e	ω (rad/s)	ω_h (rad/s)
Case 1	1	10	1000	100
Case 2	0.2	10	1000	100
Case 3	0.2	100	1000	100

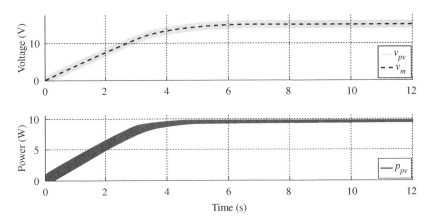

Figure 8.26 Performance of maximum power point tracking using extremum seeking with the configuration of Case 1.

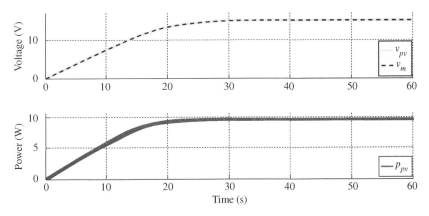

Figure 8.27 Performance of maximum power point tracking using extremum seeking with the configuration of Case 2.

Figure 8.27 shows the simulation result of the second case study. In contrast to the first case, the amplitude, A_p, is reduced to 0.2 V. This results in the peak-to-peak ripple of the PV output voltage being 0.4 V. However, it takes 35 s to reach the MPP due to the relatively low value of A_p.

Figure 8.28 shows the simulation result in the third study case. The peak-to-peak ripple of the PV output voltage is the same (0.4 V) as the second case study, since both share the same value of A_p. The seeking speed is significantly improved by the relatively high

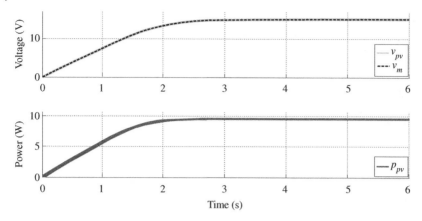

Figure 8.28 Performance of maximum power point tracking using extremum seeking with the configuration of Case 3.

value of k_e. The MPP is located after 3.5 s of tracking. In general, higher values of A_p and k_e improve the tracking speed. However, there is always an upper limit for the value of k_e, which is common for any control system when the tradeoff between the system robustness and performance is considered for practical applications. A lower value of A_p improves the steady-state performance since the ripple is reduced.

A modified version of the ES algorithm used for MPPT does not require the external perturbation signal, in contrast to the design cases shown above. The approach utilizes the inherent ripple from the switching-mode power converter as the excitation signal (Brunton et al. 2010).

8.11 Multiple Power Peaks and Global MPPT

As discussed in Section 2.2.1, the mismatch of PV modules causes multiple power peaks when all are series connected. In one case, two power peaks are noticeable in the P–V plot. Neither can represent the true available power, which is equal to the sum of the individual maximum powers of the two PV modules. The mismatch is due to a bottleneck effect caused by the under-performing PV panel. However, the case study is based on a simple shading scenario with only two PV modules in series connection. In reality, the mismatch condition can be very complicated when more than two panels are connected in series in each string. In centralized PV systems, the strings are in parallel connection to form an array. This might result in multiple peaks and complicated scenarios that are difficult to identify and predict.

The most effective way to minimize the effect of mismatch conditions is independent MPP at string, panel, and even submodule levels. The approach is commonly referred to as distributed MPPT (DMPPT), and has been broadly discussed and analyzed in Chapter 2. A DMPPT system creates an independent channel for each power contributor, with power collected at the final stage for grid injection or load consumption. The mismatch effect is ultimately eliminated or reduced according the level at which MPPT is applied.

Even though the mismatch condition prevents a centralized control system from delivering the highest amount of solar energy, significant research has been conducted to minimize the impact. The MPPT algorithm tries to find the optimal operating point, namely the global maximum power point (GMPP). For example, the case study in Section 2.2.1 shows two power peaks, one higher than the other under shading conditions. The highest point is considered the GMPP, giving the highest power output under the mismatch circumstance. Since partial shading is the main cause of such multi-peak phenomena, MPPT research mainly focuses on tracking the global peak instead of any local ones. The approach is commonly referred to as global maximum power point tracking (GMPPT).

Since MPPT algorithms are amenable optimization techniques, the particle swarm optimization (PSO) approach has been proposed for finding the GMPP under partial shading conditions. The method was originally developed in computer science, where it is treated as a multi-variable or multi-agent problem. For GMPPT, the PV array power is defined as the objective function. It is influenced by the voltage of the individual PV modules which are the agents. The agents share information with each other, leading to a unique peak in multidimensional space. Similar to the concept of hill climbing, the PSO can be considered an advanced iterative approach to manage all the agents so that they converge to the position corresponding to the global extremum. Although research has proven the effectiveness of PSO and other optimization techniques for GMPPT, practical implementation is difficult since it sensitive to the initial conditions and other settings. Considering the time-variant nature of PV power generation, practical applications are even more difficult since stability cannot always be proven under various disturbance conditions (Xiao et al. 2011).

8.12 Performance Evaluation of MPPT

The objective of the MPPT has been clearly defined as to extract the highest solar power output from each PV cell at any moment. An number of MPPT algorithms have been recently proposed to give performance superior to previous solutions. It is very important to develop a platform for fair comparison. However, comparisons are difficult since solar irradiance and ambient temperature are not controllable and not repeatable (Xiao et al. 2013).

8.12.1 Review of Indoor Test Environment

In the past, many MPPT algorithms were tested indoors. In general, with the controllable energy sources, the performance comparison of MPPT can be planned and scheduled to give a fair comparison. Based on previous studies (Xiao 2003; Xiao and Dunford 2004a), the indoor testing methods are summarized and described in the following.

1) Computer simulation
   - Always effective to quickly prove the design concept.
   - Low cost and very flexible, allowing variation of environment, loads, and grid.
   - Cannot be used to prove MPPT performance on its own.
   - Should not be confused that the hardware-in-the-loop (HIL) techniques, which are considered a simulation approach, but include some real system components.

- HIL simulation output is sometimes considered an experimental result since the HIL platform can interface with analog signals for instruments such as oscilloscopes. It should be clearly noted that the HIL system is a simulation-based platform, not an experimental one.
2) PV array simulator
  - Defined as a DC power supply which is programmable to simulate PV output characteristics (see Section 1.7).
  - Considered a moderate-cost and flexible solution for repeatable indoor tests of power-conditioning circuits and control algorithms.
  - The simulator can be flexibly programmed to represent different types of PV generators and different environmental variations.
  - The PV array simulator is a power electronic device.
  - It shares the same drawbacks as other switching-mode power supply systems, such as ripple in the current and voltage waveforms.
  - It might show self-resonance when connected to another power converter.
  - The output characteristics of PV arrays can change very quickly in response to variations in operating conditions.
  - The PV array simulator generally shows much slower responses than real PV material.
  - The constraint of dynamic response, which does not match real PV modules, should always be considered when it is used to validate fast MPPT algorithms.
3) Artificial light and temperature test chamber
  - Fully controlled environment using artificial lights can provide variable irradiance levels and regulated ambient temperature for indoor testing and evaluation.
  - Could be one of the most expensive solutions since a solar simulator with accurate solar spectrum is generally expensive and limited to small scale.
  - Difficult to test a system with significant power capacity due to the high cost of artificial light and high power consumption.
  - Furthermore, it is also a high cost to maintain chamber at constant temperature for different tests since self-heating always happens when PV modules are exposed to light.

8.12.2 Review of Outdoor Test Environments

Outdoor evaluations have great advantages since the actual MPPT behavior can be examined using real PV arrays and natural sunlight in order to avoid unrealistic effects of artificial sunlight and power electronic simulators. Furthermore, an effective MPPT algorithm performs well in real-world weather conditions rather than any pre-assumed condition. However, the accurate measurement of solar irradiance and cell temperature is difficult. Some tests intentionally choose sunny days with the assumption that the weather pattern would be repeatable. In the past, many MPPT algorithms were tested outdoors. The methods used are outlined in the following list.

1) Periodical interrupt method
  - Uses natural sunlight, which is better than artificial light.
  - Divides the testing period into small time pieces for different MPPT algorithms.
  - Usually stops one test and starts another with the assumption that the environmental condition is steady for all tests.

- Errors can be expected due to the time-variant patterns of irradiance and temperature.
- Impossible to test fast MPPT dynamics in response to conditions such as moving clouds.

2) Day-by-day test based on single PV system
    - Aims for a long-term test to evaluate the MPPT performance in response to different weather conditions.
    - A comprehensive statistical analysis is required to identify the MPPT performance, since the day-to-day weather conditions are not repeatable when the variation of both irradiance and temperature are considered.
    - The drawback lies in the long test period and the complication of statistics to prove that one MPPT algorithm is more effective than another.

3) Day-by-day testing on multiple parallel PV systems
    - Based on at least two PV systems that are identical and installed in the same condition.
    - One can be used as the benchmark for the power output comparison.
    - The two systems should be calibrated under identical conditions in order to identify any mismatches between them due to non-ideal factors in the system components.
    - The test requires multi-day evaluation and statistical analysis to identify the performance of the MPPT since no two systems are exactly identical when component tolerances are considered.
    - The approach is considered the most comprehensive evaluation method since both systems are under the same operating conditions for a long-term test.
    - Both the MPPT dynamics and steady state can be revealed for the highest power generation.

By comparing the three testing methods in natural sunlight, it has been observed that the parallel PV system is the most effective testing platform for MPPT evaluation (Xiao et al. 2013).

8.12.3 Recommended Test Benches for MPPT Evaluation

Results for two different weather patterns have been reported by Xiao et al. (2013), as shown in Figures 8.8 and 8.9. A fair comparison of one day's performance of a PV system against another is impossible since the weather pattern is not repeatable day by day. The results will be inaccurate when one PV system is used to test MPPT performance on different days. Therefore, the following introduction is based on a parallel testing configuration, with two independent PV systems tested and evaluated simultaneously under the same operation conditions and in the same natural environment.

The first test bench system we will consider is recommended for evaluating the MPPT algorithms that are usually used in DC microgrids or two-stage conversion systems. The system diagram is shown in Figure 8.29. The system is formed by a programmable DC electronic load and parallel PV power systems. The MPPT algorithms are implemented to control the two DC/DC conversion units. One can be used as the benchmark; the other can be implemented with an advanced MPPT algorithm in order to prove its effectiveness. The voltage and current at the PV and DC links should be continuously measured and recorded by a data acquisition system to assess the long-term operation.

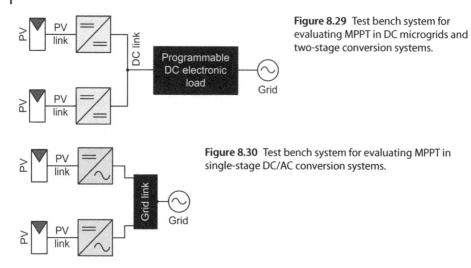

Figure 8.29 Test bench system for evaluating MPPT in DC microgrids and two-stage conversion systems.

Figure 8.30 Test bench system for evaluating MPPT in single-stage DC/AC conversion systems.

The programmable DC electronic load can create a DC link with constant voltage, which is shared by the two PV systems. It is flexible and can be programmed to have other functions, such as voltage disturbance. It should be noted that the majority of programmable DC electronic loads do not inject power into an AC grid. The grid connection is mainly for the ancillary power supply to control the load system. Depending on the application, the PV and DC/DC units can be as small as a single PV module or as large as a PV array. The PV string and array levels are important in testing the MPPT performance in response to partial shading or other mismatch conditions.

The second test bench system is used to evaluate MPPT algorithms for single-stage DC/AC conversion systems. The system diagram is shown in Figure 8.30. The system is connected to the grid directly. The test bench is especially important in testing MPPT performance for single-phase grid interconnections. The single-phase connection introduces significant double-line frequency ripples at the PV link, which limit the MPPT performance. The MPPT algorithms should be independently implemented in the two DC/AC conversion units. One can be used as the benchmark, and the other can implement an advanced MPPT algorithm for improved performance. The voltage and current at the PV link and grid link should also be recorded to enable day-by-day operation to be assessed.

8.12.4 Statistical Paired Differential Evaluation

The test systems in Figures 8.29 and 8.30 are flexible enough to test MPPT performance over any time period, from several days to several years. The comparison assumes that the parallel systems are identical. However, all PV panel manufacturers state power output tolerances ranging from $\pm 1\%$ to $\pm 5\%$. One example was discussed in Section 2.2.1 and illustrated in Figure 2.8. Even though the PV modules are the same model and were manufactured on the same date, the output characteristics are slightly different under identical test conditions. Furthermore, the components that form the DC/DC and DC/AC power converters are often non-ideal. The uncertainty in the parallel systems might therefore dominate the system output, defeating the attempt to make a meaningful performance comparison and leading to a wrong conclusion.

A calibration test should be always conducted using the same MPPT algorithm. The long-term output result reveals any differences between the two independent systems. In addition, the hardware systems can be alternated, swapping the control algorithms in order to minimize any initial mismatch effect. When a significant quantity of data has been collected, paired difference tests can be used to assess the significance of the difference between the two sets of data that are produced by the system power output (Xiao et al. 2013). The paired t-test is used to distinguish the performance of the MPPT algorithms. The t-statistic is expressed as

$$t = \frac{\overline{P_1} - \overline{P_2}}{s_{\overline{P_1} - \overline{P_2}}} \tag{8.32}$$

where $\overline{P_1}$ and $\overline{P_2}$ are the mean values of the two system power outputs. The $s_{\overline{P_1} - \overline{P_2}}$ is the standard error of the difference between the two datasets. The t-statistic has a zero-mean-normalized probability distribution function that is dependent on the number of points in the datasets.

Values of t that lie outside approximately two standard deviations of the mean (zero) are termed 95% confidence intervals, and are usually used in statistical inference as a measure of testing whether a certain parameter is significantly different from a certain hypothesized value. The correctness of the hypothesized value is called the null hypothesis. If the value of t obtained is significantly far from that value (formally outside 95% on the probability distribution), then the null hypothesis can be rejected, and it is safe to conclude that the experimentally obtained quantity is not equal to the hypothesized value with 95% confidence. The null hypothesis indicates that there is no difference between the power outputs obtained by the experimental test. It can be statistically stated as the difference between the two means being equal to zero. If the t-value obtained from (8.32) is significantly different from zero, then it can be inferred that the difference between the power outputs of the two methods is statistically significant.

The statistics toolbox of Matlab provides a function for the paired t-test. It is provided as the function $t - test(x, y)$, where x and y represent the random samples following a normal distribution. x and y must be vectors of the same length. The function can be used to conduct the t-test and assess the null hypothesis. If the result of the test returns 1, the $t - test$ indicates a rejection of the null hypothesis at the 95% confidence level. A returned value of 0 indicates a failure to reject the null hypothesis at the 95% confidence level.

8.13 Summary

For MPPT, hill climbing (HC) methods are widely used. An alternative name for this approach is perturbation and observation (P&O). The standard algorithm generally involves a tradeoff between the tracking speed and the magnitude of steady-state oscillations. The selection of the perturbation size is considered a dilemma in balancing these requirements.

The incremental conductance method (IncCond) was developed to solve the problems of the HC-based MPPT approaches. It is based on the extremum value theorem (EVT), seeking to locate the local extremum value, either maximum or minimum.

The mathematical operation of differentiation is required. However, the function of the PV output power output is not exactly known to the controller since the parameters significantly change with environmental conditions. Instead, the IncCond operation relies on a numerical approximation of the differentiation, which introduces a truncated error. Therefore, experiments show that the oscillations in the steady state cannot be removed by the IncCond method. The truncated errors of the numerical approximation mean that the IncCond operation is not a significant advance on HC-based algorithms.

It is important to choose carefully the sampling frequency and perturbation size in HC-based or EVT-based MPPT algorithms. These parameters drive the response speed and steady-state performance. The sampling frequency can be determined by system dynamic analysis, ideas introduced in Sections 6.3 and 7.8 respectively. With a dedicated voltage-regulation loop for the PV link, the system dynamics can be improved, benefiting MPPT performance. The perturbation size should be determined through consideration of the ripple that is caused by the switching-mode power interface. For practical applications, the SNR should always be investigated to decide the size. Additionally, weather variation patterns must be considered on a case-by-case basis.

A start-stop mechanism can be integrated with a conventional HC-based MPPT algorithm, bringing the advantages of simplicity and effectiveness. It can give oscillation-free operation in the steady state since unnecessary perturbations can be avoided. When the MPP changes, active perturbations can be restarted to locate the new MPP. The implementation is relatively easy and straightforward and is an ideal complement to conventional HC-based MPPT algorithms.

The step size in active tracking can be adaptively adjusted to give fast tracking in transient conditions, and low steady-state errors. In theory, this is an ideal solution since many mathematical methods can be used to drive the adaptations. The common algorithms used are the steepest decent method and the Newton–Raphson method. However, practical implementations can be difficult because the mathematical model to represent the real-time PV output characteristics is missing. When making numerical approximations of the differentiation, truncated errors have been reported when using Euler methods. It turns out that the MPP cannot be fully represented when using a numerical assumption of $\Delta P/\Delta V \approx 0$.

Centered differentiation has been introduced to minimize truncated errors and improve the approximation accuracy of the condition $dp/dv = 0$. One additional measurement is required to perform centered differentiation. Improved performance has been reported when centered differentiation is integrated with the steepest descent algorithm and the dedicated start-stop mechanism. Overall, the implementation is more complicated than using only the HC algorithm and the simple start-stop mechanism.

The real-time identification method is another MPPT approach, which utilizes a mathematical model that is constructed using polynomial equations. The parameters are identified by recursive least squares (RLS) estimation in real time, so as to follow variations in operation conditions. The Newton–Raphson method is used as the numerical solver to find the PV-link voltage, which represents the MPP. The approach is model-based and does not require a numerical differentiation used in EVT-based algorithms or the perturbation in HC-based algorithms. However, practical implementation can be difficult due to the significant computation demand required for system identification. Furthermore, the stability of the RLS solution is another concern.

Extremum seeking (ES) control is another optimization approach for MPPT, which does not require prior knowledge of the system model. The standard method injects an sinusoidal excitation signal into the control loop in order to identify the extremum point. Its effectiveness was demonstrated by simulation in this chapter. Since switching-mode power converters are commonly used for PV systems, the ES algorithm can also utilize the switching ripple as the excitation signal. However, the implementation is not as straightforward as the HC and simple start-stop mechanisms. The tracking performance has not been shown to be significantly better than others.

In general, the MPPT algorithm is not as effective in dealing with PV mismatch conditions in CMPPT systems as in DMPPT systems, even though there has been significant research and many publications in the area. The complication of mismatch conditions in PV arrays creates a problem that is difficult to solve quickly for practical applications. Up to now, a DMPPT configuration has been the most effective way to minimize power degradation when there are mismatch conditions among PV cells.

Even though significant research has been published, many of the studies involved did not provide a fair comparison to demonstrate MPPT effectiveness. An outdoor test with natural sunlight is considered the ideal condition for evaluating MPPT effectiveness. However, a fair comparison of different MPPT methods is not easy outdoors. First, the true MPP is unknown when a PV system is in uninterrupted operation. The weather conditions – irradiance and temperature – are not repeatable. It is important to create a practical bench test system to evaluate MPPT performance under identical operating conditions.

A parallel testing system using two independent PV systems is highly recommended in this chapter. However, it is impossible to construct two truly identical systems when the difference between solar modules and other non-ideal factors in the power-conditioning circuits are considered. A pairwise difference comparison is introduced in order to statistically distinguish energy harvesting efficiency of two MPPT algorithms regardless of non-ideal factors. The pairwise *t*-test formula can be applied to filter out the effect of the difference in PV panel outputs, to enable accurate quantification of the statistical significance between two MPPT methods. The proposed method is considered as a systematic approach for both short-term and long-term side-by-side comparisons of MPPT performance using natural sunlight.

Problems

8.1 Readers are encouraged to research other MPPT algorithms that have not been covered in this chapter. Compare them with the solutions presented in this book and provide comment about their pros and cons.

8.2 Use any available simulation tool, construct an MPPT block based on the HC algorithm.

8.3 Based on the developed MPPT model, implement the start-stop mechanism.

8.4 Based on an example, compare the performance with and without the start-stop mechanism. The comparison can be based on the produced energy within a predefined time period.

References

An SH and Yao K 1988 Convergent and roundoff error properties of reflection coefficients in adaptive spatial recursive least squares lattice algorithm. *Circuits and Systems, IEEE Transactions on* **35**(2), 241–246.

Ariyur K and Krstic M 2003 *Real-Time Optimization by Extremum-Seeking Control*. Wiley.

Brunton SL, Rowley CW, Kulkarni SR and Clarkson C 2010 Maximum power point tracking for photovoltaic optimization using ripple-based extremum seeking control. *Power Electronics, IEEE Transactions on* **25**(10), 2531–2540.

Du Y, Li X, Wen H and Xiao W 2015 Perturbation optimization of maximum power point tracking of photovoltaic power systems based on practical solar irradiance data *Control and Modeling for Power Electronics (COMPEL), 2015 IEEE 16th Workshop on*, pp. 1–5.

Khan W and Xiao W 2016 Integration of start-stop mechanism to improve maximum power point tracking performance in steady state. *Industrial Electronics, IEEE Transactions on* **63**(10), 6126–6135.

Liu SJ and Krstic M 2012 *Stochastic Averaging and Stochastic Extremum Seeking*. Springer Science & Business Media.

Rowland T and Weisstein EW n.d. Lipschitz function http://mathworld.wolfram.com/LipschitzFunction.html. Accessed: 13 May 2016.

Xiao W 2003 *A modified adaptive hill climbing maximum power point tracking (MPPT) control method for photovoltaic power systems* PhD thesis University of British Columbia.

Xiao W 2007 *Improved control of photovoltaic interfaces* PhD thesis University of British Columbia.

Xiao W and Dunford WG 2004a Evaluating maximum power point tracking performance by using artificial lights *30th Annual Conference of IEEE Industrial Electronics Society (IECON)*, vol. **3**, pp. 2883–2887.

Xiao W and Dunford WG 2004b A modified adaptive hill climbing MPPT method for photovoltaic power systems *IEEE 35th Annual Power Electronics Specialists Conference (PESC)*, vol. **3**, pp. 1957–1963.

Xiao W, Dunford WG, Palmer PR and Capel A 2007a Application of centered differentiation and steepest descent to maximum power point tracking. *Industrial Electronics, IEEE Transactions on* **54**(5), 2539–2549.

Xiao W, Dunford WG, Palmer PR and Capel A 2007b Regulation of photovoltaic voltage. *Industrial Electronics, IEEE Transactions on* **54**(3), 1365–1374.

Xiao W, Elnosh A, Khadkikar V and Zeineldin H 2011 Overview of maximum power point tracking technologies for photovoltaic power systems *37th Annual Conference on IEEE Industrial Electronics Society (IECON)*, pp. 3900–3905.

Xiao W, Lind MG, Dunford WG and Capel A 2006 Real-time identification of optimal operating points in photovoltaic power systems. *Industrial Electronics, IEEE Transactions on* **53**(4), 1017–1026.

Xiao W, Zeineldin HH and Zhang P 2013 Statistic and parallel testing procedure for evaluating maximum power point tracking algorithms of photovoltaic power systems. *Photovoltaics, IEEE Journal of* **3**(3), 1062–1069.

9

Battery Storage and Standalone System Design

A standalone system supplies electric power independent of an electrical distribution network. Such systems can be generally classified into two categories: those with or without significant storage, as shown in Figure 9.1.

Systems without bulk energy storage can be direct-coupled, power-conditioned or hybrid, the configurations of which are shown in Figure 9.2. The diagrams provide a basic representation, but practical systems are more complex than shown.

Direct-coupled systems are the simplest PV applications, designed to match the PV output to the load, as shown in Figure 9.2a. They are commonly used for ventilation fans and water pumps for irrigation, and so on. In most systems, control circuits are implemented to switch on and off the connection depending on the voltage the PV output, which is an indicator of the available solar power. Such systems are incapable of accurate MPPT due to their lack of power conditioning.

All direct-coupled systems can be modified to incorporate a power-conditioning unit to enhance control. Such a system is shown in Figure 9.2b, with a power conditioner included between the PV generator and the load. Power conditioning is very effective for a PV system without significant energy storage. The MPPT function can be used to maximize the PV output regardless of environmental variations. The power interface can also be controlled to meet specific load requirements, such as a constant voltage supply.

A hybrid system usually uses a conventional engine-based generator in parallel with the PV power source. Both share the same distribution channel to the common load, as shown in Figure 9.2c. The DC/AC converter produces AC power from the PV source and supplies the load through the distribution network. The PV power contribution can reduce the power generation required from the generator, resulting in a fuel saving. The PV system is usually controlled through MPPT to yield the highest solar energy harvest.

Even though the diagram in Figure 9.2c shows only one motor generator, the system can be composed of multiple generation units in parallel connection. For the best fuel-saving performance, communication is generally required to coordinate the PV output with the engine-based generators. Coordination is important when the PV power penetration becomes significant, say more than 25%. When multiple motor generators are available in a standalone system, the system can be optimized to improve efficiency. For example, running two generators each at 50% of capacity is generally more fuel efficient than operating four, each at 25% of capacity. Using real-time measurements of the load condition and PV generation, the centralized coordinator can schedule the overall generation facilities in an economical way, so as to give the best performance in terms of system efficiency and fuel saving.

Photovoltaic Power System: Modeling, Design, and Control, First Edition. Weidong Xiao.
© 2017 John Wiley & Sons Ltd. Published 2017 by John Wiley & Sons Ltd.
Companion Website: www.wiley.com/go/xiao/pvpower

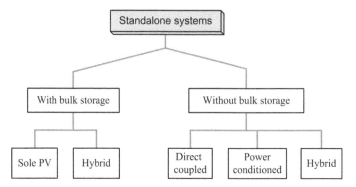

Figure 9.1 General classification of standalone PV systems.

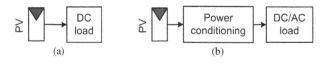

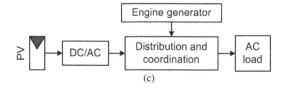

Figure 9.2 Standalone systems without bulk energy storage: (a) direct-coupled; (b) with power conditioning; (c) hybrid solution.

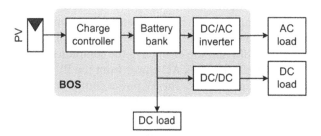

Figure 9.3 Standalone PV system with bulk energy storage.

A steady power supply is demanded by loads in the majority of standalone systems. As discussed in Section 1.9, bulk energy storage – mainly rechargeable batteries – is generally required to mitigate the intermittency of solar energy. Systems with bulk energy storage, can be divided into two groups: standalone PV and hybrid. These are shown in Figures 9.3 and 9.4, respectively. It should be noted that the diagrams are general representations, since practical systems may be more complex. Furthermore, overcurrent protection (OCP) is mandatory for any system with battery storage, even though no protection circuits are shown in the diagrams.

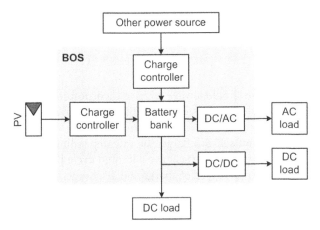

Figure 9.4 Hybrid system including PV and bulk energy storage.

As shown in Figure 9.3, the battery charging cycles should be properly maintained by the charge controller with integration of MPPT to give the highest solar energy harvest. Any extra PV power bypasses the battery buffer and supplies the loads directly. The battery bank forms an unregulated DC bus since the voltage varies by up to ±20% from the nominal voltage according to the state of charge. The unregulated bus can directly supply DC loads if they are insensitive to voltage variation. A DC/DC converter should be implemented to regulate the output voltage and supply the dedicated load if constant DC voltages at different levels are required. When AC loads are present, DC/AC conversion should be used to convert from the unregulated DC bus to AC.

A hybrid system involves use of additional power sources alongside the PV generator, as shown Figure 9.4. Additional power can come from wind turbines, fuel cells, or conventional engine-based generators. Such a system can balance the battery charge and the power from PV or other power sources since multiple charge controllers are included. The coordination can be achieved using a centralized controller or a distributed sharing algorithm implemented in each charge controller. PV power generation is expected to use MPPT to give the highest clean energy contribution. There is also an option for fuel-based power sources to bypass the battery and supply the load directly, although this configuration is not shown in Figure 9.4. A direct connection can minimize power losses in the conversion stages.

9.1 Batteries

A rechargeable battery is defined as an energy storage device that converts chemical energy into electrical energy and vice versa. People are tending to use more and more rechargeable batteries instead of the disposable counterparts due to cost savings and environmental concerns. Due to the bulk energy storage requirement in standalone PV systems, it is important to understand the different battery technologies and to select one accordingly. Battery cells are the basic electrochemical units that form a battery module and pack, as shown in Figure 9.5. A large battery power system is usually constructed from battery banks. When the battery system becomes more and more

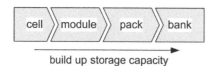

Figure 9.5 Formation of battery power systems from cell to bank.

complicated and powerful, special attention should always given to incorporate protection devices for safety purposes. For portable devices such as cell phones a single battery cell can be used.

It should be noted that all batteries exhibit self-discharge, the rate depending on the battery type and temperature. Furthermore, all batteries have a limited cycle life, their lifetime depending on the battery type, the conditions of charge and discharge, the temperature, and other conditions of use.

9.1.1 Battery Types

Common rechargeable batteries are based on lead, nickel, lithium, or sodium. It is important to understand the cell nominal voltage and characteristics and to select the correct type for the given application.

Lead-based batteries
Lead-based battery technology is mature and therefore low-cost and reliable. This is the main reason that lead-based batteries are still widely used for vehicles and back-up energy storage. They are often referred to as "lead-acid" batteries, because dilute sulfuric acid and lead are used for the electrolyte and plate, respectively. Table 9.1 summarizes the common terms and basic characteristics of lead-acid batteries. Two main categories can be considered: flooded and sealed. Because of their advantages of safety and low maintenance requirement, more and more applications are using sealed modules instead of the old flooded type. Sealed modules also include the VRLA, AGM, and gel types, which are described in the table.

The battery module, typically rated at either 12 V or 6 V, is built from lead-acid cells with a nominal voltage of 2 V. Lead-acid batteries can also be classified in another way:

Table 9.1 Common lead-acid battery types.

Type	Description
Flooded	Old lead-acid technology with liquid electrolyte inside the cell. Regular maintenance is required by adding distilled water.
Sealed	Also called "maintenance-free" battery since the cell is sealed without water. Generally considered safer than flooded batteries.
Valve-regulated lead-acid (VRLA)	The battery module includes a valve for release of hydrogen and oxygen gasses during charging.
Absorbent glass mat (AGM)	The electrolyte is suspended in a special glass mat. High efficiency.
Gel	Contains a silica type gel to suspend the electrolyte. Does not tend to sulfate or degrade as easily as wet cells.

as starting or deep-cycle batteries. A starting battery is to used provide instant power, such as in a starter motor. Such a battery is also termed as the "cranking" type. For standalone PV systems, deep-cycle batteries are commonly used. These are built to provide continuous electric power, with high capacities and long cycle lives. The commercial term "solar batteries" often refers to deep-cycle batteries. Even though lead-acid batteries are still widely used in renewable energy systems, the following drawbacks lead to the use of other battery technologies:

- low energy capacity density per unit volume
- low energy capacity density per unit weight
- slow charge speed
- limited cycle life.

Nickel-based batteries

The common types of nickel-based batteries are summarized in Table 9.2. Nickel–metal-hydride (NiMH) batteries were introduced in 1992 and are commonly used in low-cost consumer products (Powers 2000). The battery is composed of nickel hydroxide and a hydrogen-absorbing alloy. Most rechargeable NiMH batteries follow the the American National Standards Institute (ANSI) standard, and are sized as A, AA, or AAA. A standard AA size cell typically has a capacity of 1800–2500 mAh. The cylindrical forms are available to replace primary batteries, which are not rechargeable. The hydrogen-absorbing alloy is capable of absorbing and releasing hydrogen at a higher density level than the cadmium compound used in nickel–cadmium (NiCd) batteries. One important application of NiMH technology is in the hybrid vehicle, the Toyota Prius. According to a report from the National Renewable Energy Laboratory of the USA (Kelly et al. 2002), the battery system is designed to have a nominal voltage of 273.6 V and a capacity of 6.5 Ah. The system includes 38 battery modules, each formed by six cells in series.

NiCd batteries used to be considered one of the most reliable battery technologies, and have been produced since 1980 (Powers 2000). They have been widely used in the space and aviation industries. For example, they have been used in the main and auxiliary power units in the Boeing 777 airliner. The Boeing 777 has one of the best safety records in aviation history, supporting claims for the reliability performance of NiCd batteries. The battery pack used has a nominal voltage of 24 V and a capacity of 16 Ah. It is formed from 20 NiCd cells and weights 48.5 kg.

The battery shares the same structure as its NiMH counterpart. NiCd batteries use nickel oxide as the cathode, a cadmium compound as the anode and potassium

Table 9.2 Common nickel-based batteries.

Type	Description
Nickel–metal-hydride (NiMH)	The cell is rated at 1.2 V. The battery is composed of a positive plate containing nickel hydroxide as the positive electrode and a hydrogen-absorbing alloy as the negative electrode.
Nickel–cadmium (NiCd)	The cell voltage is rated at 1.2 V. The battery uses a nickel oxide and cadmium compound.

hydroxide solution as the electrolyte. They also use the ANSI standard AA and AAA sizes, but are no longer as widely available as their NiMH counterparts. They exhibit low voltage drop over the discharge period, but the energy density is no longer competitive with the latest battery technologies. Standard AA cells generally have capacities of 600–800 mAh, which is significantly lower than NiMH cells. Nickel-based batteries show a memory effect that tends to "remember" the previous operation cycle, so a relatively deeper charge/discharge cycle is required than for batteries without a memory effect. The self-discharge rate of nickel-based batteries is significantly higher than the latest lithium-based technologies.

Lithium-based batteries

Lithium-based technologies have drawn significant attention and grown exponentially due to their significant advantages, such as high energy density per unit volume and per unit weight. The typical cell voltage rating is 3.6 V, which is generally higher than other types. The self-discharge rate is lower than that of NiMH batteries and there is no memory effect. Furthermore, high efficiencies can be demonstrated in fast charge and discharge. In a lithium-ion cell, the positive electrode is usually activated by cobalt acid lithium, while the negative electrode is activated by highly crystallized specialty carbon. The lithium cobalt oxide material is ionized and the ions move to the negative electrode. During discharge, the ions move to the positive electrode and turn into the original compound.

The advantages of lithium-ion batteries mean they are widely used for transportation fleets and portable electronic devices. Table 9.3 outlines two examples of lithium-ion batteries that have attracted much media coverage in recent years. Further information about these technologies and the latest updates can be sourced from the relevant websites at www.boeing.com and www.chevrolet.com.

The Boeing 787 airliner uses lithium-ion batteries to replace their NiCd counterparts. This gives a significant improvement in terms of power capacity and high power over the Boeing 777. However, in the first year of service, the aircraft suffered a number of fire incidents resulting from the lithium-ion battery packs. As a result of the Boeing 787 incidents, in 2013, Airbus reverted to NiCd batteries for the newly developed A350 XWB.

The Chevrolet Volt is a plug-in hybrid vehicle and is an example of a lithium-ion battery application for ground transportation. The battery pack is rated at 16.5 kWh at a nominal voltage of 355.2 V. The capacity of the battery pack is not as high as other

Table 9.3 Typical applications of lithium-ion batteries.

Applications	Battery pack specification
Boeing 787	Each battery unit is configured with 8 lithium-ion cells
	Nominal voltage of 32 V, 28.6 kg weight
	More powerful but lower weight and smaller size than NiCd counterpart
Chevy Volt	Battery pack weighs 197 kg, formed from 288 lithium-ion cells
	Nominal voltage 355.2 V and 16.5 kWh capacity
	Narrow cycle of charge and discharge (60% of total capacity) is used for day-to-day operation

electric vehicles because the vehicle is also equipped with an internal combustion engine which can drive an electric generator and provide charge if required. The system is designed for a light charge and discharge cycle in order to prolong the battery life cycle. It has generally received positive reviews thanks to the selection of an appropriate battery technology and the design of the battery management system.

Lithium-based technologies are the latest battery technology, with more and more applications emerging, including utility-scale storage. Safety precautions should be always taken; the high energy density is advantageous, but is potentially hazardous. Misuse and lack of protective measures may cause battery explosion or ignition.

One special technology is the sodium–sulfur (NaS) battery, which has several important advantages: high energy density, long cycle life, low-cost materials, and high efficiency. The Japanese company, NGK Insulator Inc, is the major manufacturer for this technology (Beaudin et al. 2010). According to the company's website at https://www.ngk.co.jp, an NaS battery cell uses sulfur as the positive electrode and sodium as the negative electrode. Beta alumina ceramic is used between the electrodes. During discharge, sodium ions are released from the negative electrode and are transferred through the solid electrolyte into the sulfur at the positive electrode. The battery charging is the reverse of the discharge process, with sodium forming in the negative electrode. High operating temperatures are required for molten sodium, which limits use to stationary applications. The nominal voltage of an NaS cell is 2 V. The technology is considered suitable as an energy buffer for mitigating power intermittency in renewable electricity resources, such as solar and wind. The high operating temperature is also a concern because of the fire hazard involved.

9.1.2 Battery Terminology

A battery module is formed from series-connected cells sealed into one unit. Battery modules are mechanically and electrically assembled to form a battery pack. In large-scale systems, battery packs are grouped together to form a battery bank. A battery management system is generally required to regulate the battery system operation and protect it from damage if the capacity becomes significant; multiple cells in complex configurations can exhibit mismatches, which can be unsafe. The important terms relating to battery voltage are described in the following:

- *Nominal voltage*: can refer to the cell, module, or pack levels, and is the voltage reference for system rating and design. When batteries are used in practical systems, the terminal voltage is generally different from the nominal voltage. The terminal voltage depends on the instant load condition, state of charge, and temperature.
- *Cut-off voltage*: the least allowable voltage to discharge the battery. For rechargeable batteries, any discharge operation may result in damage when the terminal voltage is lower than the cut-off limit.
- *Open-circuit voltage*: measured when no charge or discharge is taking place. It is commonly considered as an indicator of battery capacity. Compared to the nominal voltage, a high value indicates high remaining SOC. However, the absolute value also depends on the temperature.
- *Float voltage*: voltage level for the battery charger to maintain a trickle charge when the battery has been fully charged. The voltage level is generally defined to maintain the battery capacity against discharge and avoid overcharging the battery.

The terms for energy and capacity are as follows:

- *Nominal battery capacity* is generally defined as the total amp-hours available. If a battery capacity is 1 Ah, without considering power loss, it indicates that the nominal capacity can be fully discharged in 1 h if the discharge current is 1 A. On the other hand, in theory, the battery can be fully charged from empty when a current of 1 A is applied for 1 h.
- *C-rate* is the rate at which a battery is charged or discharged relative to its rated capacity. If a battery capacity is 1 Ah, a 1C discharge rate means that the discharge current is 1 A. A C/4 charge rate indicates the charge current is 0.25 A. In theory, the time to fully charge from empty is 4 h when C/4 is applied.
- *State of charge (SOC)*: is an expression of the present battery capacity as a percentage of the rated capacity.
- *Depth of discharge (DOD)* refers to the percentage of battery capacity that has been discharged over the rated capacity.
- *Nominal energy* is an important term to define the energy capacity of batteries. It can be derived from the nominal voltage (V) and capacity (Ah). The measurement unit is Wh or kWh. Its value is important to size standalone PV systems with consideration of the load profile.
- *E-rate* is the rate at which a battery is charged or discharged relative to its nominal energy.

Other important terms are:

- *Charge cycle* is defined as the process of charging a rechargeable battery to a high capacity and discharging to a low capacity. In a PV power system, it is usually counted day by day since the charge happens during daytime and discharge is at nighttime.
- *Deep cycle* for a rechargeable battery is when it is charged to more than 80% of the SOC and then discharged to less than 20% of the SOC. Since the cycle life depends on the DOD, frequent deep-cycle operation generally causes fast aging and a short life span.
- *Recommended charge current* is the ideal current at which the battery is initially charged (to roughly 70% of SOC) under a constant charging scheme before transitioning to constant voltage charging. Temperature should be always considered when determining the recommended charge current.
- *Recommended charge voltage* is the voltage at which the battery is allowed to be charged up to the full capacity. The cycle charge includes both the regulation of current and voltage. Temperature is generally required to precisely determine the recommended charge voltage.
- *Cycle life* is the number of discharge-charge cycles the battery can experience before it fails to meet specific performance criteria. The evaluation and estimation of cycle life should be based on specific charge and discharge conditions, such as the charge rate, discharge rate, DOD, temperature, and humidity. It is generally considered that the higher the DOD per cycle, the lower the cycle life.

9.1.3 Charging Methods

Charging is a process to restore a discharged battery to the required capacity. It is important to use a proper charging method and maintain proper charge cycles for long battery lifetime. Battery charging methods are in three categories:

- cycle charging
- compensating charging
- float charging.

The standard cycle charge usually starts with a constant current, and then a constant voltage. Cycle charging is designed to charge the battery from a low percentage of SOC to a higher level or full charge. It starts with a constant-current charge cycle and allows the voltage to rise, as illustrated in Figure 9.6. It is commonly referred to as the bulk stage, since most capacity is recovered during this cycle. The maximum charging current is usually set to prevent any significant temperature rises, would result in fast aging and early degradation. The C-rate is commonly used as a measure of the charging current. When the voltage reaches the upper level, the charging cycle becomes a constant-voltage operation. This is also referred to as the "absorption stage". The battery capacity will be recovered to the full level at this stage when the charge current reaches a significantly low level, as shown in Figure 9.6.

Compensating charging is also called trickle charging, and is applied to maintain the battery capacity against self-discharge. Float charging is used where the battery is in parallel connection with the load. In PV power systems, solar power is used to supply the load during the day, while any surplus power is used to charge the battery. A strict charging cycle for constant current is difficult to maintain in the float charging process due to the variation of load conditions. Both trickle and float charging are considered constant-voltage charging methods since the battery voltage must be maintained at a constant level.

Table 9.4 shows the typical ratings to charge lead-acid batteries. It should be noted that the ratings are for a constant temperature of 25°C and are presented only for reference purposes. Accurate ratings should follow the manufacturer's recommendations, along with consideration of environmental conditions. The setting can be calculated according to the nominal voltage and battery capacity. For a battery module rated as 12 V/100 Ah, to perform cycle charging, the recommended charge current is 40 A and the voltage is 14.4 V.

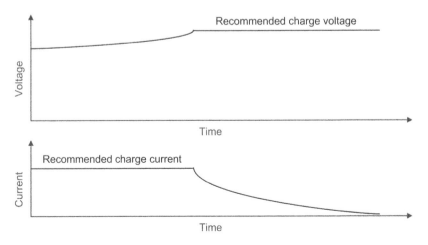

Figure 9.6 Cycle charge with constant current and constant voltage: top, charging voltage; bottom, charging current.

Table 9.4 Typical settings for charging lead-acid batteries.

Charging method	Recommendation
Constant current for cycle charge	Recommended current is 0.4C or lower. The lower the rating, the longer the charging time.
Constant voltage for cycle charge	Recommended voltage is 120% of the nominal voltage.
Trickle charging	Recommended voltage is 115% of the nominal voltage.
Float charging	Recommended voltage is 115% of the nominal voltage.

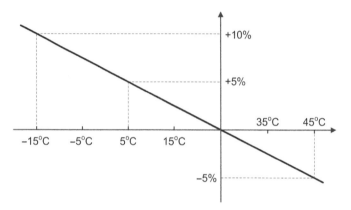

Figure 9.7 Charging voltage with temperature compensation.

Ideally, the battery should be stored or operated in a environment in which the ambient temperature is conditioned. High temperatures accelerate aging, while low temperatures constrain capacity. Temperature compensation should be considered for the charging voltage when the battery temperature varies significantly and can be sensed. The objective is to either prevent fast aging if the battery temperature too high or to guarantee sufficient charging when the temperature is lower than required. The general rule is to increase the charging voltage if the temperature is lower than 25°C and decrease it when temperature is higher than 25°C. For lead-acid batteries, it is recommended to have a compensation slope to adjust the charging voltage according to the temperature. Figure 9.7 shows a reference chart for temperature compensation with a slope of −0.25% per degree. For example, if the nominal charge voltage is 14.4 V for a 12-V battery module, the charging voltage should be adjusted to 15.8 V if the temperature is 15°C below zero. When the temperature becomes 45°C, the charging voltage is set to 13.7 V according to the compensation curve. It should be noted that different battery manufacturers might recommend different values for temperature compensation.

The settings in Table 9.4 are considered a general reference. Battery temperature is considered as the main factor that affects aging and damage. The charging rating can be more accurately defined and automatically adjusted if the battery temperature can be monitored as it charges. Without temperature sensing, the battery charging cycle is usually defined in a conservative way in order to avoid any unexpected temperature rises that would cause fast aging and damage. Rapid charging can be performed when the battery temperature is either monitored or regulated. The charging current and voltage can be adaptively adjusted in real time in response to temperature variations.

For example, the charging current for rapid charging can be regulated up to 1C for NiMH batteries when the ambient temperature is higher than 0°C but lower than 40°C, according to the recommendation of Panasonic. The rapid charging should be turned off if one of the following conditions is detected, at which point trickle charging is activated:

- the NiMH cell voltage reaches the extreme level of 1.8 V
- the cell voltage starts to drop by up to 10 mV
- the cell temperature increases too fast; up to 2°C/min
- the cell temperature reaches the upper limit of 40°C
- the rapid charge time is recorded up to 90 min.

Panasonic recommends a two-stage cycle charging for its lithium-ion batteries. It starts with a constant-current charging cycle and the voltage is allowed to rise. When it reaches the upper voltage limit, the charging cycle becomes a constant-voltage operation, which maintains the battery voltage at a constant level. The battery capacity is considered to be fully recovered at this stage. The typical settings for lithium-ion batteries are summarized in Table 9.5. The charge should be stopped if the charge current decreases to 0.1C. It is considered a malfunction if the charging time is longer than 720 min and the current is still higher than 0.1C. Charging should be stopped and an error should be signaled by the charge controller.

If the initial battery voltage is only 80% of the nominal voltage, the battery is usually considered as deep discharged. To avoid any damage caused by rapid charging, a slow charging process should be used, in which the charge current is set to 0.1C or less. It should be noted that the recommended ratings for various charge methods is only a general reference; the manufacturer's recommendation should be always followed because of the variety of different battery technologies.

9.1.4 Battery Mismatches and Balancing Methods

Similar to the construction of PV modules and strings, battery modules and packs are formed from multiple cells in series connection, so as to reach the required voltage level. Ideally, all battery cells in one battery pack are identical and share the same SOC and electrical characteristics. However, mismatches happen along each string since battery cells can be different, for various reasons, such as manufacturing defects, temperature gradients, and uneven aging. Between 1% and 10% of mismatches are due to manufacturing defects (Rehman et al. 2014).

During charging and within one string, a stronger cell gains SOC faster than a weaker one. During discharging, the SOC of the weaker cell decreases faster than the

Table 9.5 Typical settings for charging lithium-ion batteries.

Charging method	Recommendation
Constant current for cycle charge	Recommended current is up to 0.7C. The lower the rating, the longer the charging time.
Constant voltage for cycle charge	Recommended voltage is 117% of the nominal voltage.
Charge temperature	Between 10°C and 45°C.

stronger one. Due to the unbalanced charging and discharging, the SOCs of the cells are not equally distributed. The weakest cell becomes a bottleneck along the string. The repeated charging and discharging eventually brings about early-stage failure, first of the single cell and then propagating to the whole battery pack.

During either charging or discharging, the mismatch generally causes a voltage difference across battery cells since they are connected in series and share the same string current. The term "equalization" is commonly used for the method to balance the cell differences in term of SOC levels and voltages. The process is especially important for high-capacity battery systems, so as to ensure safety, reliability, and a long lifetime for the system. The equalization approaches are conventionally divided into two main categories: passive equalization and active equalization (Daowd et al. 2011). However, the classification causes confusion since passive balancing methods also include the switched shunt resistor method, which can also be considered an active approach. In this book, a new classification is defined: heat dissipation methods and energy buffer methods, as illustrated in Figure 9.8. This generally provides a clear framework for understanding the methodologies used for battery balancing.

Equalization based on dissipation elements includes three different methods:

- zener diode
- fixed shunt resistor
- switched shunt resistor.

These are shown in Figure 9.9.

The zener diode is selected to match the rated cell voltage at 100% SOC. When the voltage is higher than the rated voltage of the zener diode, the internal loop conducts to dissipate energy through the resistors. The fixed shunt resistor method is simple and based on the principle that higher voltages result in higher self-discharge when resistances are equally applied. The technique is also commonly used for balancing the voltage for capacitor banks of stacked capacitors. However, their drawback is the continuous energy losses.

Voltage sensing and control are required for active methods, such as the switched shunt resistor approach, as shown in Figure 9.9c. In normal operation, the switches are open to allow the cells to be balanced by themselves. When an abnormal voltage

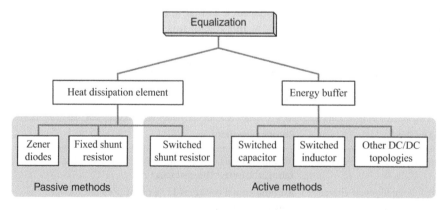

Figure 9.8 Typical battery cell equalization techniques.

Figure 9.9 Balancing circuits based on heat dissipation: (a) zener diode; (b) fixed shunt resistor; (c) switched shunt resistor.

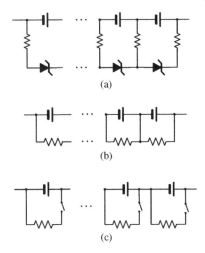

is detected across an individual cell, the individual switch can be closed to maintain the cell voltage through dissipating electricity into heat. The switches are usually transistors rather than mechanical based relays.

The passive methods are simple, low cost, and robust. They are shown in Figures 9.9a and 9.9b. Their disadvantage lies in the unstoppable power loss of the shunt resistors since any excessive charge is dissipated as heat. The implementation of zener diodes can reduce the continuous losses of the resistor network. However, zener diodes do not show an abrupt voltage breakdown, but instead a soft reverse knee. Temperature effects mean that additional losses due to imprecise matching between the zener diode breakdown voltage and the battery cut-off voltage cannot be avoided. Switched shunt resistors generally exhibit smaller losses than the fixed solution. However, active methods increase the circuit and control complexity compared to passive approaches. Significant losses cannot be avoided when action is taken to balance battery cells.

A short-term energy buffer can be used with DC/DC conversion to balance the differences between battery cells. In other words, the excess or deficient energy between the cells can be balanced by storage components instead of heat dissipation. This is expected to be more efficient than topologies based on heat dissipation. The energy buffer can be either a capacitor or an inductor, as summarized in Figure 9.8. The common DC/DC topologies without galvanic isolation, include the switched capacitor and switched inductor topologies. Isolated topologies, such as flyback, can also be used to equalize individual cell voltages. Active switching becomes essential in order to balance the voltage differences between battery cells. It should be noted that the method using an energy buffer is not completely loss-free since ancillary power is needed for sensing, control and active operation.

One idea is to use the switched capacitor topology for the equalization, as shown in Figure 9.10 (Daowd et al. 2011). The circuit includes N battery cells, and is implemented with $N - 1$ capacitors and N single-pole–double-throw (SPDT) switches. When all SPDT switches are at the A position, the capacitors are connected in parallel with the cells from 1 to $N - 1$. When the switches are at the B position, the capacitors are connected to the cells from 2 to N. The continuous switching between A and B aims to balance the voltage between any two adjacent battery cells. If the voltages of adjacent

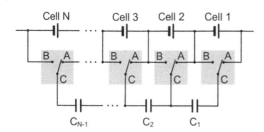

Figure 9.10 Balancing circuits using switched capacitors.

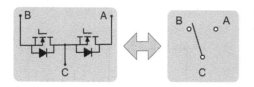

Figure 9.11 Semiconductor circuit for the single-pole-double-throw switch.

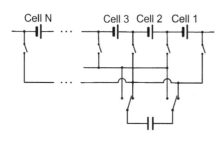

Figure 9.12 Balancing circuit based on switched capacitors.

cells are different, the capacitor is charged by the higher-voltage cell and discharged by the lower-voltage cell. Depending on the inequality level and the capacitance, in theory the overall battery pack can be equalized in voltage by the continuous switching and propagation.

As shown in Figure 9.11, the SPDT switch can be constructed from two power MOS-FETs instead of mechanical SPDT relays (Kimball et al. 2007). In the circuit, the A terminal should be always connected to positive due to the body diode inside the MOSFET. The equalization does not require either a sophisticated control algorithm or voltage sensing of each battery cell. The operation can made flexible for active balancing operations. It should be noted that the topology shown in Figure 9.10 was granted a US patent in 1998 (Pascual and Krein 1998).

Another topology using one capacitor for the switching operation is shown in Figure 9.12. Voltage sensing is essential for the operation since the controller needs to identify the voltage differences among the battery cells. The balancing operation can target the cells showing the highest and lowest voltages, making the equalization more efficient. A start-stop mechanism can be used to stop the operation if the imbalance is not significant.

The topology shown in Figure 9.10 can easily be implemented using repeated switching operations without voltage sensing, but requires more capacitors for balancing. It should be noted that the topology is also flexible and can produce fast balancing if voltage sensing and centralized control are available for coordination. The topology shown in Figure 9.12 reduces the capacitor count into one, but requires voltage sensing and relies on a centralized controller for coordination.

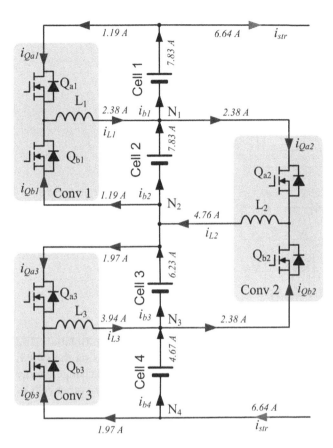

Figure 9.13 Balancing circuits based on switched inductor.

Figure 9.13 shows an example of the switched inductor topology used for balancing mismatches among four battery cells during discharge. Three differential processors are required and these are labeled Conv 1, Conv 2, and Conv 3. The analysis is based on a continuous conduction model at steady state. It shows that the output current of an individual cell can be different from the others. Even though the string current has a value of 6.64 A, the current output in Cells 1 and 2 is 7.83 A, but 6.23 A for Cell 3, while Cell 4 outputs only 4.67 A. The unequal distribution can be used to equalize the unbalanced cells through control of the switching duty cycle.

When the switching duty cycle of three converters is controlled to be 50%, the voltage values of all the cells are kept equal. The differences in the current outputs represents the imbalance among the battery cells. The strong cells, such as Cell 1 and Cell 2, contribute higher currents than the weak cells in the string. The operation is similar to the MIDPP function used for balancing PV mismatches, as introduced in Section 2.4.4. The switched inductor circuits provide a bypass current route to distribute the power output and equalize the cell voltage and capacity.

The switched inductor configuration can be applied in another way using the flyback topology, as shown in Figure 9.14. The inductor is formed by the magnetic inductance

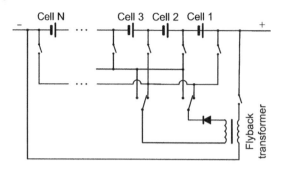

Figure 9.14 Balancing circuits based on flyback topology.

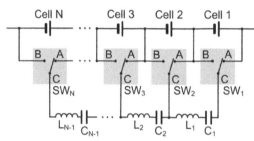

Figure 9.15 Balancing circuits based on switched capacitor with resonance.

inside a flyback transformer, which also provides the required galvanic isolation. The winding-turn ratio is determined by the number of battery cells, N. Using the switch configuration, the terminals of an individual cell can be selected to exchange energy with the overall battery pack. Any weak cell can be specially charged from other cells in the pack.

Other DC/DC conversion topologies are also available for balancing networks. For example, the switched capacitor topology, shown in Figure 9.10, can be modified by adding magnetic components, as shown in Figure 9.15. The storage elements include both inductors and capacitors, which result in resonance in the circuit. Soft switching can be achieved by the resonant solution. Alternative balancing topologies are also available, and these have been discussed by Daowd et al. (2011).

9.1.5 Battery Characteristics and Modeling

Chemical reactions are complicated, so it is hard to develop a general and accurate model of the electrical characteristics of different batteries. The battery can be considered a variable voltage resource, with the steady-state value affected by the SOC and the rate of charge or discharge. Figure 9.16 illustrates the typical response of the battery voltage, V_{bat}, to a step change of load current, I_{bat}. It shows a voltage drop in response to a step increase of the discharge current. The short-term steady state is when the rate of charge and discharge dominate the variation of V_{bat} and the SOC is relatively steady over a short period.

A simple battery model can be represented by a Thévenin equivalent circuit, which is formed by a voltage source in series connection with a resistor, as shown in Figure 9.17. The model can be configured to simulate the steady-state value of the voltage in response to the variation of the discharge current, as expressed in

$$V_{bat} = V_{OC} - I_{bat}R_{bat} \tag{9.1}$$

Figure 9.16 Transient response of battery voltage to step change of discharge current.

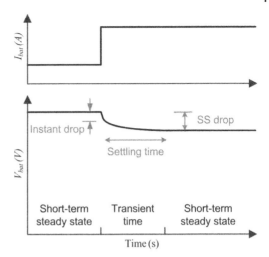

Figure 9.17 Simplified Thévenin model for battery modeling.

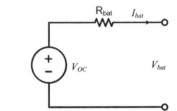

Figure 9.18 Simulink model for the simplified Thévenin battery model.

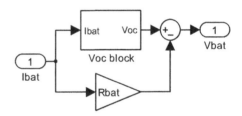

A positive value of I_{bat} indicates that the battery operation is dominated by the discharge. The voltage across the equivalent series resistance (R_{bat}) causes the terminal voltage V_{bat} to be higher than V_{OC} when charge is conducted, but lower when discharge is dominant.

A simulation model can be constructed using Simulink, as shown in Figure 9.18. The model includes one output, which is the instant battery voltage, V_{bat}. The input is the battery current, I_{bat}. It should be noted that R_{bat} is not counted for the loss of SOC in the simulation model.

The center of the simplified model is the voltage source, shown as V_{OC}. The value of V_{OC} and R_{bat} are considered constant in the short-term steady state since the SOC and temperature is relatively steady. In a long run, the value changes with battery capacity in terms of either the SOC or the DOD.

An example is used to demonstrate battery modeling and simulation of the voltage source. The specific battery module is the BK-10V10T, which is a Panasonic NiMH product. The nominal voltage and capacity are 12 V and 90 Ah. At a constant temperature of 25°C, the discharge characteristics are given by the product datasheet. Figure 9.19 shows

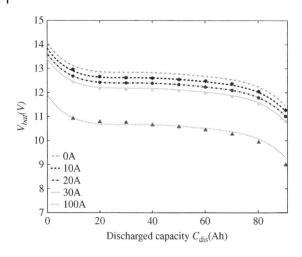

Figure 9.19 Battery voltage versus discharged capacity for model output and product data.

Table 9.6 Polynomial curve fitting parameters for battery pack BK-10V10T.

P_5	P_4	P_3	P_2	P_1	P_0
-7.1626×10^{-9}	1.7234×10^{-6}	-1.6205×10^{-4}	0.0072	-0.1526	14.06

that the battery voltage reduces with increasing discharge capacity, C_{dis}. A higher value of C_{dis} indicates a lower voltage, by following the curves. The top curve indicates the voltage variation in response to a 10 A discharge current. The bottom curve represents the discharge when the current is 100 A.

Figure 9.19 shows that the variation of the battery voltage corresponding to the discharged capacity and the discharge current. When the discharge current is high, the voltage drop is also high. This is caused by the resistance, R_{bat}. For example, when the discharge capacity is 40 Ah, battery voltages of 12.602 V and 12.386 V correspond to discharge currents of 10 A and 20 A, respectively. Considering that the difference is caused by R_{bat}, the resistance can be estimated as $(12.602 - 12.386)/10$, which is 21.6 mΩ.

When the measured data of V_{bat} are available, the dataset of V_{OC} can be derived from

$$V_{OC} = V_{bat} + I_{bat}R_{bat} \tag{9.2}$$

Based on the voltage curve of the 20 A discharge rate, polynomial curve fitting can be used to derive the relation between V_{OC} and the discharged capacity, C_{dis}. A fifth order polynomial can be used to represent the function. For the BK-10V10T module, the polynomial function is derived as

$$V_{OC} = P_5(C_{dis})^5 + P_4(C_{dis})^4 + P_3(C_{dis})^3 + P_2(C_{dis})^2 + P_1C_{dis} + P_0 \tag{9.3}$$

with the parameters as shown in Table 9.6.

The function is plotted in Figure 9.19, illustrating the correspondence between V_{OC} and C_{dis}. The discharge current is shown as 0 A for the open-circuit condition. For comparison, it is also plotted with the battery voltage, V_{bat}, in response to different levels of discharge current: 10 A, 20 A, 30 A, and 100 A. The data can be derived from the function for V_{OC}, applying (9.1) for the various discharge currents. The measurement data

Table 9.7 Polynomial parameters for modeling battery module BK-10V10T.

P_{S5}	P_{S4}	P_{S3}	P_{S2}	P_{S1}	P_{S0}
42.2942	−98.3961	88.7769	−40.3893	10.2942	11.4802

Figure 9.20 Open-circuit voltage (V_{OC}) versus SOC: model output and BK-10V10T product data.

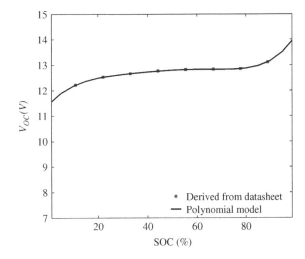

sourced from the product specification are also marked in Figure 9.19 for comparison. The evaluation demonstrates agreement between the simulation model output and the product data.

For a general representation, the discharge capacity, C_{dis} can be converted into the SOC, which is the normal way to represent the battery capacity. For the BK-10V10T module, the polynomial function is derived as

$$V_{OC} = P_{S5}SOC^5 + P_{S4}SOC^4 + P_{S3}SOC^3 + P_{S2}SOC^2 + P_{S1}SOC + P_{S0} \tag{9.4}$$

with the parameters as shown in Table 9.7.

Figure 9.20 shows the function of (9.4) in comparison with data derived from the product datasheet. The plot is based on the relation between the value of V_{OC} and SOC which is general for simulation. It shows that the value of V_{OC} increases with the level of SOC in a nonlinear manner. The output of the polynomial function is a good match for the product data.

Figure 9.21 shows the simulation block for the output of V_{OC} in response to the level of SOC. The integral block inside the simulation model calculates the variation of the SOC. The initial value can be set in the integral block to represent the original capacity, before the simulation starts. It should be noted that the unit of the parameter $C - total$ is a pre-calculated value with units of ampere seconds (As) instead of ampere hours (Ah). The polynomial block produces the value of V_{OC} using (9.4). The Simulink model should be implemented inside the Voc block in the Thévenin model, as shown in Figure 9.18 to simulate the characteristics of the battery voltage in response to real-time variation of the battery current regarding to charge or discharge.

The simple Thévenin model discussed above is effective for simulating long-term battery operations and reflects the battery voltage according to the variation of SOC and

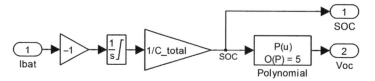

Figure 9.21 Simulink model for the voltage source to output V_{OC} based on the level of SOC corresponding to the battery current.

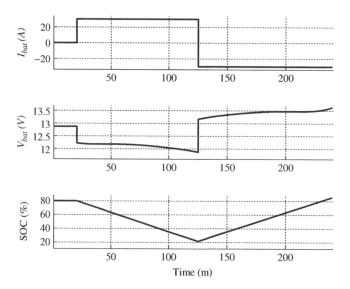

Figure 9.22 Simulation of four-hour charge and discharge of BK-10V10T battery module.

battery current. A four-hour operation of the BK-10V10T battery module is simulated and shown in Figure 9.22. The initial SOC is set to be 80%. The battery voltage is the same as the open-circuit voltage, V_{OC}, since the battery current is zero at the start. The discharge current is set to be 30 A at 20 min. The SOC and V_{bat} drops accordingly to follow the output of the simulation model. At 120 min, the mode is changed from discharge to charge, with a current of 30 A, which is equivalent to C/3. Both the SOC and voltage start to recover and show increasing trends. This demonstrates the effectiveness of the simulation of the characteristics of the battery voltage in response to real-time variations of the battery current in either charge or discharge mode.

The transient response, which is indicated in Figure 9.16, cannot be represented by the simplified Thévenin model that is discussed above. A modified Thévenin model utilizes resistor–capacitor (RC) circuits with one additional resistor and one capacitor, labeled as R_{nl} and C_d in Figure 9.23. The resistances, R_{bat} and R_{nl}, give rise to a voltage difference between the V_{OC} and the terminal voltage V_{bat} in steady-state conditions. The presence of C_d introduces the dynamic behavior in the battery that is shown in the Laplace function

$$V_{bat}(s) = V_{OC} - I_{bat}(s) \left[\frac{R_{bat} + R_{nl} + R_{bat}R_{nl}C_d s}{1 + R_{ni}C_d s} \right] \tag{9.5}$$

Figure 9.23 Modified Thévenin model for batteries.

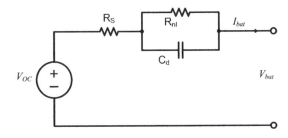

Figure 9.24 Simulink model for the modified Thévenin battery model.

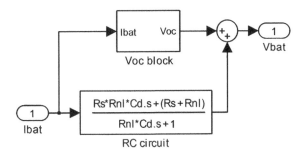

The value of I_{bat} is defined as positive when the rate of discharge is higher than the discharge rate. The RC-based model generally gives a better representation of the chemical reaction phenomena when the battery current instantaneously changes.

The simulation model can be constructed by following (9.5) and is shown in Figure 9.24. The modified Thévenin model is equivalent to the simplified model when the parameter C_d is set to zero. The Voc block is the same as in the polynomial-based model shown in Figure 9.21. With the help of the product data, the sum of R_S and R_{nl} can be estimated from the voltage drop in the steady state. When the battery voltage response is tested by applying a pulse current, the instantaneous voltage drop, as shown in Figure 9.16, can be used to estimate the value of R_S. The resistance of R_{nl} can then be identified because the total resistance is known. The capacitance C_d can also be estimated by measuring the settling time, as shown in Figure 9.16. The settling time of an RC network is roughly calculated as $4R_{nl}C_d$.

Figure 9.25 illustrates the dynamic response of the battery voltage relative to the pulsed current variation. The model is constructed for a battery module rated at 12 V and 90 Ah. The model parameters are: $R_S = 5.4$ mΩ, $R_{nl} = 16.2$ mΩ, and $C_d = 33$ F. It should be noted that the parameters were not derived from the BK-10V10T battery model, since the experimental data are unavailable for the parameterization. The waveform and parameters are solely for demonstration purposes, in order to demonstrate the modified Thévenin model. The SOC decreases from the initial level of 80% in response to the pulsed discharge current. The waveform of the battery voltage V_{bat} demonstrates a fast transient response between steady states.

Even if all parameters in the Thévenin models are considered as constant, it should be noted that the values of R_{bat}, R_{nl}, and C_d are influenced by the SOC. The resistances R_{bat} and R_{nl} generally decrease with an increase of the SOC. However, the capacitance C_d is very nonlinear, and is difficult to predict accurately through a mathematical function.

A more complicated Thévenin-based circuit known as the "run-time-based battery cell" model aims to give a more accurate representation (Chen and Rincon-Mora 2006;

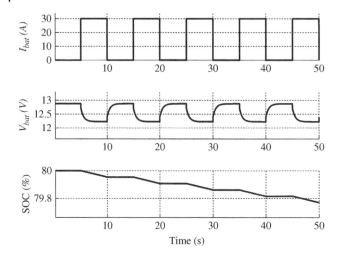

Figure 9.25 Simulation result of the dynamic response of the battery voltage for the impulsed current variation.

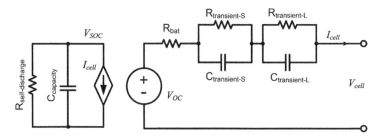

Figure 9.26 Comprehensive Thévenin model for battery modeling.

Kim and Qiao 2011). The equivalent circuits are illustrated in Figure 9.26, and include two sections. Since one section of the model is constructed with RC circuits, it is sometimes called the comprehensive Thévenin model.

The left-hand section is designed to track the status of the SOC, with the capacitor representing the charge stored in the battery cell and the resistance characterizing self-discharge. The current source I_{cell} represents the charge/discharge current of the battery cell.

The right-hand section is developed to simulate the I–V characteristics and transient response of the battery cell. The circuit follows the same format as the Thévenin models, including both the voltage source and the RC circuits. The controlled voltage source, V_{OC}, represents the open-circuit voltage and corresponds to the instantaneous SOC. The parallel RC circuits are divided into two groups to characterize the short-term and long-term transient responses of the battery cell, and are shown as transient-S and transient-L in the circuit.

There is always a tradeoff between the battery model's simplicity and accuracy. A more complicated model generally requires more experimental data for estimating the parameters. It also requires more computational power for simulation. In this section, three modeling approaches are discussed and summarized and the findings are summarized in

Table 9.8 Comparison of battery models.

Modes	Simple Thévenin	Modified Thévenin	Comprehensive Thévenin
Steady state	Yes	Yes	Yes
Transient response	No	Yes	Yes
Run-time feature	Yes	Yes	Yes
Self-discharge	Optional	Optional	Yes

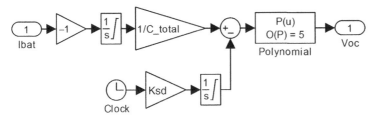

Figure 9.27 Simulink model for the voltage source to output V_{OC} based on the SOC and self-discharge rate.

Table 9.8. All models are capable of simulating the battery output voltage with respect to the charge or discharge current and can predict the SOC. The two complicated models, namely the modified Thévenin and comprehensive Thévenin, can simulate the transient response when the battery current changes suddenly.

The comprehensive model has a specific mechanism for self-discharge, which is not included in other Thévenin models. However, a circuit-based implementation might not be necessary since it is easy for other models to incorporate self-discharge using a mathematical implementation. Figure 9.27 shows that the Voc block can be implemented with a self-discharge feature using an integral function with time. Ksd is the self-discharge rate, which varies the SOC. It should be noted that accurate prediction can be difficult since the self-discharge rate is nonlinear and is significantly affected by temperature. For accurate prediction of self-discharge, the nonlinearity and temperature information should be considered.

It is known that all parameters in the equivalent circuits change not only with the SOC, but also with the rate of charge or discharge, temperature, aging, and so on. These factors can be incorporated only if the relevant information is accurately and reliably acquired. The model parameters are also affected by the frequency that is commonly introduced by switching-mode power converters. Since electrochemical reactions are complicated, it is difficult to develop a general modeling approach to accurately predict the battery output in specific conditions. If there are no accurate and reliable data for parameterization, it should be straightforward to use the simplest model instead of complicated counterparts. It is always important to define the simulation objectives and choose the simplest model that will meet the requirements, thus avoiding unnecessary complication. For example, the charge and discharge operations usually take hours, and so the slow variation of the voltage and SOC might be more of interest than the fast dynamics. In this case, the simplified Thévenin model can do the job without any need for high-powered computing.

9.1.6 Battery Selection

It should be noted that NiCd batteries can only be used where there is a simple requirement for steady runtime voltage and reliability. NiCd cells are not as low-cost as their lead-acid counterparts, but the low capacity density prevents their use in the majority of applications. In PV systems, the battery implementation can be either for stationary energy storage or in mobile applications. Lead-acid, NiMH, lithium-ion, and NaS batteries are all widely available for such uses. The following tips can be considered for selecting among them.

- *Lithium-ion technology* is suitable for different scales and applications, has high performance in terms of capacity density, and is becoming cheaper due to recent mass production. Detailed sensing and protection should always be used to guarantee safe operation, especially for systems with significant storage capacity.
- *Lead-acid batteries* can be selected for either vehicular or stationary implementations. Their advantages lie in the maturity of the technology and their low cost. The drawbacks are the low capacity density per unit weight and volume.
- *NaS technology* has potential for large-scale stationary energy storage due to its high operating temperature. However, the supply is limited to a single source, which is different from the other technologies.
- *NiMH batteries* can be used for small-scale PV systems for either vehicular or stationary implementations when the power density is not critical. The technology is generally low cost, but has shorter life cycles, higher self-discharge rates, a memory effect, and is inefficient in comparison with lithium-ion batteries.

9.2 Integrating Battery-charge Control with MPPT

Charging is the process of storing solar power in discharged batteries. A suitable charging method should be used to charge them efficiently and prevent any damage. To stop the PV generation from overloading the charging capacity, the charging cycle should be maintained with an MPPT function.

When the voltage and current of the battery reach their charging-cycle limits, MPPT should be stopped in order to reduce PV power generation. Instead, the control action should shift the operating point of the PV generator in the open-circuit voltage direction or into the voltage-source zone. It has been reported that the system dynamics of the voltage-source zone gives better damping performance than the current-source zone (Xiao et al. 2007). The PV I–V curve has been divided into three operating zones: current-source zone, power-source zone, and voltage-source zone, as shown in Figure 6.2. It is recommended to operate the PV output in the power-source and voltage-source zones and to avoid the current-source zone.

Figure 9.28 illustrates the integration of MPPT and cycle-charge control, in terms of constant voltage and constant current regulation. For battery-charging applications, the battery voltage and current are sensed to determine the operation status. If both are under their defined limits, MPPT is used to deliver the highest charging power. If either limit is reached, the control variable moves the operating point away from the MPP, in the direction of the open-circuit voltage. For the case study, $\pm 1\%$ hysteresis is applied to the regulation for both battery voltage and battery current. The hysteresis error can

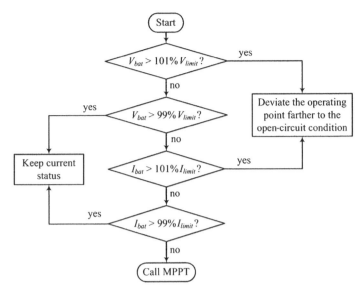

Figure 9.28 Integration of maximum power point tracking for battery charge control.

be other values depending on the practical requirements. When the sensed voltage and current are within their predefined tolerances, the controller maintains the steady-state condition.

9.3 Design of Standalone PV Systems

This section introduces the design, integration, and simulation of standalone PV power systems. The systems may or may not include significant energy storage. "Significant" energy storage means that the storage capacity is sufficient to mitigate the intermittency of PV generation. Switching-mode power supplies contain energy storage units, such as inductors and capacitors, to constrain switching noise, which is considered insignificant in comparison with batteries.

9.3.1 Systems without Significant Energy Storage

Most DC loads can be supplied with a certain range of DC voltage. A direct-coupled system is the simplest solution and can be designed to match the PV module with the load demand. An example is used to demonstrate the design procedure. A solar-powered ventilation system is required to provide airflow up to 700 CFM (cubic feet per minute). A cooling fan is selected to meet the airflow requirement, the specifications for which are shown in Table 9.9. A PV module should be selected to match the supply voltage range and the power requirement.

To match the voltage range from 18 to 30 V, as shown in Table 9.9, a PV module is constructed from 48 crystalline solar cells. The size of the solar cells is 6 inches, with a current rating of 7.20 A. A suitable PV module is selected, and the specification is shown in Table 9.10. The rated MPP of the PV module is close to the rated voltage and current required by the cooling fan. The peak power at STC is 170 W, which is slightly lower than

Table 9.9 Specification of Orion 12HBXC01A cooling fan.

Term	Rating
Model number	NMB F225A4-092-D0530
Nominal voltage	24 V
Nominal current	7.20 A
Voltage range	18–30 V
Rated power	172.8 W
Rated air flow	763 CFM

Table 9.10 Specification of Invensun i170-48P PV module.

Basic information			
Manufacturer	Model	Cell material	Dimension
Invensun	i170-48P	Poly-crystalline	1316 mm × 995 mm × 40 mm

Electrical performance at STC					
Cells	P_{MPP}	I_{MS}	V_{MS}	I_{SCS}	V_{OCS}
48	170 W	7.14 A	23.81 V	7.72 A	28.8 V

Temperature coefficients		
α_T	β_T	γ_T
0.06%/°C	−0.35%/°C	−0.45%/°C

α_T, β_T, and γ_T are temperature coefficients for correcting the PV module output current, voltage, and power, respectively.

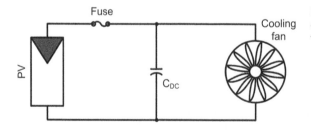

Figure 9.29 System diagram of direct coupled PV system for ventilation.

the rated power of the load, 172.8 W. The system can safely operate at 170 W, without being overloaded.

The system is shown in Figure 9.29. A DC-link capacitor gives a start-up current and maintains a steady voltage at the DC link. The fuse is rated as 15 A, in line with the recommendation in the PV module datasheet.

Although direct coupling is the simplest solution, it is very difficult to achieve an exact match between the off-the-shelf load and the PV generator when variation of environmental conditions is a factor. In many cases, the system design is customized for either the load or the PV unit. Simulation for case studies can be conducted only if the I–V characteristics of the PV module and the load are available. However, the manufacturer of the cooling fan does not provide such information. Experimental testing therefore becomes important.

It should be noted that many cooling fans monitor the input voltage and apply autostart and stop based on the voltage level. If such a function is not supported or does not fit the PV output curve, an additional control circuit should be included to support the start and stop function with a hysteresis loop. In this case, the voltage window of the load is 18–30 V. The voltage threshold for start can be set to 24 V or higher to ensure that the PV module can support enough power at the start point. Without the hysteresis setting, the system can start and stop frequently in the early morning since the irradiance level is not sufficient to support the nominal operating power even though the open-circuit voltage is high enough.

Without power-conditioning circuits, the MPPT function cannot be implemented. This hints that the PV module mostly operates away from the peak power point due to the variation of environmental conditions. Therefore, more and more recent PV systems are including power-conditioning units, to allow use of MPPT and flexibility in design and operation. A direct-coupled system can be redesigned to integrate a dedicated power conditioner.

One important application is proposed by Mascara NT, a French company, which supplies highly efficient off-grid systems for seawater desalination. The concept is simple, since fresh water is only produced when solar power is available. The system stops at night or when the weather is bad. The "storage" is fresh water instead of any electrical format. Even though no significant battery storage is required, power conditioning is included in the conversion stage between the PV section and the load. Reverse osmosis is used for high-efficiency desalination. Solar power drives the pump motor and supplies ancillary devices. A DC bus is formed from multiple PV strings in order to directly supply the desalination unit.

9.3.2 Systems with Significant Energy Storage

When energy storage is required, system design generally follows six steps, as shown in Figure 9.30 and described in the following steps:

1) The load profile should be clearly specified, including the daily energy consumption and the rated voltage. The energy consumption is measured in Wh or kWh for a 24-h period. The nominal load voltage is the reference for rating the battery pack and solar generation unit voltages.

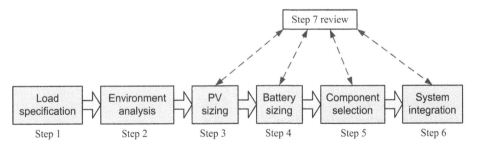

Figure 9.30 A design procedure for PV-battery systems.

2) The location should be thoroughly examined to derive information about the level of solar irradiance and ambient temperatures. It is known that solar irradiance is directly correlated with PV power generation. High temperatures can degrade power generation. It should be noted that the analysis is based on rough estimates since the exact weather cannot be accurately predicted.

3) The rule for sizing the PV generation unit is based on the daily load consumption and the lowest solar generation period. PV generation should be more than the load consumption over a 24-h period. The PV generation unit can be slightly oversized to deal with uncertainty.

4) The voltage of the battery pack should be specified to match the load voltage rating in order to avoid a high ratio of voltage conversion. The battery capacity should be sufficient to mitigate the power variation between daytime generation and nighttime load consumption. The possibility of a deep charge/discharge cycle – more than 60% – should be minimized since it accelerates battery aging. The capacity of the battery pack should generally be oversized by 20–30% to cope with charge and discharge losses.

5) The system should be designed according to the specifications of the load, PV generation unit, and battery pack. DC/DC and/or DC/AC converters might be used according to the load requirement. Protection and disconnect functions should be included. A complete schematic should be presented for review.

6) The design document should include a single line diagram of the system structure, showing all component ratings and details.

7) The final step is to review the whole system specification and design. The design can be revised if any outstanding issues are raised.

The design process can be demonstrated by an example. The load specification is shown in Table 9.11. The 24-h energy consumption, E_{load}, is calculated to be 3200 Wh using (9.6). According to the voltage rating, a battery pack rated at 48 V should be used to meet the acceptable voltage-variation range.

$$E_{load} = 140 \times 8 + 130 \times 16 \tag{9.6}$$

Step 2 is to evaluate the environmental conditions at the installation location. The information collected is shown in Table 9.12. It should be noted that it is impossible to predict any weather condition accurately. Extreme weather is not included in this design process due to its unpredictability. Since the highest temperature is not significantly higher than STC, the power degradation caused by high temperatures can be neglected in this example.

Table 9.11 Specification of DC load.

Term	Rating
Nominal voltage rating	48 V DC
Acceptable voltage range	42–56 V
Averaged power consumption from 9am to 5pm	140 W
Averaged power consumption at other hours	130 W

Table 9.12 Weather information for PV power generation.

Term	Rating
Average irradiance in 8 h period year round	600 W/m^2
Low irradiance in 8 h period on rainy days	200 W/m^2
Average high temperature over year	25.9°C
Average low temperature over year	8.1°C

Step 3 is to size the PV power generator unit. The solar power generation is expected to be sufficient even though the average irradiance is only 200 W/m^2 on rainy or cloudy days. The daily energy generation should be more than 3200 Wh, which is the daily load requirement. The rated power of the PV unit should be calculated from the 8-h generation limit in a 24-h base period, corrected for any estimated losses. The minimum PV power capacity is calculated to be 2500 W, with a 20% loss expected in the power conversion:

$$P_{STC} = \underbrace{\frac{1000}{200}}_{\text{irradiance}} \times \underbrace{3200}_{\text{load}} \times \underbrace{\frac{1}{1-20\%}}_{\text{loss}} \times \underbrace{\frac{1}{8}}_{\text{8-hour}} \tag{9.7}$$

Step 4 is to size the battery pack to store excess energy produced by the PV unit and supply the load at night. It is estimated that there will be 16 h without any solar energy input in a 24-h period. According to the load profile, the minimum energy capacity of the battery is 480 Wh, with a cycle between 0% and 100%. It is commonly recommended to operate the battery in a light cycle, from 40 to 90%, for long operational life. Therefore, the nominal energy capacity of the battery pack is rated as 4160 Wh:

$$E_{bat} = \underbrace{\frac{1}{90\%-40\%}}_{\text{battery cycle}} \times 16 \times 130 \tag{9.8}$$

Based on the requirements for its capacity, the battery capacity in C rate terms can be rated as 87 Ah from

$$C_{bat} = E_{bat}/V_{b-nom} \tag{9.9}$$

where V_{b-nom} refers to the nominal voltage of 48 V in this case. The battery pack can be formed from four 12-V battery modules in series connection. The BK-10V10T battery modules are used to form the battery pack, since each is rated at 12 V and 90 Ah.

A solar charge controller is used to maintain the battery charging cycle and maximize PV generator output. The FLEXmax 60 used is manufactured by Outback Power Inc. The key specifications are shown in Table 9.13. Further information can be found on the product website at www.outbackpower.com.

The voltage rating of the PV string should match the input voltage window of the charge controller, which should not be higher than 150 V in cold conditions. The voltage at MPP should be higher than the open-circuit voltage of the battery pack at 100% SOC. In this case, the voltage window is from 56 to 150 V in order to construct the PV string.

Table 9.13 Key specification of the charge controller–FLEXmax 60.

Term	Rating
Nominal battery voltage	Programmable for 12, 24, 36, 48, or 60 V
Maximum output current	60 A
Maximum input voltage	150 V
Voltage conversion	Step down

Table 9.14 Specification of PV module Q.PLUS BFR-G4.1.

Basic information			
Manufacturer	Model	Cell material	Dimensions
Q Cells	Q.Plus BFR-G4.1	Multi-crystalline	1000 mm × 1670 mm × 32 mm

Electrical performance at STC				
Cells	P_{MPP}	I_{MS}	V_{MS}	I_{SCS}	V_{OCS}
60	280 W	8.84 A	31.67 V	9.41 A	38.97 V

Temperature coefficients		
α_T	β_T	γ_T
0.04%/°C	--0.29%/°C	−0.40%/°C

α_T, β_T, and γ_T are the temperature coefficients for correcting the PV module output current, voltage, and power, respectively.

The Q.Plus BFR-G4.1 solar module is selected for this application. It is produced by Hanwha Q Cells GmbH. The key specifications at STC are shown in Table 9.14. Detailed and updated information is found at the product website at www.q-cells.com.

Nine PV modules are required for this application, since each is rated at 280 W. The total power capacity is 2520 W, which is higher than the fundamental requirement of 2500 W at STC. The modules can be configured into three strings, each with a number of PV modules in series (N_{series}) of 3. The maximum DC voltage is calculated as the sum of the rated open-circuit voltage of each string, with correction for the lowest expected temperature:

$$V_{pvdc}(\max) = K_T \times 3 \times 38.97 = 116.91 K_T \tag{9.10}$$

where K_T is the correction factor. The installation location is expected to become as cold as 8.1°C, and K_T can be as determined from the temperature coefficients listed in Table 9.14. Using (9.11) it can be calculated as 1.17. Therefore, the maximum DC voltage is rated as 137 V, lower than the 150 V required by the charge controller. At the STC, the string voltage at MPP is 95 V. Even though the MPP voltage is lower when the temperature is higher than 25°C, the voltage rating at MPP should be always higher than the minimum requirement of 56 V in order for the step-down voltage conversion that is required by the battery-charge controller.

$$K_T = 1 + \beta_T(8.1 - 25) \tag{9.11}$$

Figure 9.31 is the single line diagram for the standalone system configuration. The specifications of the charge controller and PV module are shown in Tables 9.13 and 9.14. The information about other key components, the battery pack, PV source circuit, DC combiner, DC disconnect, and DC circuit breakers, are summarized in Tables 9.15–9.19 and 9.20, respectively. Together with the climate information for the installation site, the diagram and tables should be included in the final design document. It should be noted that the 48-V battery system requires a higher voltage rating for the circuit components interconnected with the battery voltage bus because of the expected battery voltage variation.

Even though AC loads are common in many standalone systems, these can be equivalent to DC loads since a DC/AC power interface is required to draw DC power from the system. DC/AC conversion systems have been discussed in various books on the subject of power electronics. Therefore the subject is not covered in this book. For a

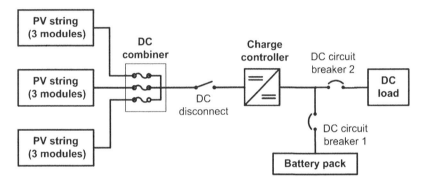

Figure 9.31 Single line diagram of standalone PV system with battery storage.

Table 9.15 Battery pack configuration and specification.

Term	Rating or description
Configuration	4 battery modules in series connection
Battery pack voltage rating	48 V
Battery pack nominal capacity	90 Ah
Battery module model	Panasonic BK-10V10T
Battery type	Nickel metal hydride (NiMH)

Table 9.16 Configuration and specification of PV source circuit.

Term	Rating or description
PV array configuration	3 PV strings in parallel connection
PV string configuration	3 PV modules in series connection
Maximum array voltage rating	137 V
Maximum array current rating	35.3 A

Table 9.17 Configuration and specification of DC combiner.

Term	Rating or description
Configuration	No less than three inputs
Rating of string protection fuse	15 A
Voltage rating	Not less than 137 V

Table 9.18 Specification of DC disconnect.

Term	Rating or description
DC disconnect current rating	Not less than 44 A
DC disconnect voltage rating	Not less than 137 V

Table 9.19 Specification of DC circuit breaker 1.

Term	Rating or description
Amperage rating	90 A
Voltage rating	Not less than 56 V
Configuration	With integrated DC disconnect

PV–battery system, it is always good practice to minimize the use of AC loads due to the loss and complications of DC/AC conversion. More and more modern equipment and appliances are using DC supply.

9.4 Equivalent Circuit for Simulation and Case Study

The equivalent circuit for simulation can be derived from the system design in Figure 9.31. Figure 9.32 illustrates the equivalent circuit for analysis and simulation. The simplified Thévenin model is used, and includes the battery pack voltage V_{OC} and the equivalent series resistance R_{bat}. The dynamics of the battery link shows interaction with the capacitor, C_{BAT}, which is the equivalent capacitance across the battery pack. All DC loads are simplified to draw current from the battery link, which is denoted as i_{load}. A buck converter is the charge controller that transfers the PV power to the battery and supplies the load. The inductor current of the buck converter is denoted i_L. The circuit dynamics are

$$i_L + i_{bat} = C_{BAT}\frac{dv_{bat}}{dt} + i_{load} \tag{9.12}$$

and

$$R_{bat}C_{BAT}\frac{di_{bat}}{dt} + i_{bat} = i_{load} - i_L \tag{9.13}$$

with the assumption of a constant voltage, V_{OC}, in the short-term steady state.

Table 9.20 Specification of DC circuit breaker 2.

Term	Rating or description
Amperage rating	5 A
Voltage rating	Not less than 56 V
Configuration	With integrated DC disconnect

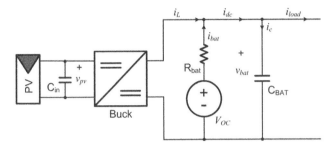

Figure 9.32 Equivalent circuit of standalone PV system with battery storage.

The battery-link dynamics can also be represented by the transfer function:

$$i_{bat}(s) = \frac{i_{load} - i_L}{R_{bat}C_{BAT}s + 1} \tag{9.14}$$

In the steady state, the battery current, i_{bat}, is the error between the load current i_{load} and the inductor current of the buck converter i_L. The aggregated capacitance across the battery terminals causes the dynamics at the battery link.

9.5 Simulation Model to Integrate Battery-charging with MPPT

MPPT using the HC algorithm was introduced in Section 8.2 and formulated for simulation, as illustrated in Figure 8.6. With the integration of battery-charging cycles, a control flowchart for the PV-powered battery charger can be developed, as shown in Figure 9.28. The Simulink model should integrate the MPPT and the regulation of the battery voltage and current.

Figure 9.33 illustrates the simulation block in Simulink. The bottom section shows the HC operation for tracking the MPP. The MPPT operation can update the control variable only if both the battery voltage and current are within their limits, which are indicated by 0.99Vlimit and 0.99Ilimit. The system is operated in MPPT mode.

If either the battery voltage or current reaches its predefined limit, the system enters the charging-cycle mode. Either the battery voltage or current is regulated to avoid battery overcharging. When the value of either the battery voltage or current is within a 2% tolerance range, the model output Xnew is maintained constant by setting the variation step to zero. However, if either the battery voltage or current reaches their upper limits, denoted 1.01Vlimit and 1.01Ilimit, respectively, the control action will move the

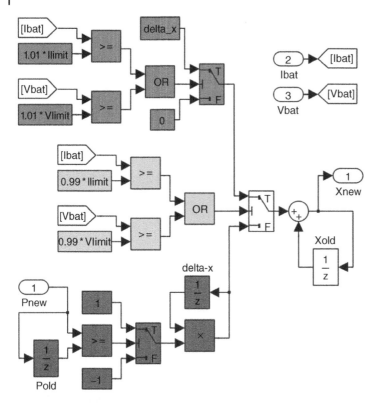

Figure 9.33 Simulink model for integrating MPPT and battery-charging cycles.

operating point in the direction of the open-circuit voltage by updating the value of Xnew. In this case, Xnew is the PV output voltage. Therefore, increasing its value brings it closer to the open-circuit condition. It should be noted that the 2% tolerance, from 99% to 101%, is just as an example for this case study. Other values can be assigned to maintain the battery voltage and current within a limited range.

9.6 Simulation Study of Standalone Systems

A simulation model for the designed system is needed to assess its response to changes of environmental conditions and load variations. The simulation system is based on the case study and design in Section 9.3.2.

9.6.1 Simulation of PV Array

The PV cell parameters can be estimated using the specification of the PV module shown in Table 9.14. The PV cell voltage is derived from the module output voltage and the number of cells in series connection. Based on the ideal single-diode model (ISDM), the ideality factor of the PV cell is determined as 1.6882.

The PV array that was developed in Section 9.3.2 has a 3×3 configuration of nine PV modules. Therefore, the PV array output characteristics at STC are represented by the

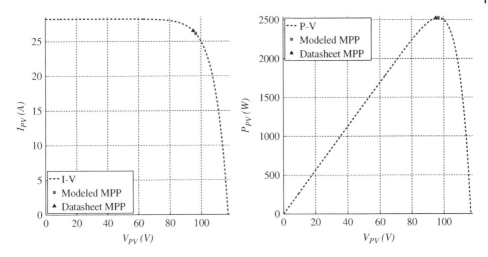

Figure 9.34 Output characteristics of the 3 × 3 PV array.

I–V and P–V curves shown in Figure 9.34. With balanced operation of nine PV modules, the MPP of the array is indicated as 2525 W and is at (96.66 V, 26.12 A).

9.6.2 Short-term Simulation

Figure 9.35 shows the overall model for simulating the standalone system including the PV array, DC/DC buck converter, battery link, and battery pack. The blocks for the solar irradiance, cell temperature, and the load current represent subsystems to be programmed to vary as predefined conditions. It should be noted that the sign of the battery current is defined as positive when it is extracted from the battery pack. Modeling of the buck converter used as the PV-side converter was introduced and developed in Section 5.1.2. The model is capable of capturing the dynamics during each switching cycle. For the case study, the specification of the buck converter is as shown in Table 9.21. The same parameters are used for the medium-term and long-term simulations described in later subsections.

The system control is represented by the blocks for the MPPT charger, PID controller, and PWM generator. The PID controller and PWM generator are those developed in Section 7.8.3. Based on the nominal operating condition and circuit parameters, as shown in Table 9.21, a small-signal model can be derived using the modeling process in Section 6.3.2. The model has a damping factor of 0.26 and an undamped natural frequency of 1.08×10^3 rad/s. Affine parameterization is used to design the feedback controller to regulate the PV-side voltage in order to follow the MPP. As a result, the PID controller for the voltage regulation of the PV link is synthesized and expressed as

$$C(s) = -0.0017 - \frac{8.4736}{s} - \frac{6.8485 \times 10^{-6}}{2.8745 \times 10^{-4}s + 1} \tag{9.15}$$

The MPPT charging block is implemented as shown in Figure 9.33. The MPPT parameters are summarized in Table 9.22. The tracking frequency should always be determined by the dynamic analysis of the PV-link voltage-regulation loop. Without the start-stop mechanism, as introduced in Section 8.6, a 1-V ripple is expected to

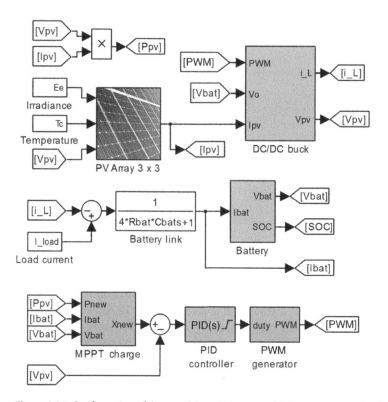

Figure 9.35 Configuration of the standalone PV system with battery storage for short-term simulation.

Table 9.21 Specification of buck converter circuit.

Term	Rating or description
Switching frequency	50 kHz
Inductance	L = 470 μH
Input capacitor	C_{in} = 470 μF
Nominal input voltage	95 V
Nominal output voltage	48 V

Table 9.22 Specification of the MPPT algorithm.

Term	Rating or description
MPPT algorithm	Hill-climbing method
MPPT tracking frequency	200 Hz
Perturbation step size	$\Delta V = 1$ V

appear in the PV-link voltage due to the active perturbation of the HC algorithm. The equivalent capacitance across the battery link is set to $C_{bat} = 33$ μF for the case study.

To avoid battery overcharging, the upper limit of the battery voltage is set to 55 V, which represents the series connection of 40 NiMH cells in the case study. Since a 2% ($\pm 1\%$) tolerance is applied, the band for the voltage limit is between 54.45 and 55.55 V. When the battery voltage is higher than 54.45 V, active tracking of the MPP is stopped and charge control is taken to maintain the battery voltage within the upper limit of the tolerated range.

The simulation model of the buck converter, as shown in Figure 5.6, is computationally intensive due to the fast switching frequency of 50 kHz. The sampling time for simulation should be at a level of nanoseconds to capture the switching dynamics and reveal the detailed transient responses up to the switching frequency. The model is suitable for the total simulation times in seconds in order to prove the design concept of the converter circuit and monitor switching ripples, since the battery voltage is relatively steady over short time periods. The detailed simulation model is inefficient for either medium- or long-term simulations for demonstration of the variation of battery SOC and voltage during charging and discharging. Therefore, the simulation results are neglected for this case study since they have been covered in Section 5.1.2.

9.6.3 Medium-term Simulation

The averaged model was introduced in Section 6.3.2. It is based on the assumption that the buck converter is in continuous conduction mode. The expression in (6.17) can be further derived as

$$i_L = \frac{1}{L} \int (dv_{pv} - v_{bat})dt \tag{9.16a}$$

$$v_{pv} = \frac{1}{C_{in}} \int (i_{pv} - di_L)dt \tag{9.16b}$$

where d is the switching duty cycle and the control variable. Other variables and parameters refer to the definitions in Figure 9.32.

Figure 9.36 illustrates the simulation blocks for the averaged model of the DC/DC buck converter. The inputs include the duty cycle, the output voltage, and the PV injection current, which are labeled duty, Vo, and Ipv, respectively. The output variables are the inductor current and the PV terminal voltage. The control variable for the averaged converter model is the value of the duty cycle instead of the PWM pulsed signals.

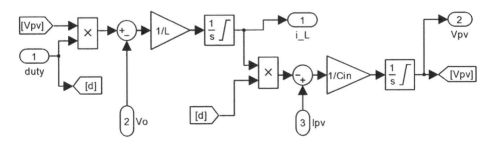

Figure 9.36 Simulink model of the averaged synthesis of the buck converter.

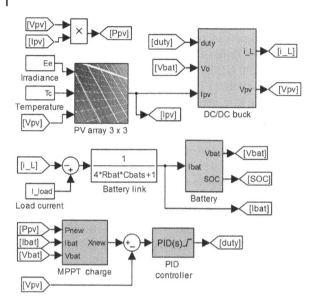

Figure 9.37 Configuration of standalone PV system with battery storage for medium-term simulations.

The switching ripples will not be expected in the simulated waveform of the PV-link voltage and the inductor current.

Figure 9.37 illustrates the block diagram when the averaged model is used to represent the DC/DC buck converter. The key difference from the switching model in Figure 9.35 is that the input of the DC/DC buck converter model is the value of the duty cycle. The high-frequency PWM signal and switching operation are neglected in the averaged model to improve simulation speeds. The control parameters are kept the same as shown in Tables 9.22 and (9.15). The battery link is represented by the transfer function (9.14). The battery model was as developed in Section 9.1.5 and shown in Figure 9.18.

The simulation result of the case study is illustrated in Figure 9.38. The initial SOC of the battery pack was set to be 50%. A load current of 2.8 A was constantly extracted from the battery link. Before 0.2 s, the irradiance was 400 W/m^2 and the cell temperature was 25°C. Since the battery voltage is lower than the limit of 54.45 V, the MPP is tracked at 0.09 s during the start-up period and maintained by the steady-state tracking operation. The voltage ripple caused by the active perturbation of the HC algorithm is noticeable. The SOC gradually increases in response to the charging current. At 0.2 s, the solar irradiance steps up from 400 to 800 W/m^2. The PV output power significantly increases, causing an increase of the charging current. The increase of the SOC becomes faster than before. Since the SOC is still relatively low, the battery voltage is still below the upper limit. The MPPT operates continuously so as to inject the highest power into the battery link.

Figure 9.39 illustrates the simulated waveform for another case study. The initial SOC of the battery pack is set to be 80%. Again, a constant load current of 2.8 A is extracted from the battery link. Before 0.2 s, the irradiance is 400 W/m^2 and the cell temperature is 25°C. Since the battery voltage is lower than the limit of 54.45 V, the MPP is tracked at 0.09 s and is maintained by the steady-state tracking operation for the highest solar

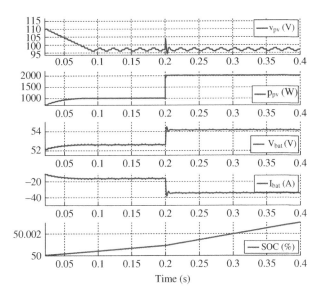

Figure 9.38 Simulated waveforms showing maximum power point tracking for battery charging.

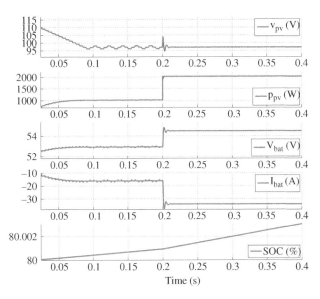

Figure 9.39 Simulated waveforms showing the transition from maximum power point tracking to battery voltage regulation.

energy harvest. The SOC gradually increases in response the charging current. At 0.2 s, the solar irradiance steps up from 400 to 800 W/m². The PV output power is significantly increased, causing the increase of the charging current, which is represented as negative in value. Since the SOC is at a relatively high level, the battery voltage reaches the predefined limit due to the equivalent resistance of the battery pack. The active tracking for MPP is stopped, as can be seen from the disappearance of the tracking ripple.

The charging control maintains the battery voltage within the upper limit range. The charging continues, and the SOC increases within the simulation time period of 0.4 s.

Figure 9.40 shows the simulated waveform for the third case study. The initial SOC is 80%, and the load current is a constant 2.8 A. Before 0.2 s, the irradiance is 800 W/m² and the cell temperature is 25°C. Since the battery voltage reaches the limit of 54.45 V, the system uses battery voltage regulation to maintain the charging cycle. The PV output voltage is maintained at a level higher than the voltage of the MPP in order to maintain low output power. At 0.2 s, the solar irradiance steps down from 800 to 400 W/m². The PV output power significantly decreases, causing a decrease of the charging current. Since the battery voltage becomes lower than the limit of 54.45 V, active MPPT is started. It is noticeable that the MPPT operation lowers the PV output voltage to a lower level for the highest power output under an irradiance level of 400 W/m². The ripple appears again on the waveform of the PV-link voltage. This is caused by the perturbation operation of the HC-based MPPT.

A sudden load change can cause a disturbance to the battery current. It might trigger a transition between MPPT and battery voltage regulation. The variation can also represent a change of the PV output power. Therefore, a case study for load variation is not presented in this section.

The voltage-regulation loop of the buck converter exhibits the highest frequency at the kilohertz level. This requires the simulation sampling time to be at the microsecond level in order to capture the fastest dynamics. For this reason, the averaged model is not efficient for simulation of long-term operations, where there are significant changes of battery SOC and voltage in response to changes in environmental conditions and load. Simulations might result in significant demands for memory and computing speed, which is impractical on standard personal computers.

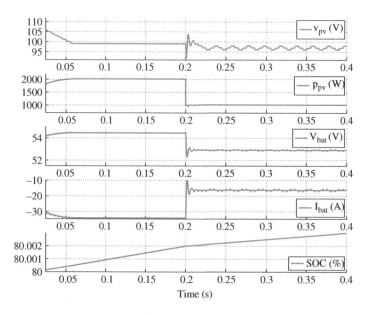

Figure 9.40 Simulated waveforms showing the transition from the battery voltage regulation to maximum power point tracking.

9.6.4 Long-term Simulations

The computer configuration for the simulation cases in this book is shown in Table 9.23. This is considered as a suitable platform to test the simulation efficiency.

It commonly takes hours to charge a battery from low capacity to full. For a long-term simulation, the fast dynamics of the buck converter and the battery link can be neglected. The equilibrium of the buck converter from input to output in continuous conduction mode without the consideration of losses can be expressed as

$$i_L = \frac{i_{pv}}{d} \tag{9.17a}$$

$$d = \frac{V_{bat}}{v_{pv}} \tag{9.17b}$$

where d is the switching duty cycle and the control variable. Other variables and parameters refer to the definitions in Figure 9.32. The dynamics caused by the capacitor, inductor, and switching operation are ignored.

The model of the standalone PV system with battery storage for long-term simulation can be constructed as shown in Figure 9.41. The model simply shows an energy correlation between the PV array and the battery pack. The buck converter is represented by (9.17). The voltage of the PV link is the direct output of the MPPT charge controller with the assumption that the PV array voltage follows the command much faster than the MPPT dynamics. The control functions for MPPT and the regulation of the battery voltage and current are the same as for the short- and medium-term simulations. The highest frequency in the simulation model becomes the MPPT frequency, which is 200 Hz. Therefore, the sampling frequency for simulation can be set to 2 kHz in order to simulate the system efficiently.

Figure 9.42 shows the simulated waveform for a two-hour system operation. The initial SOC of the battery pack is set to be 50%. Again, the load current is 2.8 A, constantly extracted from the battery link. For 40 min, the irradiance is 400 W/m^2 and the cell temperature is 25°C. Since the battery voltage is lower than the limit of 54.45 V, the MPP is tracked and maintained for the highest solar energy harvest. A voltage ripple appears in the waveform of the PV-link voltage, v_{pv} due to the perturbation of the HC algorithm. The SOC gradually increases in response the charging current.

After 40 min, the solar irradiance changes from 400 to 800 W/m^2. The PV output power significantly increases, causing an increase of the charging current. Since the SOC is at a medium level, the battery voltage is still below the voltage limit. Therefore, MPPT operates for 38 min and gives a fast increase of the SOC. After 68 min, active MPPT is stopped because the battery voltage rises to a level of of 54.45 V. The system

Table 9.23 Specification of the simulation computer.

Term	Specifications
Computer model	Dell Precision T1650
Operating system	Windows 10 Pro, 64-bit
Processor	Single Intel® i7-3770 core, 3.4 GHz
Installed memory (RAM)	16 GB

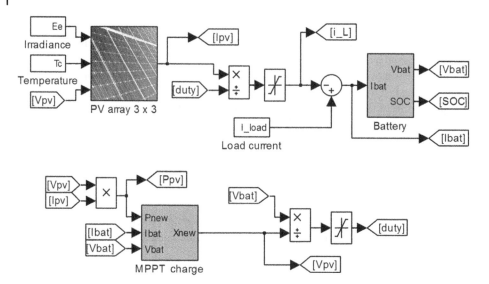

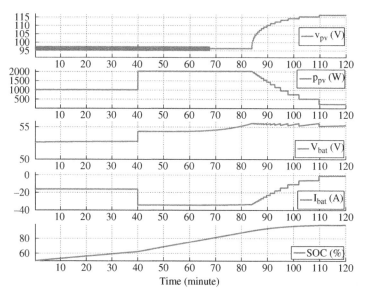

Figure 9.41 Configuration of standalone PV system with battery storage for long-term simulation.

Figure 9.42 Simulated waveforms showing the two-hour operation.

is maintained at the same charge condition until the battery voltage reaches the upper level of 55.55 V. After 84 min, the charge controller takes action to increase the PV-link voltage from the MPP. The deviation of the MPP lowers the PV output power and the charge current, even though the PV array is potentially capable of reaching the higher power level. The battery voltage is regulated for continuous charging until the battery is fully charged, which can be sensed by a reduction of the battery charging current. PV generation is maintained to supply the load power consumption. Using the simplified model, the total simulation time is 7 min for two hours of system operation showing

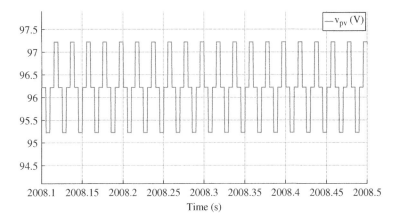

Figure 9.43 Simulated waveform of the PV-link voltage, illustrating the details of MPPT.

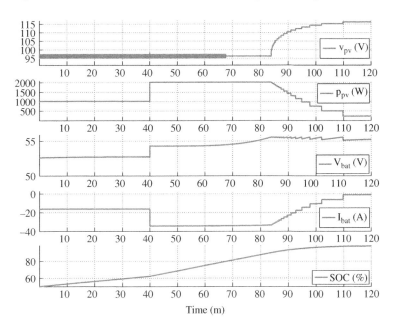

Figure 9.44 Simulated waveforms for the two-hour operation.

the detailed operation of the MPPT algorithm and charging cycle regulation. Figure 9.43 shows a zoom-in on the waveform of the PV-link voltage, v_{pv}. It reveals the perturbation details around the MPP in the early stages due to the operation of the HC algorithm.

In the simulation model, as shown in Figure 9.41, the highest frequency is represented by the HC algorithm, which is in discrete-time format. In contrast to simulations to capture continuous signals, this allows the sampling frequency to be set at up to twice the frequency, which is 400 Hz. For the same two-hour period of operation discussed above, the simulation time is reduced to 63 s, rather than the 7 min in the previous study. The simulation result is shown in Figure 9.44, which is visually identical to the waveform in Figure 9.42, which reflects the cycle charge and MPPT operation.

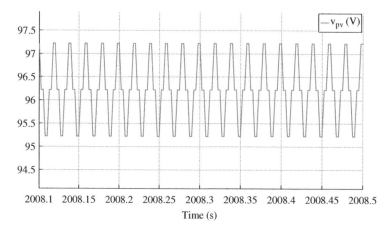

Figure 9.45 Simulated waveform of the PV-link voltage illustrating the MPPT detail.

Figure 9.45 is a zoom-in on the waveform of the PV-link voltage, v_{pv}. It clearly shows the perturbation around the MPP due to the operation of the HC algorithm, but the details are not as accurate as the waveform in Figure 9.43. If the waveform details are not critical, the reduced sampling frequency can significantly improve the simulation efficiency for long-term simulation studies.

9.6.5 Very-long-term Simulations

The proposed method is capable of simulating even longer-term operations with the simplified model and a reduced sampling frequency. When environmental data, such as irradiance and temperature, are available, the simulation can be used to predict the PV power system output in voltage and current terms. Using the same environmental data as the study by Xiao et al. (2013), the long-term simulation can be extended to eight-hour operation.

Figure 9.46 shows the simulated waveforms for the irradiance, cell temperature, PV-link voltage, battery voltage, PV output power, battery current, and battery SOC. It took 7.5 min to simulate the eight hours of system operation. The simulation result can be explained as follows:

- The irradiance increases from the early morning to its highest point in the middle of the day and decreases in the afternoon. The temperature also changes to its highest point at the middle of the day and decreases to lower level in the evening. The temperature variation affects the PV-link voltage representing the MPP, which is high at low temperatures and low at high temperatures. The solar irradiance significantly affects the PV output power.
- The initial SOC of the battery pack after the overnight discharge is 30%. The battery voltage is relatively low at the start point.
- MPPT operates at the beginning of the system control operation so as to inject the highest available power into the battery link. Ripples appear in the waveform of the PV-link voltage, indicating the active perturbation from the HC-based MPPT algorithm. The simulation shows that the system tracks the MPP at each moment and regulates the PV string voltage to follow the MPP.

- The control mode is switched to battery voltage regulation after the first 3.1 h of operation, until the battery voltage reaches the first upper limit of 54.45 V. The PV output and charge current continuously increase to respond to variations in solar irradiance.
- The system is maintained in the steady state for 15 min while the battery voltage is between 54.45 V and 55.55 V. The MPP of the PV array is no longer tracked during this period.
- Battery voltage regulation starts when the battery voltage reaches the upper limit of 55.55 V. The PV-link voltage is controlled to rise and deviate from the MPP. The PV output power and charge current are reduced by the deviation of the MPP, even though the PV array has the potential for higher power output.
- The regulation of the battery voltage is maintained for another 3 h until the solar irradiance drops to a low level in the late afternoon. The PV array output is regulated to maintain the load requirement and the float charge of the battery pack. The SOC is maintained at 100%.
- In the last hour of operation, the irradiance is close to zero. The battery is switched from charge to discharge mode. The load current is at the predefined value of 2.8 A. The battery voltage and SOC decrease accordingly. The operating mode will last until the next morning when solar radiation becomes available.

The simulation of the eight hours of operation shows the major drawback of stand-alone systems. The system is significantly oversized to accommodate bad weather conditions and low irradiation. However, solar energy is simply wasted on sunny days, as shown in this case study. The simulation result shows that the system takes the full power generation from the PV array for the first three hours. Then the control action curbs the PV power output in order to maintain constant voltage for the battery pack. Therefore, grid-connected systems represent a great advantage since the electrical network can usually absorb the PV power without limiting the generation.

9.7 Summary

Energy storage has high potential to mitigate the intermittent power generation of renewable resources. Several battery technologies were introduced at the beginning of this chapter, including those based on lead, nickel, lithium, and sodium–sulfur. The important terms for battery technologies were also introduced. Each technology has unique characteristics for practical applications. Therefore, the selection of the battery was briefly discussed.

Battery technology is still not ideal for widespread use since batteries can be expensive and have limited lifetimes, serious environmental problems, and safety concerns. Lead-acid batteries use sulfuric acid and lead, which are hazardous. Other materials used in batteries, such as nickel, lithium, cadmium, alkalis, and mercury can also contaminate soil and water. Furthermore, fire incidents involving both lithium-ion and NaS batteries raise safety concerns when these technologies are used in high-power-density batteries. Sensing and protection circuits should be carefully designed to guarantee the safe operation of battery power systems.

Battery mismatch and equalization were introduced due to their importance for reliable and long-lifetime operations. A new classification was presented in order to provide

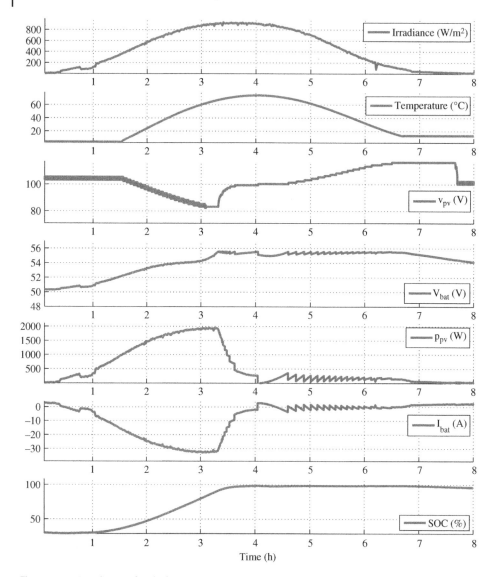

Figure 9.46 Simulation of eight-hour operation of standalone system.

a clear framework for understanding existing methodologies for battery balancing. Even though the case study for equalization was demonstrated mainly at cell level, the same principles can be applied to balance battery modules.

In general, battery modeling is difficult since the parameters are nonlinear and time-variant with many factors, such as the SOC, temperature, age, and frequency. One important development in this chapter concerning battery characteristics and modeling is equivalent circuits at different complication levels. Practical simulation models with the flexibility to represent the battery voltage and the state of charge (SOC) and the option to represent self-discharge were described. Simulations based on Thévenin circuits were built in Simulink and demonstrated for practical implementation. A practical

NiMH battery pack was used to demonstrate the analysis, modeling, simulation, and verification.

The design process of standalone PV systems was discussed. Since most standalone systems contain battery storage, the integration of battery charging and MPPT is described. The algorithm is simple and easy to implement. The design cases for the standalone system with and without battery storage were introduced separately. The direct-coupled system without massive storage represents the simplest standalone PV system, but has significant disadvantages. The typical system is coupled with battery storage and power conditioning to mitigate the intermittent power generation and ensure a steady power supply overnight.

Hybrid system design is not covered, due to the complications for dedicated applications. The knowledge of battery and standalone systems that has been presented in this chapter is useful to design efficient, safe, and reliable hybrid systems.

The simulation of a standalone system is based on one practical design example. The Simulink models for the controller and power interface are developed at different levels for the simulation study: short term, medium term, long term, and very long term. The simulation objective should be always defined clearly to avoid complexity in modeling and inefficiency in simulation. It is always difficult to develop an efficient model, which can simulate not only the long-term changes, but also the transient details of fast dynamics. A simulation of eight hours of system operation shows MPPT, battery voltage regulation, and variation of SOC in response to changes in solar irradiance and cell temperature. The case study neglects the fast dynamics of the switching and PV voltage regulation. The dynamics at the battery link are also ignored for the very-long-term simulation. Without loss of generality, the simulation models are also presented to capture the fast dynamics, including fast switching in the DC/DC converter, but only for short-term simulations due to computational constraints.

Problems

9.1 It is recommended to duplicate the simulation results presented in this chapter. The process is valuable in becoming familiar with the principles of system specification, design, component selection, simulation modeling, system protection, system integration, and verification.

9.2 Use any available simulation tool to construct a battery model based on the presented Thévenin circuits.

9.3 Find a practical battery module or cell to derive the model parameters.

9.4 Verify the simulation model output with the product data.

9.5 Design a PV–battery system for standalone applications.
  a) Construct the simulation model to represent the system operation, including the blocks for the PV, battery, MPPT charge controller, and power interface.
  b) Simulate the circuit over a short time, to reflect any transient responses that can be caused by sudden changes of load or solar irradiance.

    c) Simulate the circuit over a medium time period to reflect variations of the battery voltage and capacity.

    d) Simulate the system over the long term to reflect transitions between MPPT mode and cycle charging mode.

References

Beaudin M, Zareipour H, Schellenberglabe A and Rosehart W 2010 Energy storage for mitigating the variability of renewable electricity sources: An updated review. *Energy for Sustainable Development* **14**(4), 302–314.

Chen M and Rincon-Mora GA 2006 Accurate electrical battery model capable of predicting runtime and I-V performance. *Energy Conversion, IEEE Transactions on* **21**(2), 504–511.

Daowd M, Omar N, Bossche PVD and Mierlo JV 2011 Passive and active battery balancing comparison based on MATLAB simulation *2011 IEEE Vehicle Power and Propulsion Conference*, pp. 1–7.

Kelly K, Mihalic M and Zolot M 2002 Battery usage and thermal performance of the Toyota Prius and Honda Insight for various chassis dynamometer test procedures *17th Annual Battery Conference on Applications and Advances Long Beach, California.*

Kim T and Qiao W 2011 A hybrid battery model capable of capturing dynamic circuit characteristics and nonlinear capacity effects. *Energy Conversion, IEEE Transactions on* **26**(4), 1172–1180.

Kimball JW, Kuhn BT and Krein PT 2007 Increased performance of battery packs by active equalization *2007 IEEE Vehicle Power and Propulsion Conference*, pp. 323–327.

Pascual C and Krein P 1998 Switched capacitor system for automatic battery equalization. US Patent 5,710,504.

Powers RA 2000 Sealed nickel cadmium and nickel metal hydride cell advances. *IEEE Aerospace and Electronic Systems Magazine* **15**(12), 15–18.

Rehman MMU, Evzelman M, Hathaway K, Zane R, Plett GL, Smith K, Wood E and Maksimovic D 2014 Modular approach for continuous cell-level balancing to improve performance of large battery packs *2014 IEEE Energy Conversion Congress and Exposition (ECCE)*, pp. 4327–4334.

Xiao W, Dunford WG, Palmer PR and Capel A 2007 Regulation of photovoltaic voltage. *Industrial Electronics, IEEE Transactions on* **54**(3), 1365–1374.

Xiao W, Edwin FF, Spagnuolo G and Jatskevich J 2013 Efficient approaches for modeling and simulating photovoltaic power systems. *Photovoltaics, IEEE Journal of* **3**(1), 500–508.

10

System Design and Integration of Grid-connected Systems

The key components of grid-connected systems were introduced in previous chapters. The PV output characteristics and mathematical models were described in Chapter 4, the power-conditioning circuit in Chapter 5, modeling and control solutions in Chapters 6 and 7, and MPPT functions in Chapter 8. This chapter discusses the integration of all the components and control functions to formulate a complete grid-connected PV system. The safety standards, guidance, and regulations were briefly described in Chapter 3, and can be used to direct the design process. Simulation of the overall system is presented to verify the effectiveness of the design in meeting the performance expectation. The system integration generally follows seven steps, which are illustrated in Figure 10.1.

First, the system capacity should be clearly defined with reference to the installation location, area available, and other features. If there is a constraint on the installation area, a high-efficiency PV technology should be considered. Crystalline-based PV cells are usually more efficient than their thin-film counterparts. Otherwise, the important factors are the price per watt and the PV product quality. Certified products should always be used to ensure safety. The AC voltage and frequency ratings can match the utility standard and local grid code. The selection of single-phase or three-phase depends on the specification of the point of the distributed resource connection (PDRC). Three-phase AC is preferable to single-phase if the interface is readily available and the system capacity is reasonably significant. A balanced three-phase interconnection does not cause significant ripple at the DC side. In special cases, a three-phase system can be formed from three single-phase grid-connected inverters thanks to the modular nature of PV generators. One example was introduced in Section 1.12.

When the system specification has been clearly defined, the next step is to choose key system components. The grid-connected inverter is considered the center of PV power systems. Its selection should be based on the system capacity, the electric rating at the PDRC, and the certification requirement. It should be always kept in mind that a PV system is highly modular, which is different from conventional power generation systems. A high-capacity system can be formed from multiple, hundreds, or thousands of grid-connected PV inverters. For example, a 10-kW system can be formed from two 5-kW PV inverters. A 10-MW power plant can be constructed from twenty 500-kW PV inverters. The accumulated capacity of the inverters should be equal to or higher than the overall system power rating. If a grounded system is required, the inverter should be able to support galvanic isolation. Ancillary transformers can be installed for galvanic isolation if the inverter does not support this function.

Photovoltaic Power System: Modeling, Design, and Control, First Edition. Weidong Xiao.
© 2017 John Wiley & Sons Ltd. Published 2017 by John Wiley & Sons Ltd.
Companion Website: www.wiley.com/go/xiao/pvpower

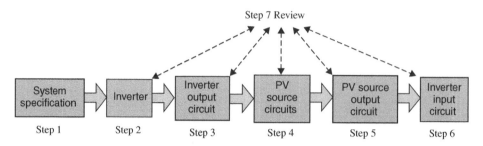

Figure 10.1 A design procedure for grid-connected PV systems.

Based on the selected inverter, the inverter output circuit can be sized and designed. The rating of the AC disconnects and devices for overcurrent protection (OCP) can follow either the recommendation of the inverter manufacturer or the local standards for electricity. When a low-voltage network is connected, Article 690 can be followed, since this standard provides general safety guidance for grid interconnections. Other countries can use it as the reference if the local electric code does not have specific requirements for PV power systems.

Based on the DC input voltage rating of the inverter, the PV panels can be selected to give the required power capacity. The input voltage window of the inverter is the guideline for deciding how many PV modules are connected in series to form a string. The temperature coefficient should always be considered when determining the maximum DC voltage in the PV source circuits, as described in either the product specification or the electric code. The parallel connection of the strings builds the PV array capacity to match the inverter capacity. It should be noted that it is sometimes difficult to match the formulated PV array with the exact power rating of the inverter. For example, if the PV module is rated as 270 W, it is impossible to construct a system of exactly 10 kW. The capacity of PV systems is rated at the STC. When the system is in operation, the system output power heavily depends on the weather conditions, which might significantly differ from STC. Therefore, it might not be meaningful to match the power rating exactly to the system specification. A certain tolerance in the capacity should be allowed for. The rating and selection of cables for outdoor installation can follow the introduction in Section 3.3.5. It is very important to identify the cabling for outdoor use. It must have the appropriate UV or sunlight resistance. If the cable is not flame resistant, it should not be installed inside conduits.

The next step is to design the PV source output circuit, which usually includes the DC combiner boxes, OCP devices, and cables. This process is based on the rating of the DC circuit voltage and current. Article 690 provides a detailed sizing requirement if the system is installed in the USA. It is commonly referred to as the Rule of 1.56 for the ratings of the PV output circuit, as discussed in Section 3.3. Other countries might have different rating requirements. Article 690 can be used as the guideline if the local electric code does not have requirements for sizing of DC components.

The inverter input circuit is also based on DC rating, and usually includes the DC disconnects and cables. Article 690 requires the Rule of 1.56 to be used when the system is installed in the USA and this standard can be a reference for installations in other countries.

The process of review and redesign can be one or several cycles until the designed system meets the overall system specification and the detailed rating of each component. This chapter discusses the design of the DC parts of the system; the AC parts are covered in other books, and is common in electrical engineering.

10.1   System Integration of Single-phase Grid-connected System

The design procedure can be demonstrated to design a practical PV system for single-phase AC grid interconnection. The system specification is briefly summarized in Table 10.1. The system capacity and specification usually fit the roof-mounted application for residential houses or apartment buildings. The expected temperature of the lowest and the highest at the installation location is also included for proper DC voltage rating.

10.1.1   Distributed Maximum Power Point Tracking at String Level

Based on the interconnection specification, the grid-connected inverter can be selected to meet the requirements for voltage, frequency, and capacity. One suitable product is the ABB-PVI-6000, which has of IEC-62109 certification by TUV.[1] All details can be found from the website at www.abb.com. The product datasheet indicates that the system uses two-stage conversion, which was introduced at the beginning of Chapter 5 and in Chapter 7. Two PV-side converters (PVSCs) are included in the inverter unit, and provide independent MPPT at the string level. One important parameter is the maximum AC output current, which is specified as 30 A. This value can be used to specify the AC disconnect and AC circuit breaker according to the Rule of 1.56 or other electrical codes. The rating can also follow the inverter recommendation of the contributory fault current, which is 40 A. Therefore, the system is designed to include the inverter and the relevant output circuit, as shown in Figure 10.2. The line diagram shows the system structure of the AC section and also includes all component ratings.

Table 10.1 Specification of a grid-connected system.

Term	Specifications
AC grid connection type	Single phase
Grid frequency	50 Hz
Rated AC grid voltage	230 V (RMS)
Rated power at STC	6 kW
Lowest temperature in winter	−35°C
Highest temperature in summer	+25°C

---

1 TUV stands for *Technischer Überwachungsverein*, the Technical Inspection Association in Germany.

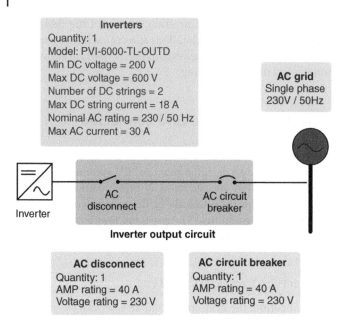

Figure 10.2 Line diagram of inverter output circuit.

The specification of the inverter has the requirements for the DC section, which includes the PV source circuit, PV source output circuit, and inverter input circuit. Various PV modules can be chosen to form the PV source circuit, which includes PV strings. The specific model selected is the HiS-M250RG, which is manufactured by Hyundai Heavy Industries. The PV module is certified by VDE and UL for ratings of 1000 V and 600 V respectively. The PV module specifications are shown in Table 10.2. The information is based on the datasheet that can be downloaded from

Table 10.2 Specification of PV module HiS-M250RG I.

Basic information					
Manufacturer	Model	Cell material	Dimension		
HHI	HiS-M250RG	Multi-crystalline	998 mm × 1640 mm × 35 mm		

Electrical performance at STC					
Cells	P_{MPP}	I_{MS}	V_{MS}	I_{SCS}	V_{OCS}
60	250 W	8.1 A	30.9 V	8.7 A	37.4 V

Temperature coefficients		
α_T	β_T	γ_T
0.048 %/°C	−0.32 %/°C	−0.43 %/°C

α_T, β_T, and γ_T are the temperature coefficients for correcting the PV module output current, voltage, and power, respectively.

http://www.hhi-green.com/. It should be noted that the datasheet information shown was downloaded on 7 June 2016 and may have changed since that time.

To form the 6-kW PV array, 24 HiS-M250RG PV modules are required, since each is rated at 250 W at STC. They can be divided into two strings, each with 12 PV modules in series (N_{series}= 12). The maximum DC voltage is the sum of the rated open-circuit voltages of each string after correction for the lowest expected temperature using

$$V_{pvdc}(max) = K_T \times 12 \times 37.4 = 448.8 K_T \tag{10.1}$$

K_T is the correction factor for the lowest expected ambient temperature. The installation location is expected to be as cold as $-35°C$, and the correction factor K_T can be determined from the temperature coefficients listed in Table 10.2. The correction factor is calculated from

$$K_T = 1 + \beta_T(-35 - 25) \tag{10.2}$$

and has a value of 1.19. Therefore, the maximum DC voltage is 542 V, which is lower than the upper limit (600 V) of the inverter as defined in the specification. The string voltage is calculated as 370.8 V, representing the nominal operating condition when the system is operated at STC, as calculated using $V_{MS} \times N_{series}$. In summer, the voltage at the MPP is always higher than the lower limit of the inverter requirement, 200 V, as shown in Figure 10.2. Therefore, the PV source circuits of the PV array can be considered to be valid. The PV source circuit and specification are shown in Figure 10.3, which is the line diagram of the whole 6-kW system.

Since the inverter accepts two strings, the system does not need a DC combiner box. Each string is designed and connected to the inverter through the DC disconnect. The maximum string current is calculated as $1.25 \times I_{SCS}$, which is 10.9 A according to Article 690. The DC disconnect should be rated accordingly, with consideration of the factor of 1.25. As a result, the design and rating are as shown in Figure 10.3, which includes all the essential information to construct the grid-connected PV system. Inverter manufacturers usually supply DC disconnects, from which a suitable one can be selected. Some can be integrated with the inverter enclosures.

The size of cable can be determined by estimating the loss due to the transmission distance. This can be different from case to case depending on the efficiency requirements and cable length. The minimum cable size can be found from the electrical code along with consideration of temperature correction. The cable gauge must be oversized if the installation is to be located in a hot place. All components should be appropriate to their place of installation – indoor or outdoor – to ensure their long-term reliability. All metal parts of frames and enclosures in the system should be grounded.

The system can be classified as DMPPT at the string level since each string is independently operated to achieve the highest power output.

10.1.2 Distributed Maximum Power Point Tracking at PV Module Level

PV module mismatch within each string can cause significant generation degradation in the system, as discussed in Chapter 2. For the system specification shown in Table 10.1, another configuration is available. This utilizes module-integrated parallel inverters (MIPIs). It should be noted these are commercially available, and were introduced in Section 2.4.1. One product meeting this requirement is the SMA Sunny Boy 240

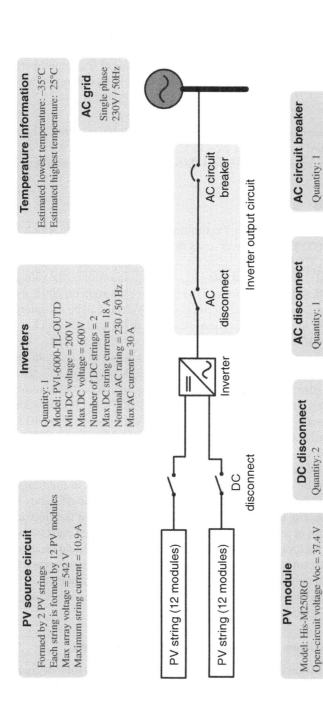

Temperature information

Estimated lowest temperature: –35°C
Estimated highest temperature: 25°C

AC grid

Single phase
230 V / 50Hz

Inverters

Quantity: 1
Model: PVI-6000-TL-OUTD
Min DC voltage = 200 V
Max DC voltage = 600V
Number of DC strings = 2
Max DC string current = 18 A
Nominal AC rating = 230 / 50 Hz
Max AC current = 30 A

AC circuit breaker

Quantity: 1
AMP rating = 40 A
Voltage rating = 230 V

AC circuit
breaker

Inverter output circuit

AC
disconnect

AC disconnect

Quantity: 1
AMP rating = 40 A
Voltage rating = 230 V

Inverter

PV source circuit

Formed by 2 PV strings
Each string is formed by 12 PV modules
Max array voltage = 542 V
Maximum string current = 10.9 A

DC
disconnect

DC disconnect

Quantity: 2
Voltage rating: 600 V
Current rating ≥ 14 A

PV string (12 modules)

PV string (12 modules)

PV module

Model: His-M250RG
Open-circuit voltage Voc = 37.4 V
Short-circuit current = 8.7 A
Max power = 250 W

Figure 10.3 Single line diagram of the grid-connected system example for single-phase-low-voltage grid interconnection.

Table 10.3 Specification of SMA-SB-240 inverter.

Term	Specifications
AC grid connection type	Single phase
Grid frequency	50 Hz
Rated AC grid voltage	230 V (RMS)
Maximum DC power	300 W rated at STC
Maximum short circuit current	12 A rated at STC
DC input voltage range	23–45 V
Rated DC input voltage	29.5 V
Peak conversion efficiency	95.8%
European weighted efficiency	95.3%

(SMA-SB-240), the key specifications of which are shown in Table 10.3. The product has been certified for installations in Europe.

The PV module specification is shown in Table 10.2. The short-circuit current is 8.7 A at STC, which is lower than the inverter limit of 12 A. With a coldest-temperature correction, the maximum output voltage is estimated as 44.5 V after applying a temperature correction of 1.19, according to (10.2); the open-circuit voltage is rated as 37.4 V at STC and the lowest temperature is estimated as $-35°C$ in the installation location. The upper limit of the inverter input is 45 V. At STC, the MPP is at 30.9 V, which satisfies the input voltage window of 23–45 V, and is also close to the rated voltage of 29.5 V.

When the key components of the MIPIs and PV module have been selected, designing the system becomes straightforward because of the highly modular nature of MIPI configurations. The 6-kW capacity is formed using 24 PV modules rated at 250 W at STC. Twenty-four SMA-SB-240 MIPIs are also required to form the power interfaces to execute MPPT at the module level and inject AC into the grid. Therefore, the system circuit can be laid out as shown in Figure 10.4, which includes the rating of each key component.

The system is divided into two circuit branches, each formed from 12 PV modules, 12 MIPIs, one AC disconnect, and one AC circuit breaker. Each branch can be implemented with an optional "Sunny Multigate" to communicate with the SMA-SB-240 units and collect and send data through the internet for remote monitoring. The manufacturer is SMA, the same as for the SMA-SB-240. The maximum efficiency of the SMA-SB-240 is 95.8%, which is lower than the 97% of the ABB-PVI-6000 string-level inverter. However, the system provides independent MPPT for each PV module, which effectively reduces power losses resulting from any PV module mismatch or partial shading. Thanks to the parallel infrastructure, individual faults can also be isolated without affecting other healthy devices.

For cabling, SMA provides specific cables and adapters for the DC connections between the PV modules and the MIPIs and supplies AC cables for the parallel connections between MIPIs. Other wiring can follow the local electrical code.

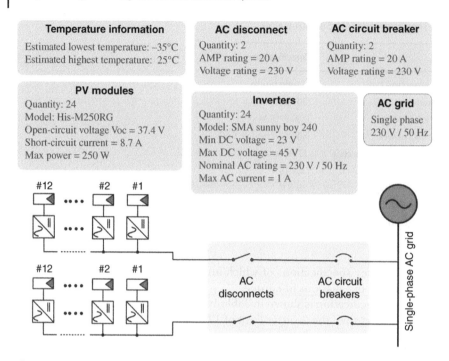

Temperature information

Estimated lowest temperature: –35°C
Estimated highest temperature: 25°C

AC disconnect

Quantity: 2
AMP rating = 20 A
Voltage rating = 230 V

AC circuit breaker

Quantity: 2
AMP rating = 20 A
Voltage rating = 230 V

PV modules

Quantity: 24
Model: His-M250RG
Open-circuit voltage Voc = 37.4 V
Short-circuit current = 8.7 A
Max power = 250 W

Inverters

Quantity: 24
Model: SMA sunny boy 240
Min DC voltage = 23 V
Max DC voltage = 45 V
Nominal AC rating = 230 V / 50 Hz
Max AC current = 1 A

AC grid

Single phase
230 V / 50 Hz

Figure 10.4 Line diagram of grid-connected system with MIPIs.

10.2 Design Example of Three-phase Grid-connected System

The design procedure can be demonstrated using the construction of a 500-kW PV system with three-phase grid interconnection. The specification is shown in Table 10.4. The temperature information for the installation location is also included.

The grid-connected inverter should be chosen according to the requirements for voltage, frequency, and capacity. One suitable product is the ABB-PVS800-57-0250kW-A, which has certification from VDE and BDEW and is rated at 250 kW. Each inverter supports two DC inputs. Other details can be found from the website at www.abb.com.

The 500-kW system can be constructed from the two grid-connected inverters with a common coupling for MV grid interconnection. The product datasheet shows the system is constructed using one-stage conversion. One important parameter of the

Table 10.4 System specification of grid-connected system.

Term	Specifications
AC grid connection type	Three phase through LV/MV transformer
Grid frequency	50 Hz
Rated power	500 kW
Lowest temperature in winter	10°C
Highest temperature in summer	50°C

grid-connected inverter is the nominal AC output current, which is specified as 485 A. This value can be used to specify the AC disconnect and AC circuit breaker according to the Rule of 1.56 or the local electrical code.

The specification of the inverter is used to design the DC section, which includes the PV source circuit, PV source output circuit, and inverter input circuit. The Hyundai Heavy Industries PV module, model number HiS-M250RG, can be used for this application. It is certified by VDE and UL at ratings of 1000 V and 600 V respectively. The PV module specification is shown in Table 10.2.

To form the 250 kW PV subarray for each grid-connected inverter, 1000 HiS-M250RG PV modules are required; each is rated at 250 W. The inverter datasheet shows that the input DC voltage should be higher than 450V, but always lower than 900 V. The voltage window determines the number of PV modules connected in series to form the PV string. The initial design uses 20 PV modules in series connection to form each string.

The maximum DC voltage is calculated as the sum of the rated open-circuit voltage of each string with correction for the lowest expected temperature using

$$V_{pvdc}(max) = K_T \times 20 \times 37.4 = 748 K_T \tag{10.3}$$

where K_T is the correction factor based on the lowest expected ambient temperature. Since the lowest temperature of the installation location is expected to be $+10°C$, the correction factor of K_T can be determined from the temperature coefficients listed in Table 10.2. This gives a value of 1.05 from

$$K_T = 1 + \beta_T(10 - 25) \tag{10.4}$$

Therefore, the maximum DC voltage is 784 V, which is lower than the upper limit of the inverter (900 V).

The string voltage is rated at 618 V to represent the nominal operating condition when the system is operated at STC. This is calculated from $V_{MS} \times N_{series}$. In summer, the voltage can be lower than the rated value since the ambient temperature is expected to be as high as 50°C. The lowest voltage to represent the MPP can be evaluated using the temperature coefficients and expressed as

$$V_{pvdc}(min) = [1 + \beta_T(80 - 25)] \times 20 \times 30.9 \tag{10.5}$$

If the PV cell temperature is estimated to be 80°C, the lowest voltage is 509 V, which is higher than the lower limit that is required by the grid-connected converter. Considering the output characteristics of PV modules, it is very important to verify the MPP voltage during hot summer days and the open-circuit voltage for the coldest winter temperatures in order to meet the input requirements of the grid-connected inverter. The line diagram for the system, with its four PV subarrays and two centralized inverters, is shown in Figure 10.5. The system is coupled to the MV network through an LV/MV transformer. The design document should include the specification for the PV source circuits, PV output circuit, inverter input circuit, inverter, and inverter output circuit. These are summarized in the following paragraphs.

The PV source circuits (PVSCs) include four independent PV subarrays. Each PV subarray is formed from 500 PV modules configured as 25 strings. Each string is formed from 20 PV modules in series connection. The specifications of the PV module are summarized in Table 10.2. Each has 250 W power capacity. Table 10.5 summarizes the PV array design configuration and parameters.

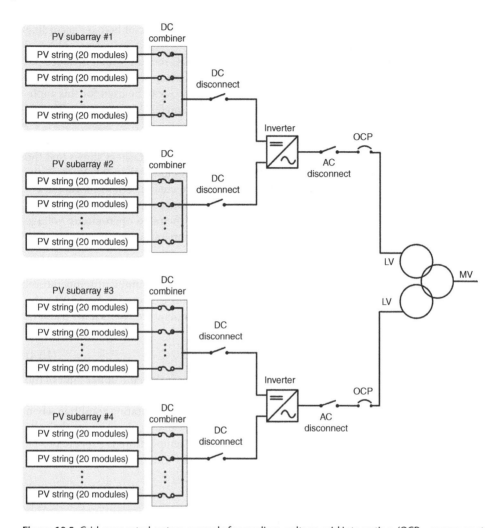

Figure 10.5 Grid-connected system example for medium-voltage grid integration. (OCP = overcurrent protection.)

Table 10.5 Design specification of PV source circuits.

Term	Design specifications
Subarray	Each includes 25 PV strings
String	Each is formed by 20 PV modules in series connection
Maximum string voltage	784 V (after temperature correction using the PV module datasheet)
Maximum string current	10.9 A (from NEC Article 690)

Table 10.6 Design specification of PV combiner boxes.

Term	Design specifications
Input strings	Not less than 25
Voltage rating	> 784 V
Maximum string current	10.9 A (from NEC Article 690)
Fuse rating	Not lower than 12.6 A for each string (from NEC Article 690)

Table 10.7 Design specification of DC disconnects.

Term	Design specifications
Voltage rating	> 784 V
Current rating	Not less than 316 A (from NEC Article 690)

The PV output circuit is formed from four DC combiner boxes with the OCP devices. Each connects 25 PV strings in parallel. The specifications of the combiner box are shown in Table 10.6. One suitable product for the circuit design is the ABB-VSN600-32, which allows 32 string inputs.

The inverter input circuit requires four DC disconnect switches to serve the outputs of the four combiner boxes. Based on the specification of the PV output circuit, the rating can be defined as in Table 10.7. One available model for the DC disconnect is the ABB-OTDC320US20S, which is rated at 1000 V and 320 A.

In this study, the inverter selected is the ABB-PVS800-57-0250kW-A, for which the detailed specifications can be found on the company website at www.abb.com. The design of the inverter output circuit and the grid connection typically follow the local grid code. The drawing of the inverter output circuit in Figure 10.5, is purely for reference purposes, and can be modified according to the specific installation requirements and local electrical code. Due to the diversity of the procedures and requirement, the details are not provided in this book so as not to cause confusion.

10.3 System Simulation and Concept Proof

A model can be built to simulate the whole grid-connected system in response to changes in environmental conditions and disturbances caused by the grid.

The simulation blocks for the power train operation of a single-stage conversion system are illustrated in Figure 10.6. The output generator output is determined by environmental conditions and the PV terminal voltage. The power exchange between the injection and extraction through the DC link determines the DC-link voltage, which is the PV terminal voltage. Due to the switching operation of the DC/AC converter, the signal V_{AC} has a pulsed AC waveform. The grid link filters out harmonics and produces an AC current waveform for grid injection. The phase is synchronized with the grid voltage. Control of the single-stage power conversion was introduced in Chapter 7 and shown in Figure 7.2.

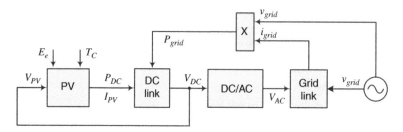

Figure 10.6 Block diagram for simulating the power train of single-stage conversion systems.

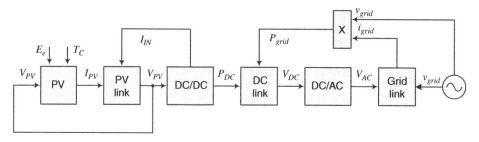

Figure 10.7 Block diagram for simulating the power train of two-stage conversion systems.

Figure 10.7 illustrates the simulation blocks and their interlink for the power train of a two-stage conversion system. The current injection and extraction of the PV link determines the variation of the PV terminal voltage, V_{PV}. The power exchange at the DC link – between the injection from the DC/DC stage and the extraction from the DC/AC stage – determines the DC-link voltage. The control diagram for the two-stage conversion system was discussed in Chapter 7 and illustrated in Figure 7.1. With the simulation block diagrams, a complete grid-connected system can be constructed and simulated. The step-by-step procedure is discussed in the following sections.

10.3.1 Modeling and Simulation of PV String

The model for a p-n junction can be used to model the PV cell, as discussed in Chapter 4. The PV cell performance can be estimated using the specification of the PV module shown in Table 10.2. The PV cell voltage can be divided by the number of cells in series in the PV module. From the ideal single-diode model (ISDM), the ideality factor of the PV cell is identified as 1.5762. Therefore, the PV module output characteristics are represented by the I–V and P–V curves shown in Figure 10.8. The deviation of the modeled MPP from the datasheet MPP is evaluated and shown as $D_{MPP} = 0.45\%$, which is acceptable for this simulation study.

Based on the 6-kW system designed in Section 10.1.1, the PV string is formed from 12 PV modules. The output is illustrated in Figure 10.9 for STC. The modeling error is accumulated and the peak power point of the simulation model is 3003.5 W, which is slightly different from the specified 3000-W power rating of each string. A Simulink model can be built to simulate the PV string output using the configuration shown in Figure 4.25. In the model, the parameter Ns represents the number of PV cells in one string, which is calculated as 720 in this system. The parameter Np is 1 since there is

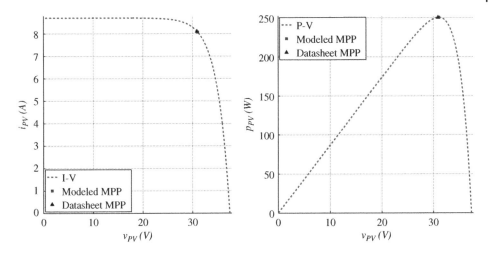

Figure 10.8 Output characteristics of the HiS-M250RG PV module.

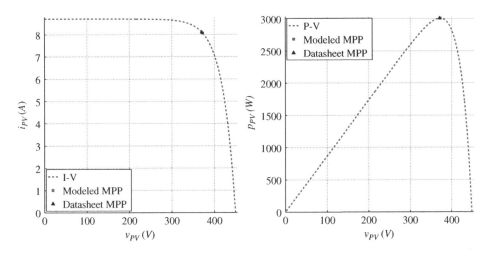

Figure 10.9 Output characteristics of the PV string that is formed by 12 HiS-M250RG modules.

no parallel connection. The model can respond to variations of solar irradiance, cell temperature, and load conditions.

It should be noted that one correction factor is ignored in the PV simulation model; the product datasheet does not provide the correction factor for the output voltage in response to the variation of solar irradiance. Therefore, the PV simulation model is slightly different from the cell model that was described in Section 4.1.3, where a polynomial function is used to represent the nonlinear correlation.

10.3.2 Modeling and Simulation of DC/DC Stage

The case study shows that the inverter accepts two PV strings that can independently track the MPP, as described in Section 10.1.1. The input voltage window is from

Table 10.8 Specification of DC/DC boost converter.

Term	Design specifications
Power rating	4000 W
Nominal input voltage	370.8 V
Nominal output voltage	600 V
Switching frequency	50 kHz
Inductance	$L = 3.5$ mH
Input capacitance	$C_{in} = 3.3$ μF

200 to 600 V. Since the manufacturer has not released detailed circuit information, the following design and simulation only mimics its operation in terms of functions and power capacity.

Two DC/DC boost converters step up the PV-link voltage to the DC-link bus voltage of 600 V for DC/AC conversion. Each PV-side converter is coupled to a dedicated PV string and operates with MPPT for maximum solar energy harvesting. The system design follows a DMPPT structure at the PV string level, as discussed in Chapter 2. The high capacitance across the DC link decouples the analysis and simulation from the DC/DC and DC/AC stages.

The DC/DC boost converter is designed and constructed using the procedure described in Section 5.1.4. Based on the PV string output characteristics and the steady-state ripple rating, the converter specification and design parameters are as shown in Table 10.8. It should be noted that the system parameters and circuit design are only for the purposes of the simulation study, and do not reflect the real parameter values for the ABB-PVI-6000. The manufacturer does not provide detailed information about the inverter circuit and control approach.

The Simulink system to simulate the DC/DC section can be constructed as shown in Figure 10.10. The system includes two PV strings each formed from 12 PV modules in series. The individual string output is independently affected by changes in solar irradiation and temperature. Two DC/DC boost converters are shown in the simulation model, and these are constructed by following the procedures described in Section 5.1.4. Both share the same output voltage that is across the DC link, but they are independently controlled in terms of MPPT and PV-side voltage regulation. The control part includes both voltage regulation and MPPT, independently operated for each string using the associated DC/DC converter.

For the nominal operating conditions shown in Table 10.8, the small-signal model is derived by following the process introduced in Section 6.3.4. The dynamic model is derived as

$$G_0(s) = \frac{-5.199 \times 10^{10}}{s^2 + 6713s + 8.664 \times 10^7} \tag{10.6}$$

with a damping factor of 0.36 and undamped natural frequency of 9.3×10^3 rad/s. Affine parameterization is applied to design the feedback controller to regulate the PV-side

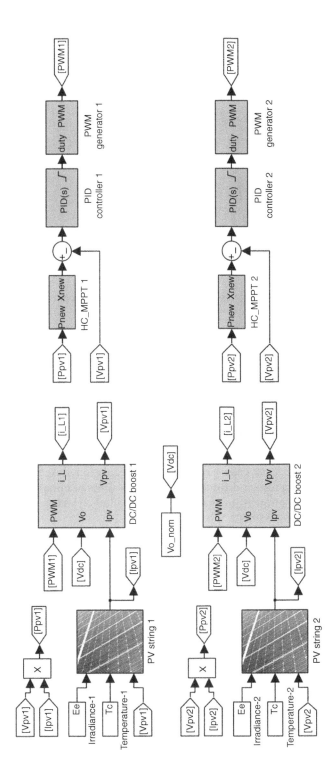

Figure 10.10 Simulink model for the DC/DC stage.

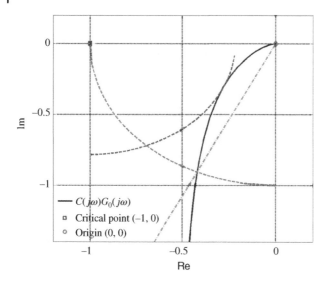

Figure 10.11 Demonstration of gain margin, phase margin, and sensitivity peak by Nyquist plot.

voltage in order to follow the MPP. As a result, a PID controller is synthesized for the voltage regulation:

$$C(s) = -0.0026 - \frac{44.33}{s} - \frac{4.62 \times 10^{-7}}{1.9184 \times 10^{-5}s + 1} \tag{10.7}$$

After the design is complete, the relative stability and system robustness should always be evaluated. From (10.6) and (10.7), the gain margin is measured as ∞, the phase margin is measured as 65.2°, and the sensitivity peak is 1.27 or 2.11 dB. The stability margins can be illustrated in a Nyquist plot, as shown in Figure 10.11. The values of the stability margins and sensitivity peak ensure robust voltage regulation. Based on the control system design, the settling time is estimated to be 161 μs at the nominal operating condition. This is the transient time from one steady state to another, and is commonly used to specify the frequency for MPPT.

The MPPT tracking frequency is conservatively set to be 2 kHz, even though the value can be higher with a settling time of 161 μs. It should be noted that the model is a piecewise linear model, while the real system always has time-variant features and nonlinearity. The algorithm is the hill-climbing method, which perturbs the PV-side voltage and detects the power variation in order to track the MPP. The step size for the hill-climbing-based MPPT is set to 1 V in this study. Figure 10.12 shows the results for the string output, using the developed simulation model and the associated control scheme.

In the simulation study, the PV cell temperatures in the two strings are intentionally set to be different: 25°C for String 1 and 30°C for String 2. The temperature variation results in a power output difference between the two PV strings, as shown in Figure 10.12. String 1 output is 3004 W in comparison 2943 W for String 2. The optimal operating voltage is 372 V for String 1 in contrast to 365 V for the MPP in String 2. The PV output current is slightly different due to the temperature difference. A 5°C variance of the PV cell temperatures is common in practical systems because it is difficult to install all PV modules

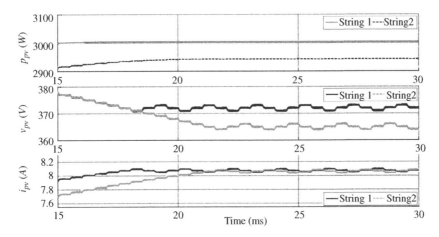

Figure 10.12 PV string output in response to the temperature difference.

under exactly equal conditions. The difference can result from the installation locations or ventilation. The simulation shows that the variance results in about 2% power degradation from STC. This agrees with a calculation based on the temperature coefficient in the product datasheet, as summarized in Table 10.2.

The simulation is designed to show that the PV cell temperature plays an important role in the power generation. It also shows the advantage of distributed MPPT at the string level, because the two strings independently track the MPPT regardless of the PV cell temperature variance. Power degradation caused by the mismatch between the two strings is therefore avoided. It also shows the advantage of two-stage conversion, namely that the DC/DC and DC/AC sections can be analyzed, designed, simulated, and evaluated independently before the final integration.

10.3.3 Modeling and Simulation of DC/AC Stage

In this stage, the system models for the DC link, DC/AC converter, and grid link are designed and constructed for evaluation. One centralized DC/AC inverter is designed to transmit the DC power from the DC link into the AC form for grid interconnection. The inverter adopts the H-bridge topology for simulation, as investigated in Section 5.4.1. The DC link uses a capacitor bank to provide an energy buffer in order to mitigate double-line frequency ripple. The grid-link section is an AC filter, which is L-type in this case, as introduced in Section 5.5. The filter should be designed to minimize the harmonic injection into the grid. Based on the PV string output characteristics and the steady-state ripple rating, the specification and design parameters for the DC/AC stage are shown in Table 10.9. Again, it should be mentioned the system parameters and circuit design are only for the purposes of the simulation study, and do not reflect the real parameters of the industrial product as indicated in the design stage.

DC-link voltage regulation is required in this stage to extract the power injected by the DC/DC stage. In this design, the small-signal model to represent the DC-link dynamics is expressed by

$$\frac{\tilde{v}_{dc}(s)}{\tilde{i}_{mag}(s)} = -\frac{25.55}{s} \tag{10.8}$$

Table 10.9 Design specification of DC/AC stage.

Term	Design specifications
Topology	H-bridge single phase DC/AC inverter
Nominal power rating	6000 W
Nominal input voltage	600 V DC
Nominal output voltage	230 V RMS value
Switching frequency	20 kHz
AC filter	L type (L_g = 5.1 mH)
DC-link capacitance	C_{DC} = 10.6 mF
Hysteresis band	± 1.5 A

which follows the modeling process that was described in Section 6.5.1. Both the feed-forward and feedback controllers are designed and implemented to regulate the DC-link voltage. The design procedures are as introduced in Section 7.10. The feedback controller is a P-type controller with a proportional gain $K_p = -10$. It is unnecessary to evaluate the stability margins since the loop is formed from a proportional controller and a model of an integral nature.

The Simulink system to simulate the DC/AC stage can be constructed as shown in Figure 10.13. The DC link takes power inputs from both PV strings. The centralized DC/AC inverter is shown in the simulation model. This is constructed following the procedures described in Section 5.4.1. The injected AC current is modulated by a hysteresis controller, which is shown as the AC modulation block. The AC current reference is formulated by the sinusoidal signal in phase with the grid voltage and magnitude determined by the control section, as shown in Figure 10.13. The control section includes both feedforward and feedback controllers, which maintain the constant voltage of the DC link and determine the value of AC current injection into the grid. The construction of the DC-link block follows the discussion in Section 5.3.

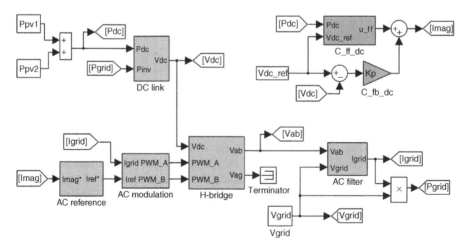

Figure 10.13 Simulink model of the DC/AC stage.

10.3.4   Overall System Integration and Simulation

The link between the DC/DC and DC/AC stages is the DC link, which is represented by the variable of the DC-link voltage. The two simulation models, as shown in Figures 10.10 and 10.13, were constructed and tested independently. They can be combined through the DC link to create the final simulation system. Figure 10.14 shows the Simulink integration of the 6-kW grid-connected PV system, including the two DC/DC conversion stages and one DC/AC conversion stage. The simulation can mimic real-world operating conditions: variations in solar irradiance and PV cell temperature and interaction with an AC grid.

One case study is created to simulate the impact of a variation of solar irradiance. Both PV strings experience an irradiance drop from 800 to 400 W/m^2 at 55 ms, while the cell temperatures is maintained at a 2°C difference between the two strings (28°C for String 1 and 30°C for String 2). The simulation results are shown in Figure 10.15. The output power of String 1 drops from 2374 to 1187 W, and the output power of the String 2 drops from 2354 to 1177 W in response to the irradiance change, as shown in Figure 10.15a. The power difference results from the temperature variance between two strings. There is also a difference in the PV-link voltages representing the MPP for each string, as shown in Figure 10.15b. The waveform shows that the DC-link voltage is maintained at the same level regardless of changes in the solar irradiance. The plot in Figure 10.15c demonstrates the grid current in response to the irradiance variation.

The simulation proves the concept of independent control of the DC/DC and DC/AC stages. The DC/DC section is controlled by MPPT and PV-link voltage regulation, implemented by the hill climbing algorithm and the PID controller, respectively. The DC/AC section maintains the DC-link voltage and modulates the AC grid current to be in phase with the grid voltage and low in harmonics. The control system is formed by feedforward, feedback, and bangbang controllers. When solar irradiance and cell temperature data are available, the model can simulate practical operations in terms of power generation and response to grid-voltage variations.

10.4   Simulation Efficiency for Conventional Grid-connected PV Systems

The computer configuration for the simulations in this chapter is as shown in Table 9.23. This was used consistently to test the simulation efficiency of different approaches.

Many studies tend to use the complete single-diode model (CSDM) to represent the PV electrical output. The transcendental nonlinear equation of the CSDM describing the PV generator generally results in slow and inefficient simulations. As recommended in Chapter 4, it is a good practice to start with simplest PV model, such as the ideal single-diode model, unless the model accuracy can be shown to be unsatisfactory. The next level is the simplified single-diode model. The CSDM is the last resort, since it is complex and difficult to compute efficiently. A good simulation system should always consider the balance between simulation efficiency and model complexity. Unnecessary complexity should be always avoided.

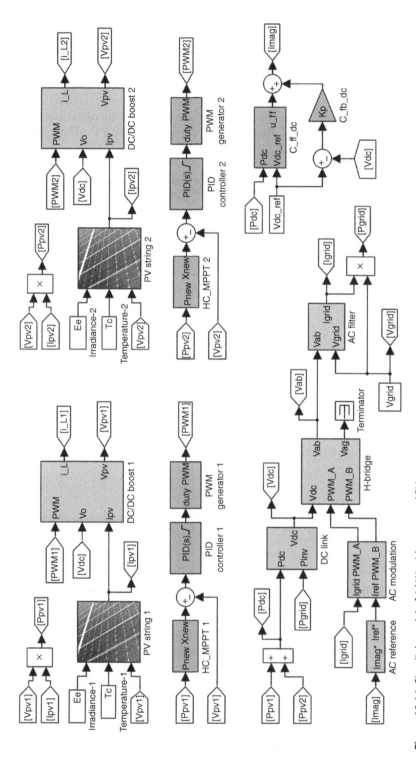

Figure 10.14 Simulink model of 6-kW grid-connected PV system.

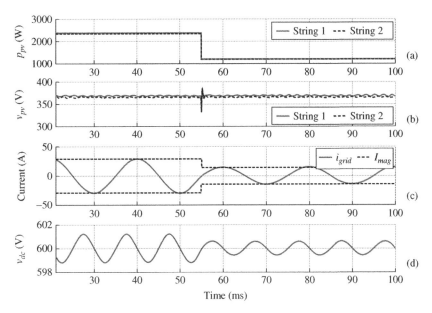

Figure 10.15 PV system simulation result in response to the variation of irradiance and temperature difference: (a) power; (b) PV-link voltage; (c) grid current; (d) DC-link voltage.

The simulation speed is mainly affected by the switching operation. For example, when the highest switching frequency is 50 kHz, the minimal sampling frequency should be 5 MHz for 1% resolution to represent the switching duty cycle. It is usually set to be either 50 MHz as the sampling frequency or 20 ns as the sampling time for simulation with consideration of higher resolution. This dramatically increases the burden of computation to achieve high-speed simulation. It takes 72 s to simulate the 100 ms of the example in Section 10.3.4 and illustrated in Figures 10.14 and 10.15. An example is the 6-kW grid-connected PV system with two PV strings and two-stage power conversion. It is difficult to adopt such a model for long-term analyses of large-scale PV systems with multiple inverters.

10.4.1 Averaging Technique for Switching-mode Converters

As discussed in Section 10.3.2, the DC/DC stage includes two boost converters that are switched at 50 kHz. They are designed and operated in continuous conduction mode. The averaged model of each converter was given in (6.32). For simulation, the equations are transformed into

$$v_{pv} = \frac{1}{C_{in}} \int (i_{pv} - i_L) dt \qquad (10.9a)$$

$$i_L = \frac{1}{L} \int [v_{pv} - (1-d)v_o] dt \qquad (10.9b)$$

where the d is the switching duty cycle. Other variables are as defined in Figure 5.16.

The Simulink model is as shown in Figure 10.16. In contrast to the switching model, the control input is the value of the duty cycle instead of the PWM signal, and the switching mechanism is removed. Other system dynamics are the same as in the switching

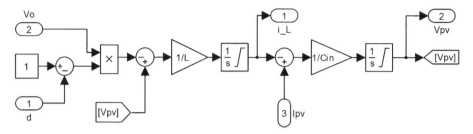

Figure 10.16 Simulink blocks of the averaged model for boost DC/DC converters.

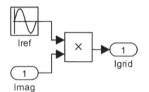

Figure 10.17 Simulink blocks of the averaged model for single-phase AC section.

model, which was developed in Section 5.1.4 and shown in Figure 5.18. Besides the control input, the model inputs also include the output voltage V_o and the PV output current i_{pv}. The main output is the PV-link voltage v_{pv}. The inductor current is considered as an internal variable, and is output for evaluation purposes.

In the DC/AC stage for a single-phase grid connection, the switching for AC current modulation is operated by the H-bridge. The basic function is to produce an AC current in phase with the grid voltage and with a controlled magnitude. The AC filter reduces the harmonics in order to keep the total harmonic distortion level lower than the upper limit, as commonly required by electrical codes and regulations. If the switching dynamics and harmonics are not the concern for the simulation study, the Simulink model for both the H-bridge and the AC filter can be simplified, as shown in Figure 10.17, to represent the AC section. The Iref block produces a sinusoidal waveform with unity amplitude and which is in phase with the grid voltage. The Imag input is the command signal produced by the control block for DC-link voltage regulation. The model is based on the assumption that the H-bridge and the AC filter always follow the control command and produce the ideal AC current waveform for grid injection.

The same method can be applied to a three-phase AC system to simplify the simulation and the representation of the AC section. The Simulink model can be built as shown in Figure 10.18. The Ias*, Ibs*, and Ics* blocks produce sinusoidal waveforms with unity amplitude, which are in phase with the three-phase grid voltage. The Imag input is the command signal produced by the control block for DC-link voltage regulation. The three-phase inverter and the AC filter are represented by the simulation block in order to show the three-phase current injection into the grid.

10.4.2 Overall System Integration and Simulation

The final Simulink model is again constructed and is presented in Figure 10.19. It includes the revised sections for the DC/DC and DC/AC stages. To simulate the model,

Figure 10.18 Simulink blocks of the averaged model for three-phase AC section.

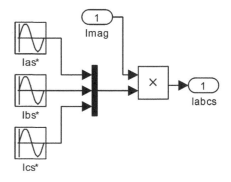

the sampling frequency should be determined by the highest dynamics in the system. Since the switching operations in both DC/DC and DC/AC stages are neglected, the dynamics can be derived from the passive components: inductors, capacitors, and resistors. The techniques involved were introduced in Chapter 6. In the present 6-kW grid-connected system, the highest dynamics can be tracked to the DC/DC stage and are expressed in the small-signal transfer function in (10.6). The undamped natural frequency is 9.3×10^3 rad/s, which is 1.5 kHz. To capture the dynamics, the sampling frequency of the simulation must be 100 times faster. Therefore, the sampling frequency is set to be 200 kHz, and this is included in the Simulink model for simulation.

The simulation keeps the same inputs as the study introduced in Section 10.3.4. The results are shown in Figure 10.20, which indicate the variance of the PV output power due to an irradiance drop from 800 to 400 W/m^2 at 55 ms, while the cell temperature is maintained with a 2°C difference between the two strings (28°C for String 1 and 30°C for String 2). The PV-link voltage is slightly different between the two strings due to the cell temperature variance. The waveform shows that the DC-link voltage is maintained at the same level regardless of the changes. Figure 10.20c shows the grid current response and its magnitude. representing the command signal, I_{mag} for current regulation. All plots follow a similar pattern to the simulation using the switching model, as shown in Figure 10.15.

The simulation can be compared with the output of the switching model by plotting the waveforms together for the same testing conditions. The simulation results are shown in Figure 10.21, which is a zoom-in plot including both models' outputs. SW and AVG indicate the output from the switching simulation model and the averaged model, respectively. Switching ripple appears in the waveforms of the PV-link voltage, the grid current, and the DC-link voltage when the switching model is used for the simulation. The averaged model captures the system dynamics but ignores the switching ripples. It should be noted that the waveforms for the DC-link voltages are almost the same in the two simulation models because the capacitance of the DC link mitigates the double-line frequency ripple. Therefore the effect of the high-frequency switching in the PVSC and GSC on the DC link is insignificant. The capacitance value is large enough to remove the high-frequency ripple that is caused by the switching operation.

It is important to use the averaged model for efficient simulation without concern for switching ripple. Using the same computer, the simulation time with the averaged model is 0.35 s, demonstrating its efficiency compared to the switching model.

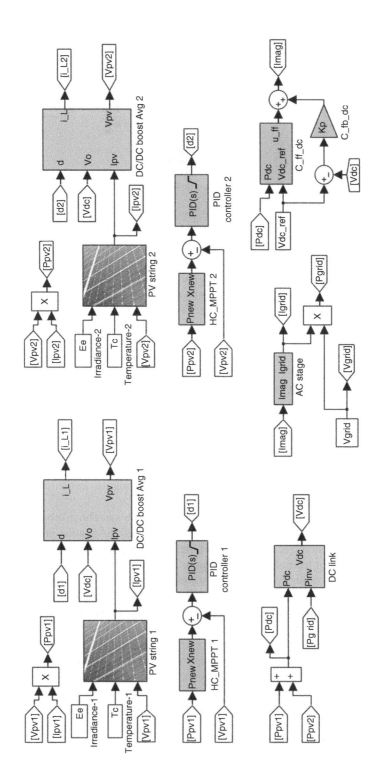

Figure 10.19 Simplified Simulink model of the 6-kW grid-connected PV system.

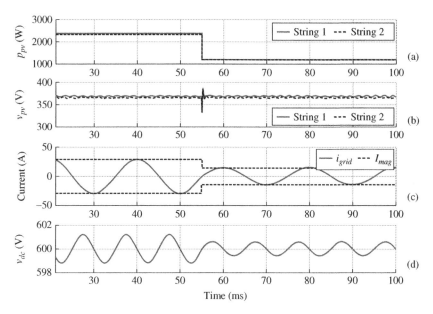

Figure 10.20 Simulation result using average model in response to variations of irradiance and temperature difference: (a) power; (b) PV-link voltage; (c) grid current; (d) DC-link voltage.

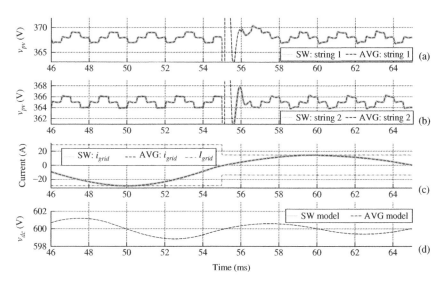

Figure 10.21 Comparison of the zoomed-in simulation results with the averaged model (AVG) and the switching model (SW): (a) PV output power; (b) PV-link voltage; (c) grid current; (d) DC-link voltage.

10.4.3 Long-term Simulation

The proposed method is efficient for simulation of long-term operations of grid-connected systems (Xiao et al. 2013). This has been verified via a comparison with experimental measurements. When environmental data, such as irradiance and temperature, are available, the simulation can be used to predict the PV power system's

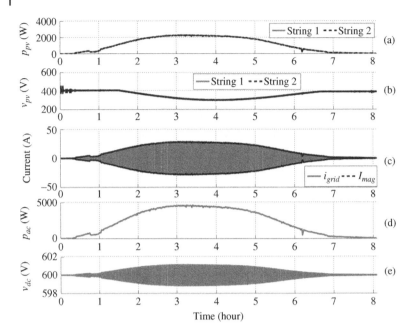

Figure 10.22 Simulation of eight-hour operation of grid-connected PV system: (a) PV power; (b) PV-link voltage; (c) grid current; (d) AC power; (e) DC-link voltage.

output in voltage and current terms. Using the same environmental data as the study of Xiao et al. (2013), a long-term simulation can be carried out, representing eight hours of operation of the design example in Section 10.3.4.

Figure 10.22 illustrates the simulated waveforms of the PV power output of Strings 1 and 2, the PV output voltage String 1 and 2, the current injected into the grid, the total injected power into the grid, and the DC-link voltage. The simulation covers the system operation over eight hours of a typical sunny day. Using the computer system in Table 9.23, the total simulation time is 12 min 21 s. The sampling frequency of the simulation is set to 20 kHz, which is sufficient to capture the fastest dynamics in the system and allows fast simulation. Therefore, the simulation result records both the steady state and the transient response according to the variation of the environmental conditions over the eight-hour period.

The simulation shows that the system tracks the MPP at each moment and regulates the PV string voltage to follow the MPP. The DC-link voltage is regulated to 600 V, at which the significance of ripples is influenced by the system power level. The grid power injection is indicated by the waveform p_{ac}, which does not count the conversion loss. The data indicate that the test is on a sunny day, where solar power increases smoothly in the morning and decreases slowly in the afternoon. To represent the real-time MPP, the waveform of the PV string voltage is high in the morning and late afternoon, but drops to a low level around noon as the cell temperature rises. It shows that the peak power level of the 6-kW-rated system reaches 4609 W, which is the peak power for that day's environmental conditions. Due to the long period involved, the plot does not show the details of the sinusoidal current and the double-line frequency ripples of the DC-link voltage. However, the envelopes show the magnitude of the grid current, i_{grid}, and the

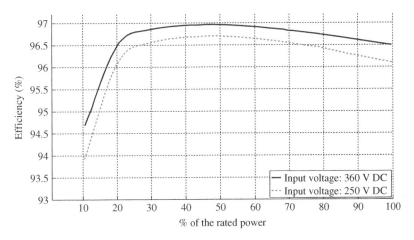

Figure 10.23 Efficiency curves of ABB-PVI-6000 PV inverter.

ripple level of the DC-link voltage, v_{dc}. Additional plots can always be produced to zoom in on these details.

To match the real system operation, the efficiency of the power conversion can be included in the simulation. This can be adopted from the inverter efficiency data. For example, the datasheet of the ABB-PVI-6000 PV inverter provides efficiency curves for the output power. The efficiency also depends on the voltage of the PV link, as shown in Figure 10.23. This information can be used in the simulation model to predict the power generation in terms of the AC form. The waveform p_{ac} in Figure 10.22 can be corrected in the real-time simulation according to the efficiency curve (Xiao et al. 2013).

10.5 Grid-connected System Simulation Based on Module Integrated Parallel Inverters

The 6-kW grid-connected system can also be constructed using 24 PV modules and 24 MIPIs, as discussed in Section 10.1.2. Even though it is a small-scale system, simulating the system operation can be very challenging due to the large number of switching-mode power units and the nonlinear nature of the PV generators (Xiao et al. 2013).

Figure 2.18a shows one of the system configurations that can be used to form a MIPI for grid interconnection. Again, it should be mentioned that the following design and parameters do not reflect specific industrial products in the design stage. The discussion here is mainly for demonstration of a simulation study of a MIPI-based system.

The MIPI system includes both a PVSC and a GSC. A DC link is required between the DC/DC and DC/AC conversion stages.

10.5.1 Averaged Model for Module-integrated Parallel Inverters

A well-known converter to act as the PVSC for DC/DC conversion is the flyback topology, which can potentially support both galvanic isolation and a high conversion ratio of voltage. The design, modeling, simulation, and control techniques involved were introduced and analyzed in Sections 5.1.7, 6.3.7, and 7.8.5. For the case study discussed in

Table 10.10 Specification of the flyback converter.

Term	Specifications	Design description
Rated DC input voltage	30.7 V	Corresponding to the MPP of the HiS-M250RG PV module
Rated DC output voltage	380 V	This is a common voltage level for DC link and DC microgrid systems
Switching frequency	200 kHz	High switching frequency can minimize the converter size
Transformer winding turn ratio	1:12	The ratio is selected to support a high conversion ratio from 30.7 to 380 V
Transformer magnetic inductance	39 μH	The inductance is rated by the inductor current ripple rating of 2 A and the switching frequency
PV-link capacitance	100 μF	The capacitance is rated for the PV-link voltage ripple of 0.2 V

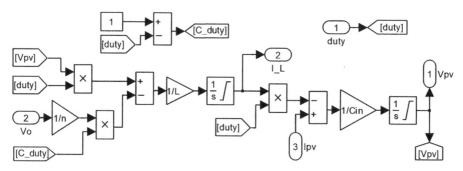

Figure 10.24 Simulink blocks of the averaged model for flyback converters.

Section 10.1.2, the flyback converter is now designed to be the PVSC, with the specification in Table 10.10.

Through the averaging technique, the averaged value of continuous signals can be used as the dynamic model for the PV-link voltage (v_{pv}) and the inductor current (i_L), as expressed in

$$\frac{di_L}{dt} = \frac{1}{L}\left[dv_{pv} - \frac{(1-d)V_o}{n}\right] \tag{10.10a}$$

$$\frac{dv_{pv}}{dt} = \frac{1}{C_{in}}[[i_{pv} - di_L] \tag{10.10b}$$

where d is the switching duty cycle and the control variable in (10.10). The output voltage, V_o, is considered constant for the DC/DC stage. The winding-turns ratio of the flyback transformer is shown as n. Based on (10.10), the Simulink model can be constructed as shown in Figure 10.24. This is based on the averaging technique and is different from the switching model in Figure 5.31. The limitation of the averaged model lies in that it is based on the continuous conduction mode of the flyback converter.

Based on the design example, the small-signal model can be derived as

$$
\begin{bmatrix} \dfrac{d\tilde{i}_L}{dt} \\ \dfrac{d\tilde{v}_{pv}}{dt} \end{bmatrix} = \begin{bmatrix} 0 & 12940 \\ -5061 & -2658 \end{bmatrix} \begin{bmatrix} \tilde{i}_L \\ \tilde{v}_{pv} \end{bmatrix} + \begin{bmatrix} 1.605^6 \\ -1.631^5 \end{bmatrix} \tilde{d}
\tag{10.11}
$$

in the state space and

$$
\frac{\tilde{v}_{pv}(s)}{\tilde{d}(s)} = \frac{-(163059s + 8121788621)}{s^2 + 2658s + 65511924}
\tag{10.12}
$$

for the transfer function. The derivation follows the procedure introduced in Section 6.3.7. The feedback controller for PV-link voltage regulation can therefore be derived as

$$
C(s) = -\frac{s^2 + 2658s + 65511924}{15.34s^2 + 926929s}
\tag{10.13}
$$

by following the design procedure in Section 7.8.5. The controller transfer function can also be transformed to the parallel PID format:

$$
C(s) = -0.0017 - \frac{70.68}{s} - \frac{1.05 \times 10^{-6}}{1.65 \times 10^{-5}s + 1}s
\tag{10.14}
$$

The verification of the averaged model can be performed by running a simulation in parallel with the switching model under the same testing conditions. The simulation result is shown in Figure 10.25, which illustrates the voltage regulation of the PV link. SW and AVG indicate the output from the switching simulation model and the averaged model, respectively.

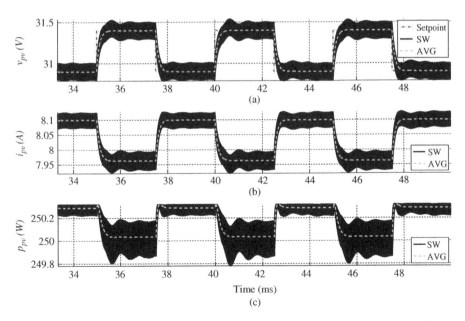

Figure 10.25 Simulation comparison between the switching model and the averaged model: (a) PV-link voltage; (b) PV output current; (c) PV output power.

Switching ripple appears in the waveforms of the PV-link voltage and the PV output current when the switching model is used. The averaged model captures the system dynamics but ignores the switching ripple. For the 50-ms simulation, it takes 0.23 s for the averaged model, but 30 s for the switching model. The averaged model neglects the high-frequency switching ripples, so the sampling time is set to 1 μs to capture the system dynamics. However, the sampling time for the switching model is set to 5 ns to record the details of the 200 kHz switching operation. The results show the effectiveness of the averaged model for simulating a grid-connected system formed by MIPIs.

For MIPIs, the basic function of the DC/AC stage is to produce an AC current in phase with the grid voltage. Since the switching dynamics and harmonics are not the concern, the DC/AC stage can be built using the averaged Simulink model, as shown in Figure 10.17. The DC-link capacitance for the MIPI system is 1 mF. The magnitude of the AC current is regulated by maintaining the DC-link voltage. The DC-link voltage regulation uses both feedforward and feedback controllers, following the approach in Section 7.10. The feedback controller adopts a P-type controller with proportional gain $K_p = -10$.

The MIPI system can be constructed in Simulink, as shown in Figure 10.26. The power train is formed by the blocks for the PV module, the flyback converter (as the DC/DC stage), the DC link (as the energy buffer), and the AC stage (for DC/AC conversion). The control functions are integrated, and include the MPPT, the PV-link voltage regulation, and the DC-link voltage regulation.

A PID controller is used for PV-link voltage regulation. The controller includes both feedforward and feedback forms for DC-link voltage regulation. These were introduced in Section 7.10. The objective of the MIPI is to maximize the PV module output and convert to AC current for grid interconnection, regardless of any environmental variations or PV module-level mismatch. As an integrated unit, the inputs of the MIPI block are the solar irradiance, the cell temperature, and the AC grid voltage. The output is the current injected into the AC grid, which is synchronized with the AC grid voltage. Other variables are considered as internal representations of the system dynamics, as shown in Figure 10.26.

10.5.2 Overall System Integration and Simulation

Figure 10.27 illustrates the modular structure of the 6-kW system formed from 24 MIPIs for simulation. Each block is independently operated for MPPT, but has the same grid voltage as the input. Each PV module can be input with different environment parameters. Each MIPI outputs the current that is injected to the AC grid. The magnitude should correspond to the difference that is caused by any mismatches among PV modules.

The following scenarios are defined for simulations using the developed model.

- The initial inputs of the first group of six PV modules is 900 W/m^2 and 25°C. The irradiance steps down to 700 W/m^2 at 255 ms.
- The initial inputs of the second group of six PV modules is 800 W/m^2 and 25°C. The irradiance steps down to 600 W/m^2 at 255 ms.
- The initial inputs of the first group of six PV modules is 700 W/m^2 and 25°C. The irradiance steps down to 500 W/m^2 at 255 ms.
- The initial inputs of the first group including six PV modules is 600 W/m^2 and 25°C. The irradiance steps up to 800 W/m^2 at 255 ms.

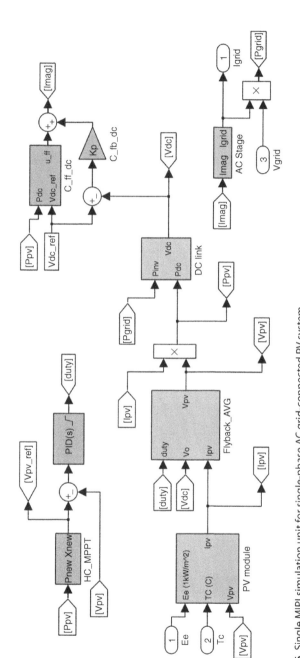

Figure 10.26 Single MIPI simulation unit for single-phase AC grid-connected PV system.

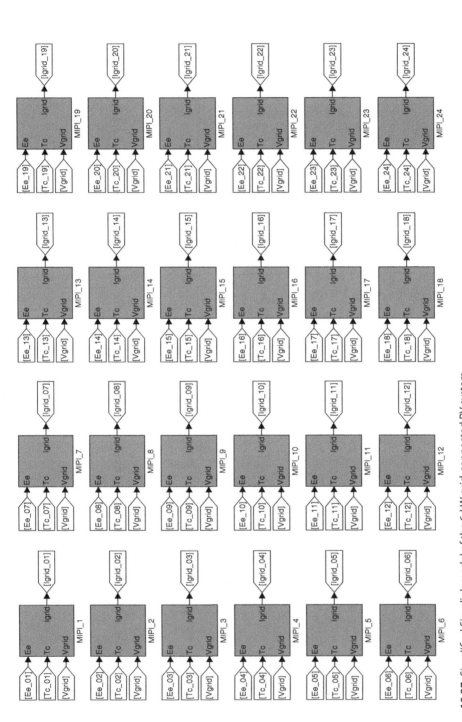

Figure 10.27 Simplified Simulink model of the 6-kW grid-connected PV system.

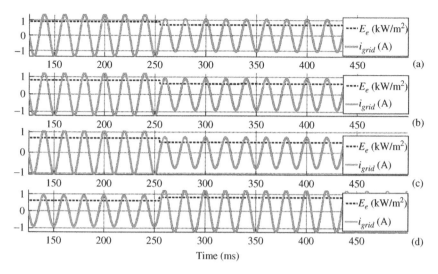

Figure 10.28 Simulation of grid-connected system with 24 module-integrated parallel inverters dealing with mismatch condition: (a) irradiance drops from 900 W/m² to 700 W/m² at 255 ms; (b) irradiance drops from 800 W/m² to 600 W/m² at 255 ms; (c) irradiance drops from 700 W/m² to 500 W/m² at 255 ms; (d) irradiance increases from 600 W/m² to 800 W/m² at 255 ms.

The AC current is plotted together with the irradiance variations, as shown in Figure 10.28. Since the grid voltage is considered to be steady, the amplitude of the AC current represents the solar energy harvesting corresponding to the environmental conditions and mismatches among PV modules. The simulation model, as shown in Figure 10.27, takes 76 s to simulate the 500-ms operation of the 24 MIPIs and 24 PV modules. Each MIPI is equipped with independent control to achieve MPPT at the module level. The same simulation study cannot be performed using switching models since the computer fails to produce any result for the predefined scenario. In comparison with the switching model, the averaged model is very efficient in simulating a distributed MPPT system with a significant number of system components (Khan and Xiao 2016).

10.6 Summary

This chapter shows the system design, integration, and simulation for grid-connected PV systems. Important components are chosen according to the system specification and performance requirements. The procedure is described and shown in a flowchart at the beginning of the chapter. Case studies of the design of practical PV systems are set out, both for single-phase and three-phase grid interconnections. When the MIPI configuration is used, the design is straightforward since the system becomes very modular. The system capacity is built by accumulation of PV modules integrated with dedicated DC/AC power interfaces. In general, the information presented in this chapter is useful for industrial design and academic study. It should also be noted that the system design approach is for reference and demonstration purposes only. Readers should always follow the regulations of the local authorities for electricity.

Special attention is given to the requirement for functional grounding. A grounded PV system is when the DC conductors, either positive or negative, are bonded to the equipment grounding system, which in turn is connected to earth. For such an implementation, galvanic isolation between the DC and AC stages should be included in the system design. A grounded system might be required by local regulators or by specific PV module manufacturers. In such systems, transformers are either integrated inside the inverter enclosure or implemented separately.

Simulations is developed to include all the components required for grid-connected PV systems with either single-phase or three-phase interconnection. The simulation proves the concept of independent control of the DC/DC and DC/AC stages, which can be analyzed and simulated separately. Techniques are introduced to simulate the grid-connected system either in detail or with fast simulation speed. The averaged PV model and the averaging technique allows simulation of long-term operation and demonstrates the essential system dynamics. This approach can also be used to simulate distributed power generation with significant numbers of power sources and converters. It is always recommended to define the simulation objective and try to use the simplest simulation model possible, so as to avoid any unnecessary complexity.

Problems

10.1 The case study and design provide sufficient information for simulation and evaluation. It is good practice to set up the case study again or to present an alternative simulation of the system and produce the same result.

10.2 A new grid-connected system can be designed according to the local regulations and weather conditions. It is good practice to follow the procedure suggested in this book. The system modeling and simulation can be performed accordingly.

References

Khan O and Xiao W 2016 An efficient modeling technique to simulate and control submodule integrated PV system for single phase grid connection. *Sustainable Energy, IEEE Transactions on* **7**(1), 96–107.

Xiao W, Edwin FF, Spagnuolo G and Jatskevich J 2013 Efficient approaches for modeling and simulating photovoltaic power systems. *Photovoltaics, IEEE Journal of* **3**(1), 500–508.

Index

Photovoltaic Power System: Modeling, Design, and Control, First Edition. Weidong Xiao.
© 2017 John Wiley & Sons Ltd. Published 2017 by John Wiley & Sons Ltd.
Companion Website: www.wiley.com/go/xiao/pvpower

Printed and bound by CPI Group (UK) Ltd, Croydon, CR0 4YY

16/04/2025

14658556-0003